The Big Book of Packaging Prototypes

RotoVision

A RotoVision Book

Published and distributed by RotoVision SA
Route Suisse 9
CH-1295 Mies
Switzerland

RotoVision SA
Sales and Editorial Office
Sheridan House, 114 Western Road
Hove BN3 1DD, UK
Tel: +44 (0)1273 72 72 68
Fax: +44 (0)1273 72 72 69
www.rotovision.com

10 9 8 7 6 5 4 3 2 1

ISBN: 978-2-88893-098-3

Art directed by Tony Seddon
Original book designs by James Campus and Fineline
This edition designed by Lisa Båtsvik-Miller
Photography by Malcolm Robertson and Simon Punter
Template designs by Three Monkeys Packaging Constructors

Reprographics in Singapore by ProVision Pte.
Tel: +65 6334 7720
Fax: +65 6334 7721

Printing and binding in China by 1010 Printing International Ltd.

The white prototypes featured in this book were programmed and produced by Three Monkey Design. The cardboard engineers and in-house graphic design team work closely together at Three Monkey to produce eye-catching, innovative packaging designs for the retail industry. This combination of skills makes Three Monkey one of the UK's leading packaging design companies.

The Big Book of Packaging Prototypes

Templates for Innovative Cartons, Packages, and Boxes

Edward Denison and Richard Cawthray

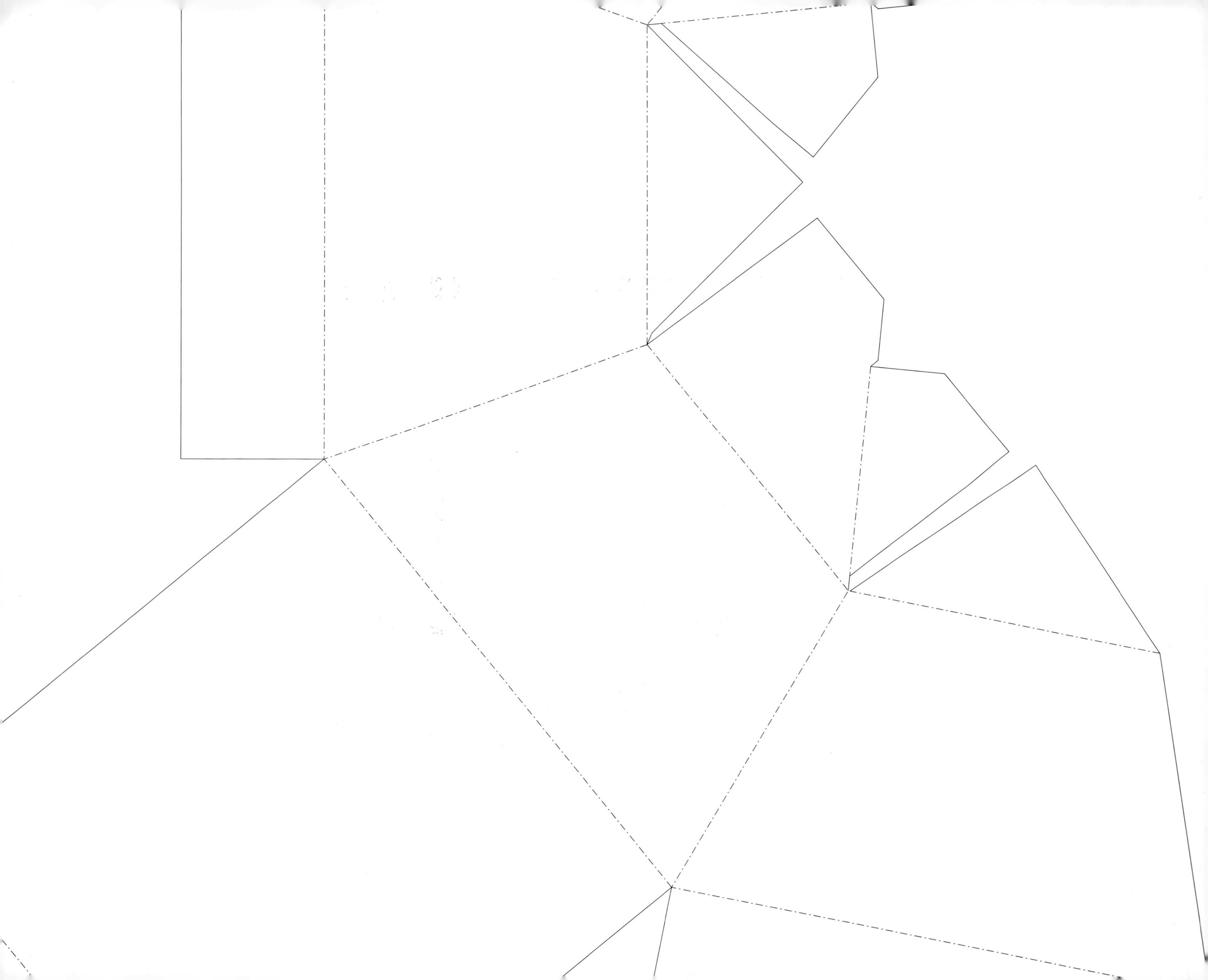

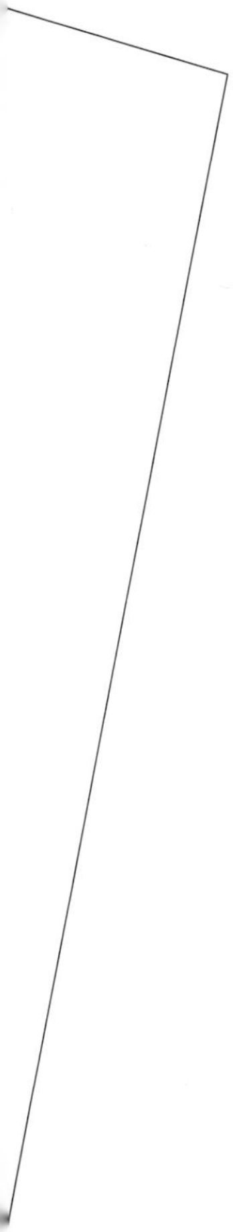

Contents

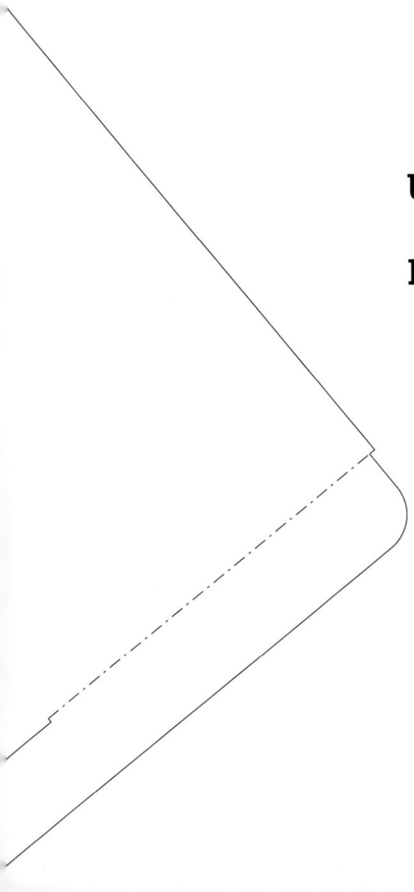

INTRODUCTION

Packaging: Insipid or Imperative?

Packaging is rarely seen as a realm of the glamorous, but in an age characterized by increasing environmental concerns, mass consumerism, and fully globalized distribution networks, it has never been so coveted, and the task of the packaging designer has never been so critical. In the past, the liberal use of materials in overpackaged or inappropriately packaged products could be seen to reflect humankind's disregard for, or relative inexperience of, the world's complex natural systems. Today, this is no longer a valid defence. We are more aware than ever before of the impact we have on the world and how the consequences of our actions, individually and collectively, can harm our own lives and the lives of others. Gone too are the days when landfills served as the sole repository of waste. Increasingly and inevitably, a "systems" view of the world is emerging in which everything is interdependent and where humankind's ultimate goal is sustainable living.

Humankind's inability to live sustainably on Earth is a fundamental flaw that has already profoundly altered the Earth's natural systems, including the atmosphere, oceanic currents, climate, and groundwater. Sustainability, in its real sense, is arguably the greatest design challenge we face. The role of design in this new realm is significantly more complex than in the past. Products and their packaging must now be designed with minimum environmental impact throughout their entire life cycle, while constantly improving their performance as a medium for providing product protection. In order for the designer to achieve these two objectives, they must have a sound understanding of the materials and processes employed in the creation, practical life, and disposal of a product or package. It is no longer sufficient for a designer to assume responsibility only for their work in isolation, when its repercussions might last for generations.

This book concentrates on a small but nonetheless imperative component of this design conundrum: packaging design. Packaging, rightly or wrongly, has long borne the brunt of public concern over environmental degradation. Perhaps this can be attributed to packaging being associated with a throwaway culture that everyone engages with and contributes to daily. Everybody disposes of precious resources in the form of waste paper, plastic, or metal packaging every day. However, while these resources are discarded or recycled, their environmental advantages are rarely considered. Indeed, a world without packaging would be a great deal more wasteful. Foodstuffs would perish before they reached the consumer, fragile electronic products would be smashed en route to retail outlets, clothes would arrive dirty and stained, liquid products would be spoiled, medicines would be contaminated, and building materials would be degraded. Packaging is an essential and beneficial component of our lifestyles in the third millennium, but it should also be viewed as one that can be constantly improved.

The designs illustrated in this book represent carton packaging, one of the fundamentals of packaging design and a sector of the packaging industry that is continuously striving for functional, esthetic, and environmental improvement. They have been selected for the varied and stimulating ways that they have responded to a wide range of packaging briefs. Cartons, by employing paper-based board to protect and often promote a product, have the capacity to deliver significant environmental benefits long into the future. They also offer remarkable scope for design creativity, providing unrivaled diversity in achieving visually and structurally flamboyant packaging solutions. However, while these designs are intended to inspire and inform designers engaged in the packaging process, the broader ramifications of packaging should always remain a foremost consideration.

The Roots of Packaging

Packaging has, over the past two centuries, burgeoned in response to an exponential rise in global commercial activity. Though packaging of some description has long been used to contain or protect products, today it is infinitely more sophisticated than at any other time in history. In a globalized world that relies on the efficient functioning of manufacturing and retail facilities located far from one another, we have become completely dependent on packaging to ensure that products withstand the rigors of global distribution networks, and that they are delivered intact to the consumer.

The origins of the modern packaging industry can be traced back to the late 18th century, when the Industrial Revolution heralded widespread changes in manufacturing. Before then, most manufacturing processes were dependent on manual labor and small-batch production, but the introduction of mechanized processes facilitated mass production, starting with a relatively small number of units manufactured daily on a production line and increasing to the millions of units per day that modern highly-automated facilities can produce.

Mechanization not only accelerated the production of all types of commodities, but also influenced their packaging. The faster the production, the greater need for packaging: supply could now anticipate and even outstrip demand, so packaging needed to be attractive as well as functional if products were to survive market competition. Packaging innovation flourished. For the first time, food could be contained in sealed and hygienic metal containers, cardboard cartons could be printed with attractive graphics and folded into striking shapes, and glass could be blown in an automated process in any number of hues. Early on, metal proved more popular than cardboard in packaging perishable goods such as cookies and confectionery, and subsequent innovations extended this demand into the containment of liquids and pressurized gases. By the turn of the 20th century, technological innovation had improved enough to permit the manufacturing of metal containers in a wide range of shapes and forms, giving rise to the first examples of novelty packaging. Today, with computerized manufacturing and advanced materials technologies, we take it for granted that such an abundance of materials, shapes, styles, and colors can be employed in packaging design.

The rapid advances in packaging technology encouraged similar developments in printing techniques, which had emerged in the early 19th century. What we now understand as "brand imagery" was becoming an important prerequisite in packaging design and had to be displayed on and reinforced through the package itself. Glass bottles, earthenware pots, metal boxes or cans, cardboard cartons, or simple paper wrappers all required a label or visual identity of some kind. The interdependent development of packaging and printing had a profound effect in cultivating branding and nurturing the idea of added value. Products previously considered too bland or too utilitarian to be given attention could now flaunt an adopted identity. Washing powders, for example, assumed evocative names, their cartons demanding attention with glaring colors and deliberately eye-catching graphics.

Printing also allowed the package to display practical information, such as illustrating pricing, contents, and instructions, as well as assisting the user in opening, resealing, or disposing of the package. These features helped to facilitate a degree of self-service and decreased the need for informed and specialized shop staff, thus contributing to the shift in the second half of the 20th century from localized independent shops selling basic ingredients, unprepared foodstuffs, and specialized products, to retail superstores offering prepackaged processed foods and a vast range of miscellaneous goods.

The role of printing and the development of specialized print techniques have had a profound effect on packaging and product advertising. Artists were first to devise graphic imagery for products. Many of these early creative illustrations evolved into established and internationally recognized brand identities. Some have developed further, defining the character of the entire corporation itself. Many of the brand images forged in the late 19th and early 20th centuries are still just as strong, and their success played a key role in forming the basis of what has since become the huge and influential industry of advertising. Advertisers rely on packaging as a key tool in the aggressive struggle for space and attention at the point of sale, and this is not always achieved by strong graphic imagery alone. The structural configuration of a package is often just as important, and can be even more effective in conferring brand identity. The ability of a packaging designer to devise a 3D form that comes to symbolize a product and creates an enduring legacy for the brand is a rare but rewarding achievement.

The Different Purposes of Packaging

Packaging design will always need to satisfy a range of objectives, both practical and conceptual. These requirements will be prioritized according to the specific conditions stipulated in the packaging brief, and will influence the purpose—and thus the form—of the package considerably. In addition, the form of a package will likely be determined by the nature of the product within. Apart from the obvious issues of size and shape, if a product's attractive appearance is its greatest asset, it might be represented on, or be visible through, the packaging; whereas for less-appealing products, the package might be used as a method of concealment or disguise. These considerations are illustrated by the contrast between (for example) an elaborately gift-wrapped box of chocolates and a budget box of breakfast cereal. Although both forms of packaging share the same fundamental purpose, they are significantly different due to the character of their respective products. Both packages are designed to contain and protect, yet they vary greatly in appearance, texture, graphics, shape, cost, and structure.

Furthermore, not all packaging is designed to be seen or used at the point of sale. Most products require extra packaging to contend with the rigors of the distribution system so that they arrive at the point of sale unspoilt and presentable. These added layers of packaging are generally termed "secondary packaging," and are designed to contain larger quantities of individual product units, each within its "primary packaging." The size and dimensions of primary packaging will govern how readily it can be stacked together in large quantities, fitted into transportation containers, and installed on standardized shelf layouts. In general, the primary packaging is that which is seen on the shop shelf, and the secondary packaging is that which is used in the transportation and distribution of the product units from the point of manufacture through to the point of sale. Depending on the product and the specific distribution requirements, there may also be a need for extra layers of packaging outside the secondary packaging (tertiary or quaternary layers).

Packaging and the Environment

Packaging and the environment are often seen as incompatible, having apparently conflicting demands. The packaging industry uses vast quantities of the Earth's resources in order to bring products safely to our homes, and the visual evidence of discarded packaging is common in the form of litter. Demands for tighter controls and more stringent regulations to curb the misuse or overuse of materials in packaging have been increasing for many years. In some countries, strict legislation to reduce or eliminate packaging materials has significantly helped to decrease dependency on packaging, while a growing international trend for locally produced goods is augmenting this process.

However, to view packaging as just a form of waste is to misinterpret its actual role. Effective packaging clearly reduces wastage of perishable items; what is questionable, however, is the quantity of material used by the packaging industry to fulfill a particular function. By improving materials technologies, manufacturing processes, and design, the packaging industry has been able to reduce significantly the quantity of materials used without compromising performance. In addition to this trend, recycling systems and technologies have improved so that the materials used in packaging can be more effectively recovered or reused. While new technologies and design innovations encourage progress within the packaging industry, the familiar exhortation remains: "Reduce, Reuse, Recycle."

The priority for the designer should be to use as little material as possible to fulfill the needs of the package. A saving in materials also results in energy savings further down the supply chain, since every unit of packaging has to be transported to the product; then, with the product, transported to the point of sale; then transported to the point of use, before finally being thrown away. The transportation cost, which uses energy and causes pollution, can be greatly reduced if the designer selects lighter or fewer materials. Although small, a slight weight reduction per unit presents a major saving over thousands or even millions of units transported every day, all over the world. Good design not only offers weight saving, but also space saving. The size and volume of a package will affect the quantity of units that can be shipped at a time. Redesigning packaging to increase the number of units that fit into a transportation container minimizes waste and maximizes efficiency. There is clearly no economic sense in packaging large volumes of air unnecessarily.

Occasionally, reducing materials is not the most efficient way to provide environmental benefits. Some packages require the use of a greater quantity of materials to make them reusable, but this is offset against the longevity of the package. The dairy and beverage industries offer a good example of the reuse method, where a high proportion of bottles are returned for refilling. Reuse is commonly regarded as more efficient and less of a burden on resources than recycling.

Recycling has become a popular banner for the championing of environmental issues in the packaging industry over the past two decades. However, the positive attributes of recycling are somewhat exaggerated. Although recycling increases the life cycle of a raw material, most materials still have a finite life span. Materials like glass and aluminum can be recycled endlessly, but plastic and carton board degrade after each use. Plastics are often contaminated during recycling; and the fiber length of carton board decreases each time it is recycled, thus reducing its strength and eventually compromising its suitability for use in packaging. Whether recyclable materials degrade or not, they all require collection, sorting, cleaning, and reprocessing, necessitating considerable energy and resources. The designer can help minimize these costs by giving careful thought to a range of design issues, such as using a single material rather than mixing materials, allowing for easy disassembly of individual components, reducing volume during collection, and making packs easy to decontaminate or clean.

As materials and resources become more scarce and therefore more expensive, and disposal becomes more costly, the pressure on designers and manufacturers to improve packaging methods and design will increase. Packaging will exist for as long as there are products that require it; but there is a fine line between the waste created by goods being damaged as a result of poorly designed or insufficient packaging, and the waste created by overpackaging. As the packaging industry strives to improve its environmental record, an observation by Sheila Clarke, Managing Director of the design agency, Packaging Innovations (UK), highlights an interesting paradox inherent in the nature of packaging:

"The rigors of the distribution system, and the lack of control over it, lead to packaging specifications which cost money and use up resources. They are geared to ensuring a very high percentage of products arrive in safe and pristine condition at their destination, despite the rigors of the journey. By definition, therefore, any product which has not been subjected to being stacked under two other pallets, has not been dropped one meter from the truck's tailboard, and has not been stored in a warehouse for the maximum period of its shelf life, is overpackaged."

Classic Packages

Packaging has become such a recognized and fundamental part of our everyday lives that wherever you live in the world, there will always be a need for packaging of some description. The following packaging examples have been chosen as a broad range of "classic" designs that have become ubiquitous in their field through continued and successful technological or design innovation.

THE TIN CAN

The storage of consumables in sealed metal cans is two centuries old, though the process relied on tin-plating techniques that had been discovered in Bohemia in the 13th century. Napoleon spurred on the innovation of conserving foods in sealed containers, when he offered a reward to anyone who could devise a successful method of preserving food for his armies. A Parisian confectioner solved this problem when he found that a sealed glass container, containing cooked food and sterilized by boiling, could preserve foods for prolonged periods. In 1810, Peter Durand patented his invention that he believed surpassed the glass canister by using instead a sealed metal can, plated with tin to prevent corrosion, that was not breakable like its glass predecessor and was also much lighter. By 1813, the first canning factory was established.

Early metal cans were made of iron and coated in tin, but later the base material switched to steel, which outperformed iron in the manufacturing process and in its quality. In the early 19th century, an individual craftsman was able to produce approximately 60 tin-plated steel cans per day using a technique that required food to be inserted into the can through a small hole in the top, which was sealed by soldering after cooking. By 1846, the invention of a can-making machine by Henry Evans had increased production to 600 cans per day. By 1900, production times were greatly improved with the invention of the "sanitary" can-manufacturing process, which allowed a more efficient method of sealing the ends. However, it was not until the 1920s, in America, that a fully automated manufacturing system was designed, and from which evolved today's highly efficient production lines that produce 2,500 cans per minute.

THE BEVERAGE CAN

Improvements in steel-can manufacturing in the mid-19th century prompted rapid progress in the packaging of consumables in metal containers. Liquid-storage techniques first started being used in 1885, when condensed milk was packaged in cans in America. By 1940, beverage cans had become a common method of packaging liquids, with beer being a major driving force in the market. In the United States and parts of Europe, beer cans were manufactured in steel and constructed in three separate parts, with a conical lid that could be sealed using a "crown" cork.

By the 1960s, the dominance of steel as a packaging material for cans was being undermined by aluminum. Frozen-juice concentrate was among the first products packaged using aluminum, which, according to consumer polls, proved a popular material among the general public. This favorable reaction encouraged further research and development in aluminum as a packaging material and helped persuade an American firm called Reynolds Metals Company to establish a division focused solely on aluminum packaging. In 1963, Reynolds Metals Company and the Dayton Reliable Tool Company invented the aluminum can with an easy-open end, revolutionizing the beverage-can market and dramatically improving sales.

The subsequent rapid development and commercial use of aluminum cans for the beer and soft-drink markets helped improve design and performance, including the introduction of the two-piece can, which by the 1980s had superseded the three-piece can completely in the UK and US beverage markets. Design innovations and manufacturing improvements have further enhanced the drinks can as a packaging device since the 1980s, with some of the most important advances occurring to meet increasingly rigorous environmental demands. One of the early developments in this field was the retained ring-pull, which proved remarkably successful and quickly replaced the highly littering detachable ring-pull.

The most significant improvements in recent years, despite fierce competition from the plastic bottle in the beverage industry, have been made in the light-weighting of cans. Enhanced manufacturing techniques and materials have brought the weight of the aluminum can down from $^3/_4$ to $^1/_4$oz (21 to 15g), and the steel tin-plate can from over 2 to under 1oz (60 to 30g) since the 1970s. These improvements have occurred through reductions in wall thickness as well as the end diameter. With modern processing plants producing over one million cans a day, and the European beverage-can industry producing over 32 billion drinks cans per year, the remarkable savings in resources are clearly evident.

The recyclability of aluminum is another key factor in its success. Recycling aluminum represents a 95% energy saving compared with producing virgin material—a much higher percentage than most other materials. The environmental cost of resource use, extraction, and transportation, and the energy required in the manufacturing process, as well as the transportation and recycling of finished and used products, have all been reduced greatly since the 1990s.

THE AEROSOL CAN

Aerosols are packaging devices that use a pressurizing agent, usually a gas propellant, to dispense a product from the container when a valve is pressed. The key innovation in aerosols was the use of a liquid that would become gas at room temperature yet remain a liquid under pressure or at low temperatures.

The idea of dispensing liquid from a pressurized container has been around for centuries. At the end of the 18th century, receptacles containing self-pressurized beverages were developed in France, and by the early 19th century, an innovation called the Regency portable fountain became the first device to use pressurized gas, in the form of carbon dioxide, to dispense carbonated beverages. By 1899, the first aerosol sprays were patented, using methyl and ethyl chloride as a propellant. In 1929, the first aerosol cans using valves were developed in Norway, providing the basis for the modern aerosol can. The aerosol had become a commercial success by the 1940s, creating an entirely new packaging medium encompassing a diverse range of uses from asthma inhalers to tomato ketchup dispensers.

The success of the disposable aerosol produced a series of design and manufacturing innovations that rely on the use of various materials including aluminum, tin plate, stainless steel, plastic, and glass. However, the most significant innovation in aerosol design around the end of the 20th century has been linked to the choice of gas used in dispensing the liquid. As is so often the case, environmental necessity forced designers to rethink the problem of aerosol packaging, as mankind's perceived problem-solving ability in one area was causing far greater problems in another. In 1974, two American scientists posited the theory that chlorine-based aerosol propellants (CFCs) were causing the depletion of the protective ozone layer in the Earth's stratosphere, causing harmful ultraviolet rays to penetrate the lower atmosphere. It was over a decade before unequivocal evidence was found to prove that the ozone layer above the North and South Poles was indeed thinning. In response, most of the world's industrialized nations signed the "Montreal Protocol" in 1987, setting out the terms for phasing out the use of CFCs in aerosols. Today, only a select few products are exempt from using CFCs, while nearly all commercial products use alternative methods of dispensing liquid products.

Manufacturing techniques for aerosol cans usually rely on two- or three-piece construction of the cylinder, although aluminum, due to its malleability, can be impact-extruded from a single ingot, allowing for considerably more inventiveness in the container's shape and form. The sheer size of the aerosol market, which exceeds 1.5 billion aerosols in the UK alone each year, is testament to the container's success as a means of packaging a wide range of products from wet sprays (such as hair spray) or foam sprays (such as shaving foam) through to dry powder for fire extinguishers.

THE GLASS BOTTLE

Glass is an ancient material, first used by the Egyptians as a packaging material in the second millennium BC. It had been used in the production of decorative ornamentation, particularly in the manufacture of beads, for thousands of years; but the manufacturing of the earliest glass containers required a process of forming concave receptacles by pressing lumps of molten glass or by coating a sand core with molten glass to form hollow containers. This first manufacturing process was succeeded by glass-blowing from the first century AD—which the Romans perfected and tried to keep a secret—until the empire collapsed, allowing the technology to spread quickly throughout Europe and the Middle East.

Glass-making flourished in parts of Europe, especially Venice, where elaborate designs and colors were used to create all manner of glass products,

including bottles and jewelry. By the 17th and 18th centuries, the innovation of the split mold allowed irregular shapes and surface decorations to be achieved in the production of bottles, including the embossing of names and product descriptions on the surface of the bottles.

High-quality glass suitable for optical lenses began with the invention of lead-crystal glass. George Ravenscroft found that by adding lead to the glass-making process the final material was not tainted by clouding and therefore had extremely high optical qualities. This innovation also augmented the use of glass as a building material, especially in large or decorative windows.

By the mid-19th century, glass had become a major innovation in the building industry, with the construction of structures such as the Crystal Palace at Britain's Great Exhibition in 1851, which spurred materials and manufacturing processes further. By the end of the 19th century, glass-blowing had become an automated process, with the invention of a machine in England that could produce 200 bottles per hour. This presented a 300% increase in output over previous techniques, but the process in turn was superseded in 1907 by an American firm that could produce 2,500 bottles per hour.

Manufacturing techniques have continued to improve throughout the 20th century, allowing modern processing plants to produce millions of bottles per day in any color and in a wide range of shapes, making it an ideal packaging material and very popular in the luxury or high-end market. In

addition to superior perceived value, glass also boasts positive environmental characteristics, since it is a stable material and also easy to reuse and recycle.

PAPER AND THE PAPER PULP CONTAINER

The origins of paper and paper pulp as a packaging material go back to the first millennium BC when they were used by the Chinese, from whom techniques for paper production spread to the Middle East and Europe. Paper was made using flax fibers and other plant matter until the mid-19th century, when wood pulp was discovered as a more effective material. Developments in manufacturing at the end of the 19th century and early 20th century allowed the mass production of paper bags. Innovations like the gusset design, gluing techniques, and printing processes improved the popularity of paper bags among the general public.

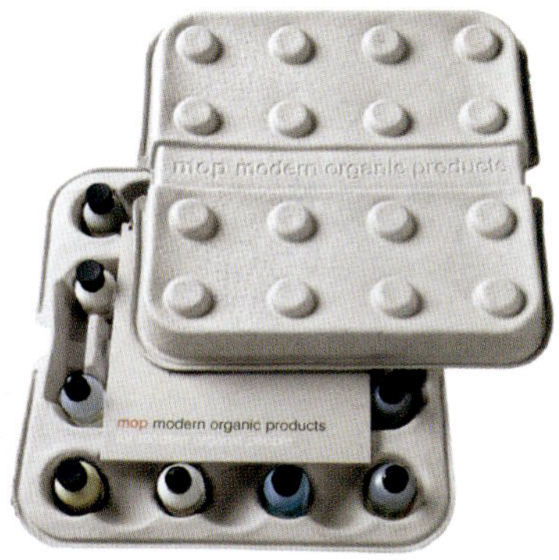

Unlike recycled paper, pulp does not require the high-quality surface and bleached appearance that paper often boasts. Pulp's rough, unrefined texture is often viewed as a positive attribute in the marketplace, evoking a sense of environmental responsibility. This perception of pulp is more than just a superficial association. With environmental degradation—including increasing fears over the production and disposal methods of petrochemical products—being one of the world's most serious problems in the 21st century, pulp offers a chemical-free packaging option that has yet to be exploited fully by designers.

Exemplary among pulp products is the egg carton, a design unsurpassed for its uncomplicated design and its performance in packaging Nature's own perfect package—the egg. Originally designed and manufactured in the 1930s, the pulp egg carton has survived many rigorous challenges from plastics-based competitors and might yet endure to symbolize the timeless appeal of paper pulp.

THE CARTON AND THE CARDBOARD BOX

Cardboard was invented by the Chinese in the 17th century and was not used as a packaging material in Europe until the 19th century. The first commercial cardboard boxes were produced in England, but the introduction of corrugated paper in the 1850s heralded new innovations in transport packaging that, by the turn of the century, started to replace wooden crates.

Developments in corrugated paper and cardboard were assisted by the proliferation of processed foodstuffs around the start of the 20th century. Major cereal-manufacturing companies were among the first to use cardboard to package their products directly in boxes sealed with a wax resin coating or wrapped in a waxed paper sleeve on which branding and advertising information was printed.

Carton and corrugated board continued to dominate the packaging industry through the 20th century until plastics were created as a mass-market material after the Second World War. From the 1950s until the 1980s, the market share of paper-based products reduced as plastics increased, but this started to shift back in favor of paper by the 1990s, in response to widespread concerns about the use of finite and nonbiodegradable resources in packaging. Concerns for the environment, also experienced in so many other fields of packaging, have been a major factor in stimulating improvements in packaging design and materials use in this specific area.

THE PLASTIC BAG

Plastics were first created in the 19th century, but it was not until after the Second World War that they became an economically and functionally viable mass-market material. The phenomenal growth of plastic since the 1960s has facilitated unprecedented innovations in the packaging industry and transformed almost every type of packaging.

However, despite the countless packaging applications that exist for plastics, none is as ubiquitous as the plastic bag. First introduced in the 1950s as a mass-produced product on a roll, plastic bags have been through many stages of development to satisfy a broad range of uses, from small sandwich bags to large refuse sacks. In the 1970s, new manufacturing processes allowed for the production of plastic carrier bags, which quickly eroded the market share once occupied by the paper bag. As the dominance of supermarkets in Europe and America increased from the early 1980s, a change in shopping habits boosted demand for bigger and stronger bags. The polythene "T-shirt" or "vest" carrier bag (with handles integral to the design, rather than attached loops) was designed in the early 1980s and has since grown dramatically to become the leading carrier bag design in the world.

Packaging manufacturers around the world have continued to refine the design of this foremost carrier bag, so that today it is lighter, stronger, and easier to manufacture and dispense than ever before. However, its unrivaled success might also account for its potential downfall. The plastic bag's omnipresence has made it a symbol of man's inability to discard waste responsibly and effectively. As litter, the plastic bag causes much anger around the world, with some countries, such as Bangladesh and Eritrea, even attempting to prohibit the use of plastic bags.

Such visible environmental problems have sullied the image of plastics in general, forcing the industry to counter claims of poor environmental performance. Indeed, plastic has very positive environmental credentials in many instances. Nonetheless, in many parts of the world, plastic bags are a conspicuous pollutant, visually and otherwise. In response to this, some countries have imposed large financial deposits on plastic bags to discourage people from throwing them away after just one use. Many supermarkets, to encourage people to use the same bag again and again, have also introduced new designs of bags that are tougher, stronger, and reusable. As materials technologies improve, it is likely that future plastic bags will be made from composite starch-based materials that biodegrade harmlessly in landfill or compost.

Today, despite the fact that tens of billions of PET soft-drinks bottles are manufactured every year, the soft-drinks industry accounts for less than half of the PET used in packaging. Due to the unrivalled popularity of PET as a packaging material, there has been a growing and urgent need to find some means of recycling this plastic, as well as finding appropriate uses for the recycled material. Today, billions of PET bottles are recycled every year around the world, feeding a wide range of other industries that manufacture anything from clothing to furniture from the recycled material.

THE PLASTIC DRINKS BOTTLE

Carbonated drinks were first invented in the 18th century, but it was not until the early 19th century that "soda" water became a bottled and marketable product. Glass bottles were used for the storage of carbonated drinks for nearly one and a half centuries before the first plastic bottles were used from 1970, but the early plastic bottles were inefficient and unreliable, with the plastic often failing and splitting due to the extreme pressures imposed by the carbonated liquid.

In 1973, the first patent was filed for the Polyethylene Terephthalate (PET) bottle, made from a form of polythene that could be mass-manufactured at a reasonable price and yet was strong enough to withstand pressurized carbonated liquid. By the end of the 1970s, the PET drinks bottle was introduced into Europe, and in America was already competing with the glass bottle. Throughout the 1980s, the PET bottle increased its share of the drinks-bottle market, expanding into other product areas such as cleaning products and cosmetics.

Different Types of Closures

The closure is an essential part of the carton, providing a temporary barrier between the product and the outside environment as well as contributing to the package's structural integrity. There are five common styles for carton closures. The following can all be further adapted to fulfill specific requirements such as tamper evidence, suitability for filling on automated assembly lines, and resealing.

A

B

C

D

TUCK-END CARTON

These closures all tuck into place and require no gluing; they can either be opened and closed many times or used once, depending on different types of fixtures such as dagger/spade, slit, and tab locks.

A] Standard Tuck-Flap Carton.

B] Slit-Lock Tuck Carton. Provides a more secure seal.

C] Tab-Lock. This provides additional protection against the lid being forced open from the inside. With slits in the tab, this design provides a level of tamper-proofing.

D] Postal-Lock. This offers a degree of tamper-proofing through the tab-ends creasing when the package is opened. Since the tabs do not tear immediately, the closure has limited reuse. A dagger lock is a variation of this design that has an arrowhead tab which tears on opening; this is not reusable, but is completely tamper-proof.

A

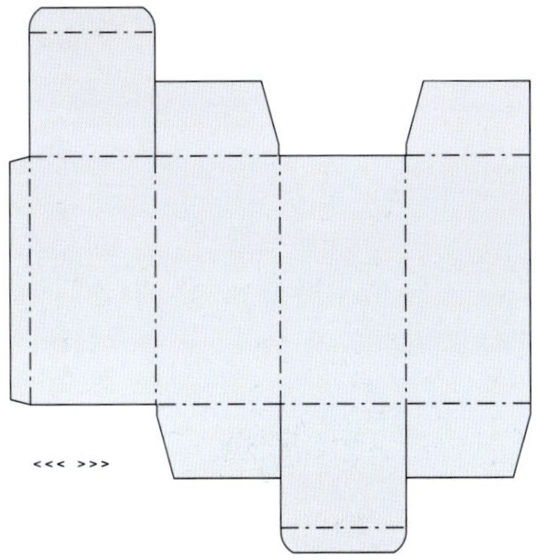

B

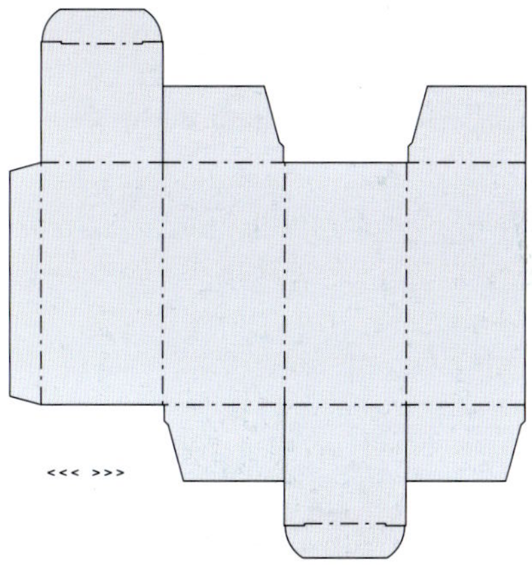

C

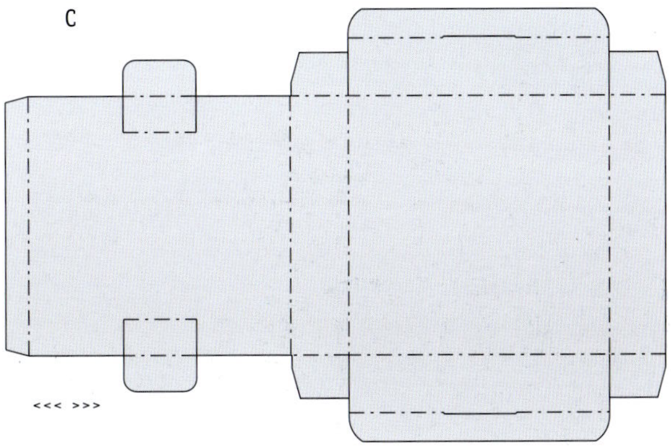

D

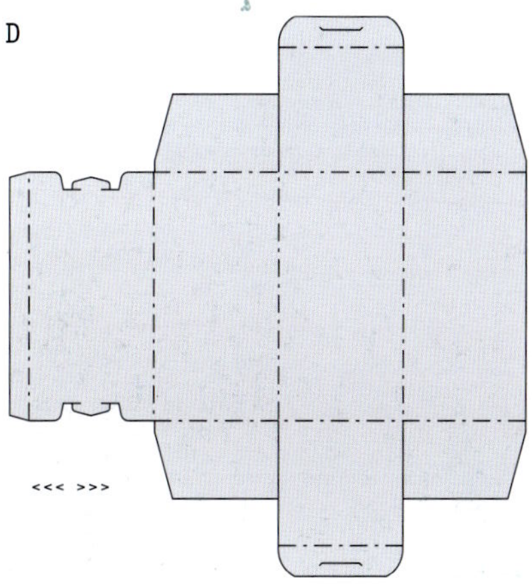

SKILLET OR SEALED ENDS

Most of the transit cartons use this type of closure, as it provides the most economical use of carton board. It also minimizes scrap removal, which is a labor-intensive and costly process. Flaps are sealed using glue or tape, which is commonly applied using an automated sealer on the production line.

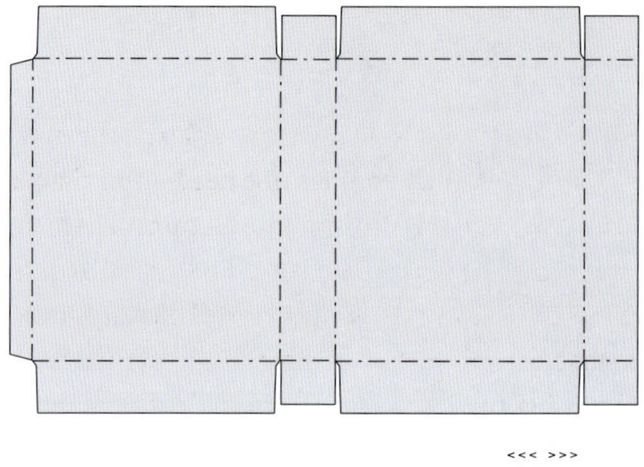

A, B] Skillet (with butting tape-sealed ends).

C] Skillet with a Partial Overlap Seal. This provides a decorative tab and lock slot.

A

B

C

TUCK-TOP CRASH-BASE CARTON

These are used increasingly when fast assembly of the carton is required. They are preglued and folded flat. For assembly, the carton needs to be opened; the base slides into position and locks when all sides meet.

The above-and-below views of this closure show it to be glued lengthways along the external sides of the carton, with the flaps from the widthways-sides glued down.
The base slides into place when the carton is pushed together, then snaps closed as the friction between the faces pushes the paperboard together (see also template below).

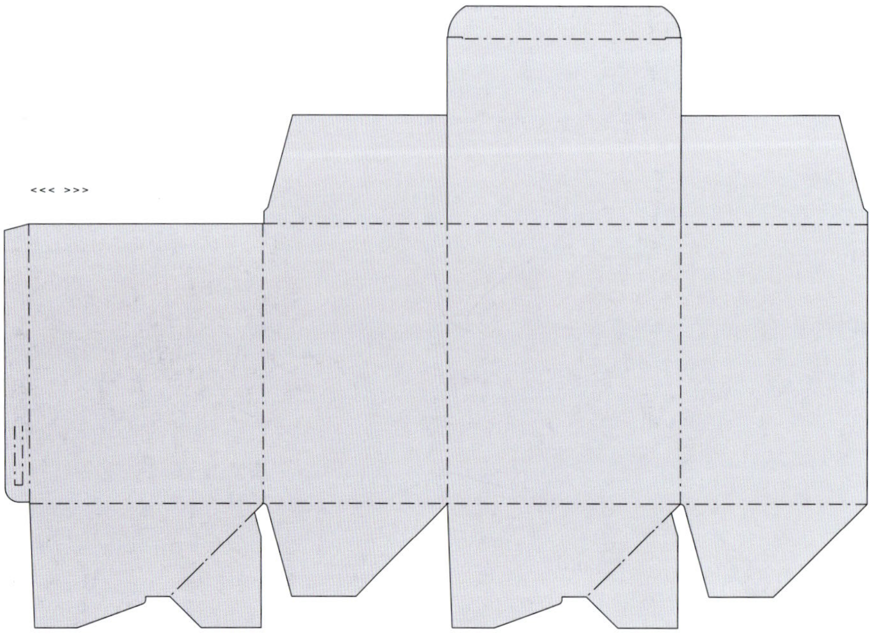

WEB CORNER TRAY

This is used for easy-erect trays without the use of glue. Because the design does not need glue, time and resources are saved during production. There is a diagonal fold across each corner that creates a web when erected; the web corners are held in place by flaps that fold down over them. The corners can be glued to give further strength if required.

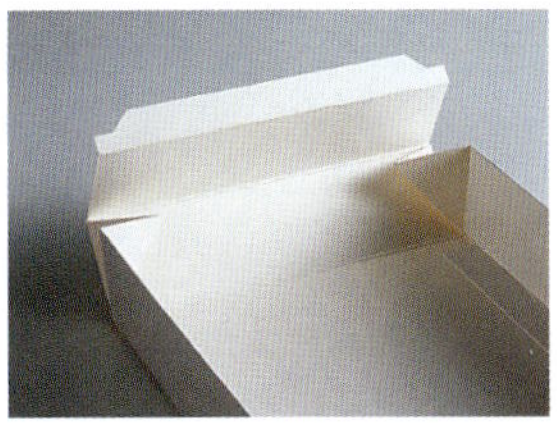

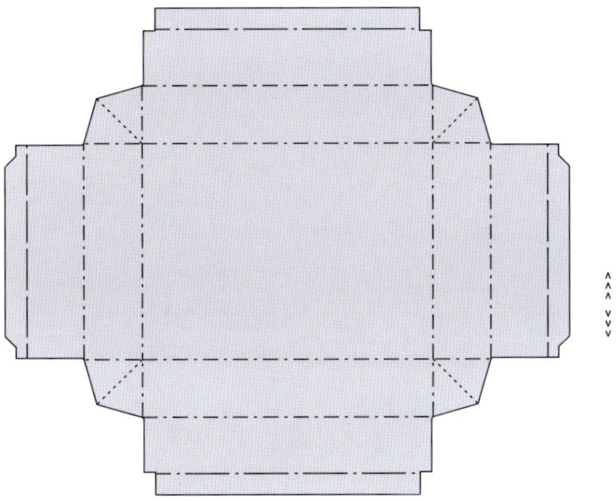

SIX-POINT GLUED TRAY WITH INTEGRAL LID

This type requires gluing for added strength and ease of assembly. The corners are preglued and the structure is erected by pulling out the sides of the tray.

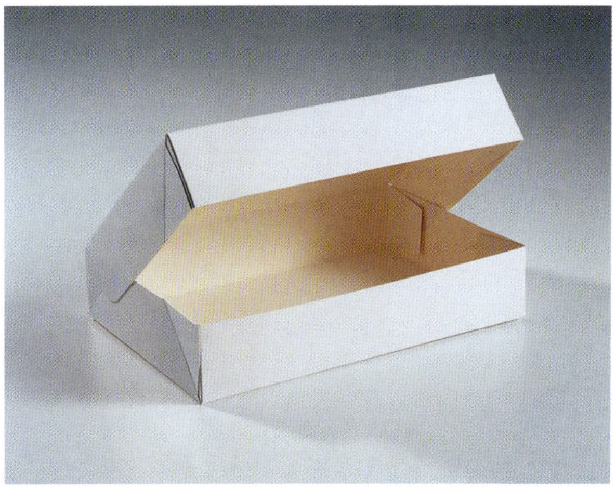

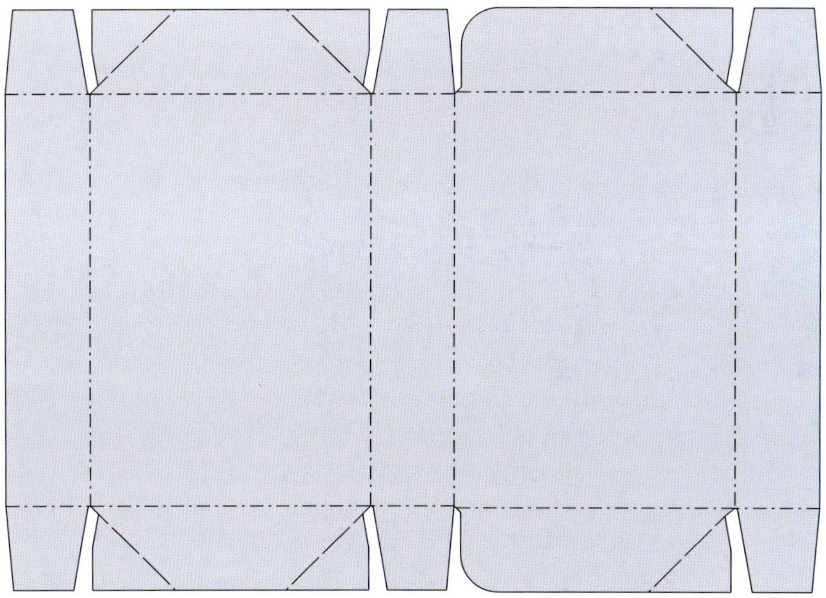

Packaging Materials

DESCRIPTION	TYPICAL USES	APPROXIMATE THICKNESS (MM)
White-back folding box board	Novelty and luxury packs such as cosmetics, confectionery, and other high-quality foods.	0.3–0.58
Folding box board reverse side cream	Food products (including frozen), medical packaging, and cosmetics.	0.35–0.65
Solid bleached board	High-quality packaging used for cosmetics and the luxury trade.	0.285–0.49
Recycled solid white-lined chipboard (minimum 75% recycled content)	Display outers and nonfood products.	0.3–0.85
Pulp board	Used for low-cost products, special promotional packs, or to emphasize an "environmental" aspect of the pack.	0.3–1
Unlined A-flute corrugated	Used for packs which need no strength, providing a layer of protection for the product.	4
Single-face A-flute corrugated	A currently fashionable pack for fast-moving consumer goods, with good crush resistance and tactile qualities.	4.2
A-flute corrugated (33 flutes per linear foot)	A pack for very fragile goods with great shock absorbency.	4.5–4.7
B-flute corrugated (47 flutes per linear foot)	High shock-absorbency packaging with optimal levels of crush resistance.	2.1–2.9
C-flute corrugated (39 flutes per linear foot)	High-level shock-absorbency packaging (greater than B-flute).	3.5–3.7
E-flute corrugated (90 flutes per linear foot)	Thinnest corrugated packaging used in instances where a narrow gauge of corrugation is required.	1.1–1.2
Double-wall corrugated (B & C-Flute)	Used to protect fragile goods and to increase the strength of cartons containing heavy objects.	5.6–6.6
Multilayered solid bleached board (waterproof lined)	Food and drinks packaging.	0.8–1
Acetate	Used to provide a barrier to touch and for safety in transport while allowing the product to remain visible.	0.3–1

Icon Key

MATERIALS

carton board

corrugated cardboard

plastic

multimaterial

waterproof board

ASSEMBLY

glue/adhesive tab

no glue

self-erecting

other fixings

SUITABLE USES

food packaging

confectionery

gift

liquid container

pharmaceutical

transit

display

other

Line Key

cut

crease

perforation

score

cut and crease

grain direction

<<< >>>

OTHER INFORMATION

one-piece design

two-/multipiece design

product visible

possible Euroslot placement

environmentally responsible design

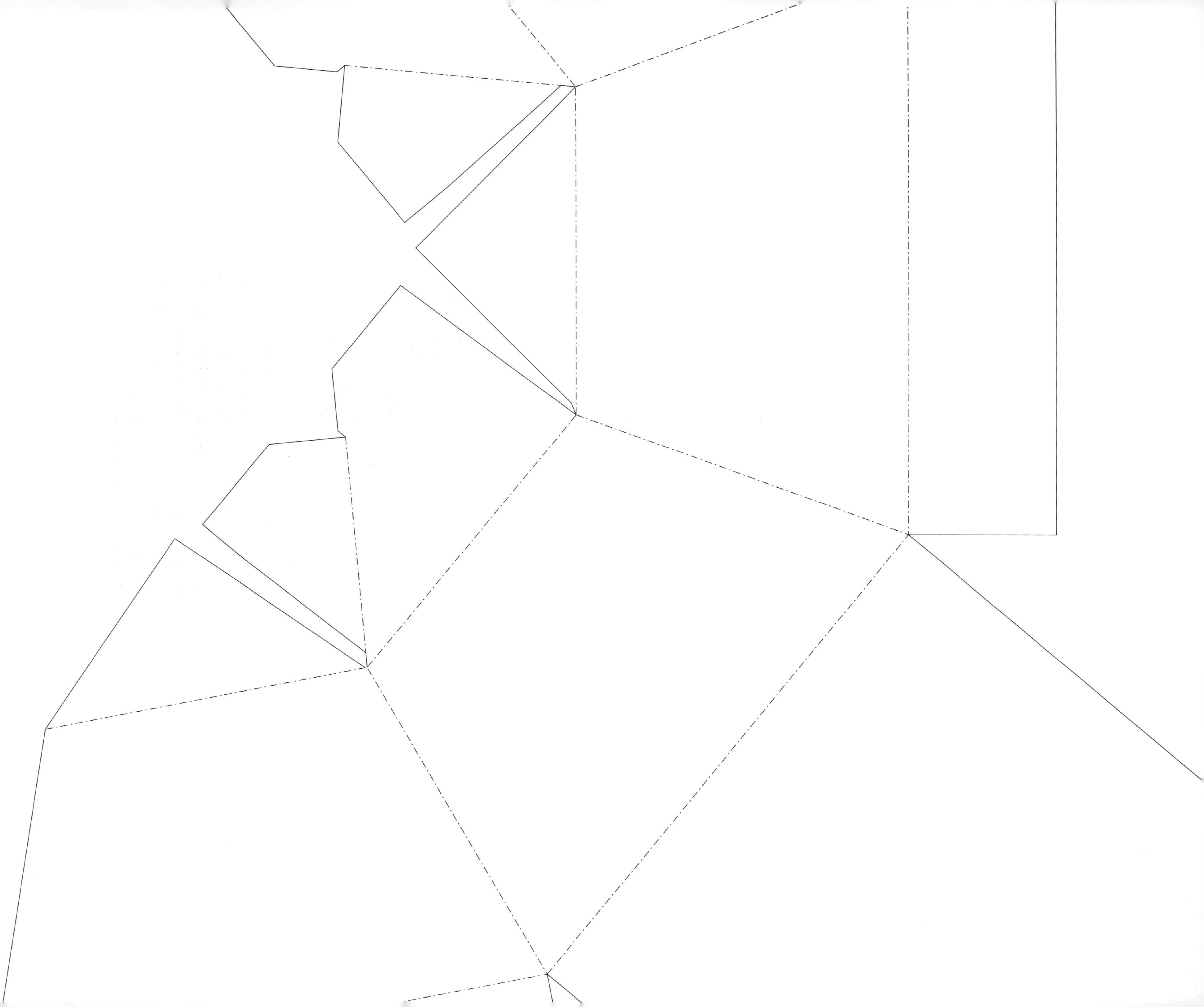

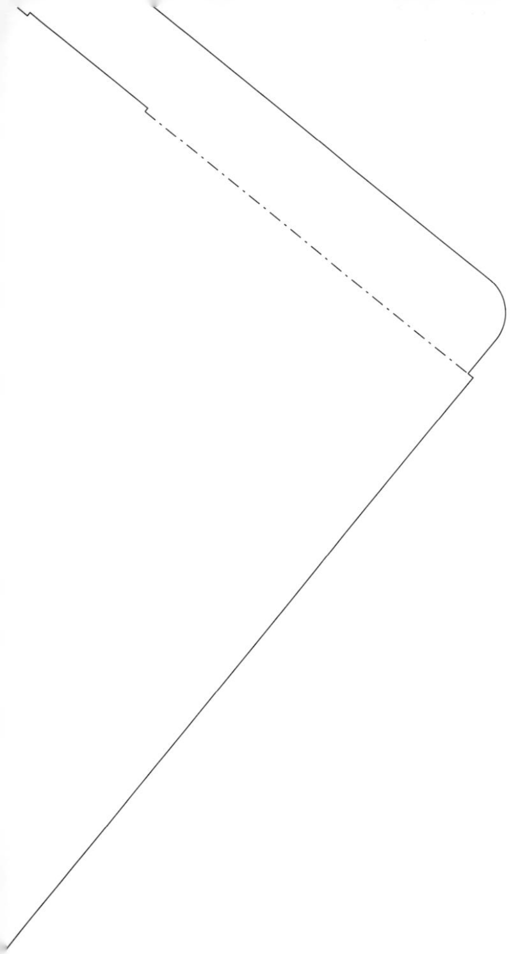

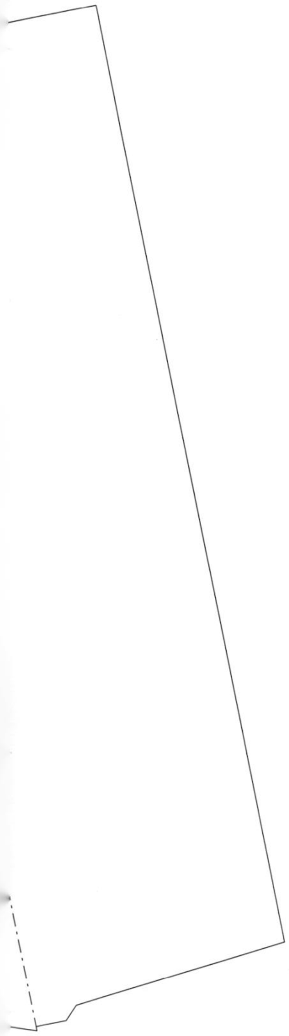

The Designs

There is no limit to the number of designs for carton packaging. Despite the proliferation of standardized templates and a preponderance of similar or successful designs available on the market, there always remains scope for improvement, innovation, and creativity. Structural design, when applied successfully, will often surpass the numerous other methods of winning customer loyalty, such as advertising, marketing, and graphic imagery, and can significantly improve competitiveness in the marketplace, as well as offer improvements in performance. A successful carton design can forge a company image in the public's eye, convey a sense of quality that becomes the envy of its competitors, and deliver functional and environmental improvements that exceed its predecessors.

The cartons featured in this book have been chosen to provide designers with a variety of samples from a wide range of applications that are intended to inspire, inform, and encourage the design process, and help stimulate future improvements in carton design. No list of carton designs will ever be exhaustive, but the principles of any design can be used in an infinite number of different applications. This book has been conceived to assist with this process and help designers arrive at packaging solutions that suit their own specific requirements.

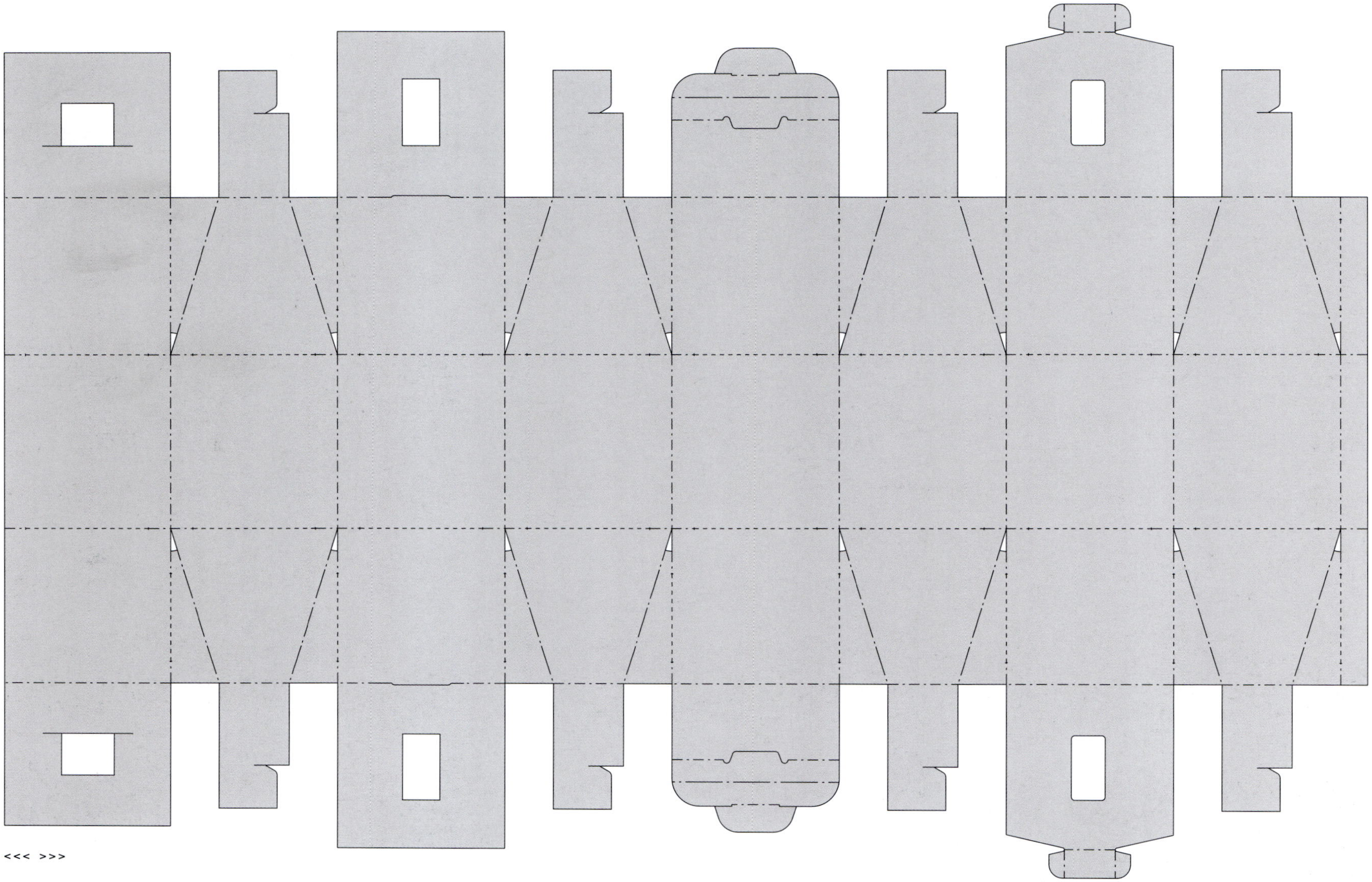

NEAR-SPHERICAL CARTON

A multiformed and perforated carton. Through a series of semicomplex folds and tucks, this template takes on the rounded form that comes close to a sphere in carton board. This is good for novelty items, and the multisided shape provides substantial opportunity for interesting graphics to be applied.

PYRAMIDAL CARTON WITH PUSH-IN CLOSURE

This pyramidal carton is similar to the following example but uses a different form of closure. Instead of a simple lid, it includes a closing mechanism on each side section. This functions by pushing down one of these top tucks and then the others, in sequence, spiralling inward.

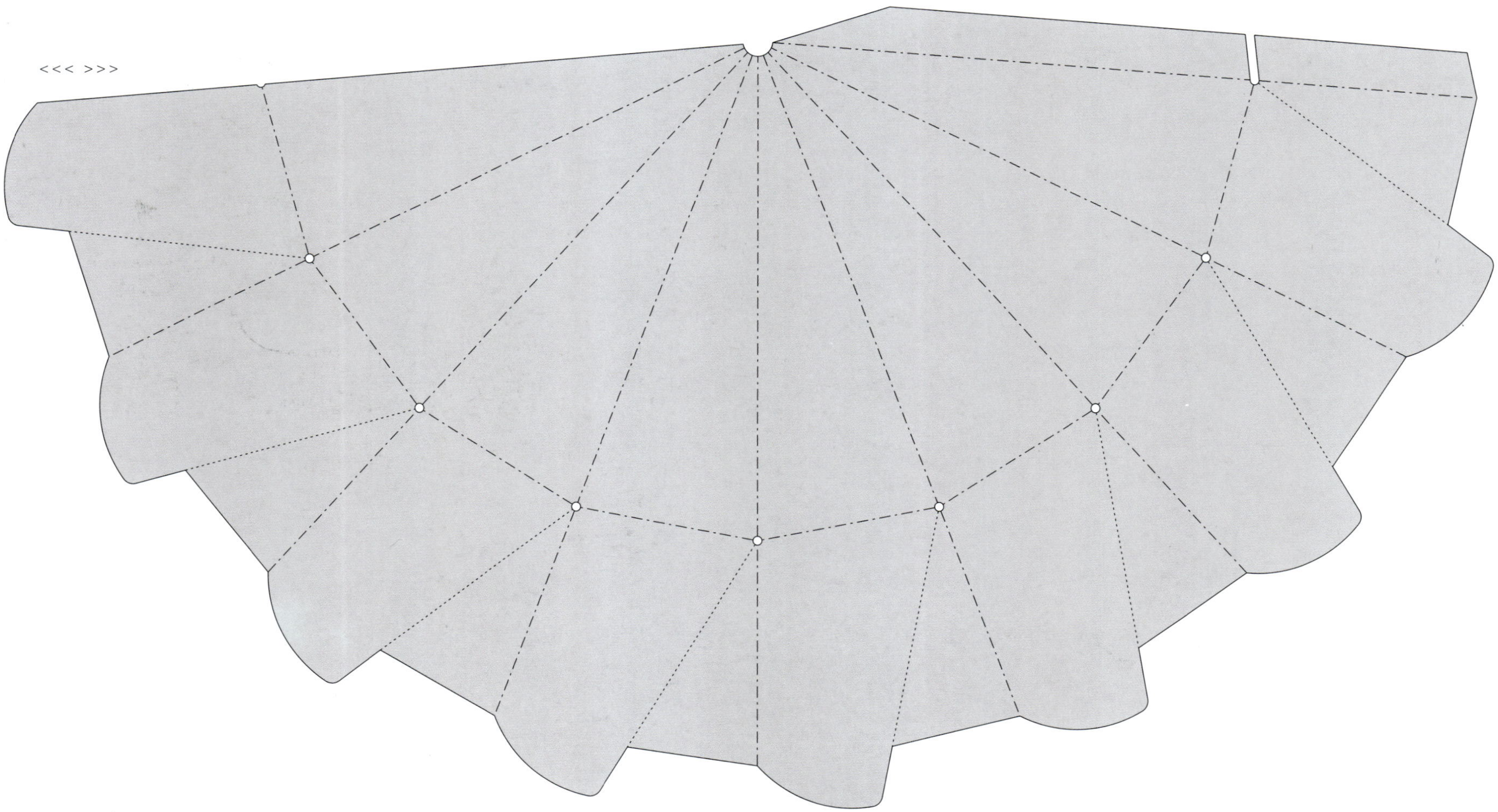

Art Institute of Pgh.

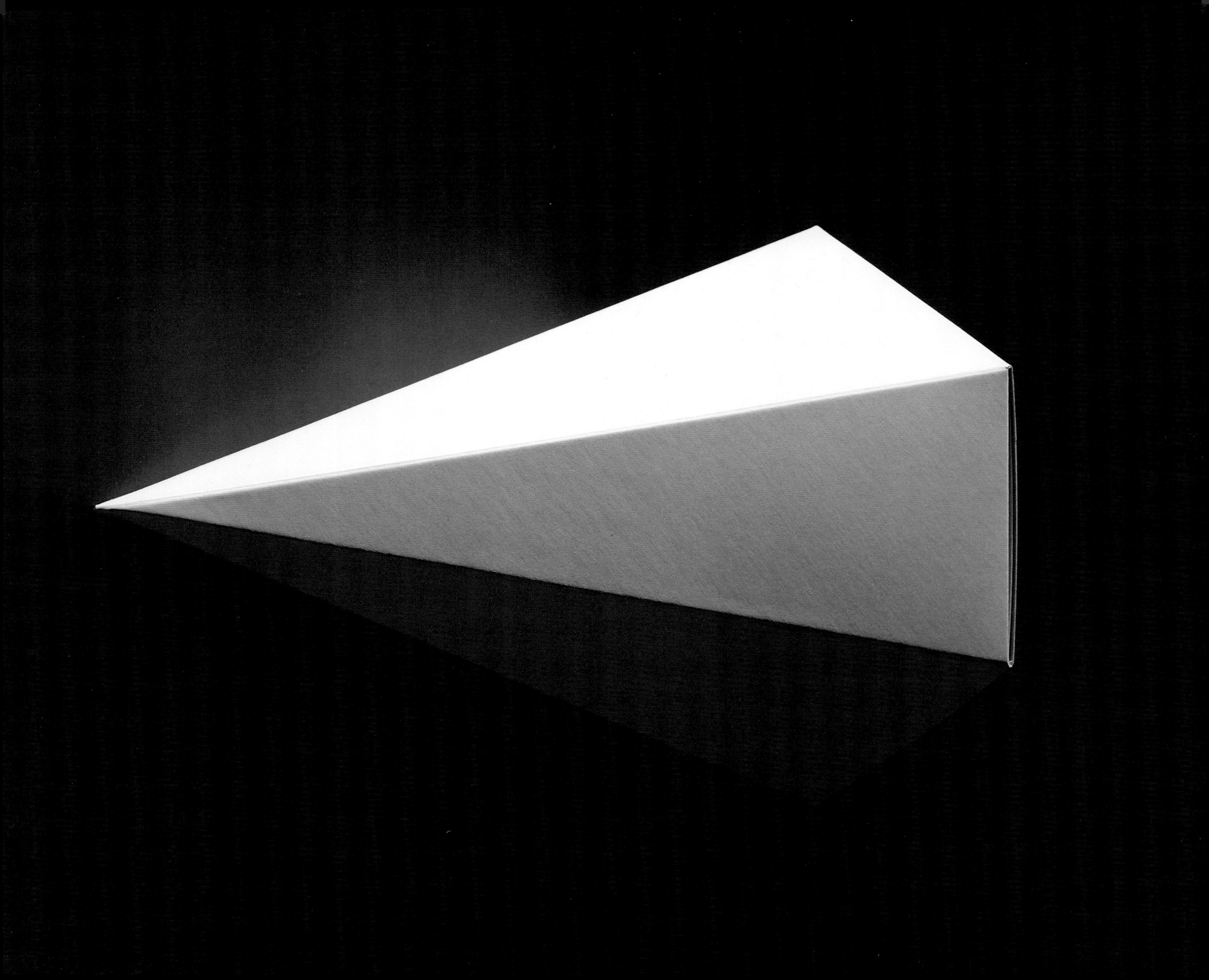

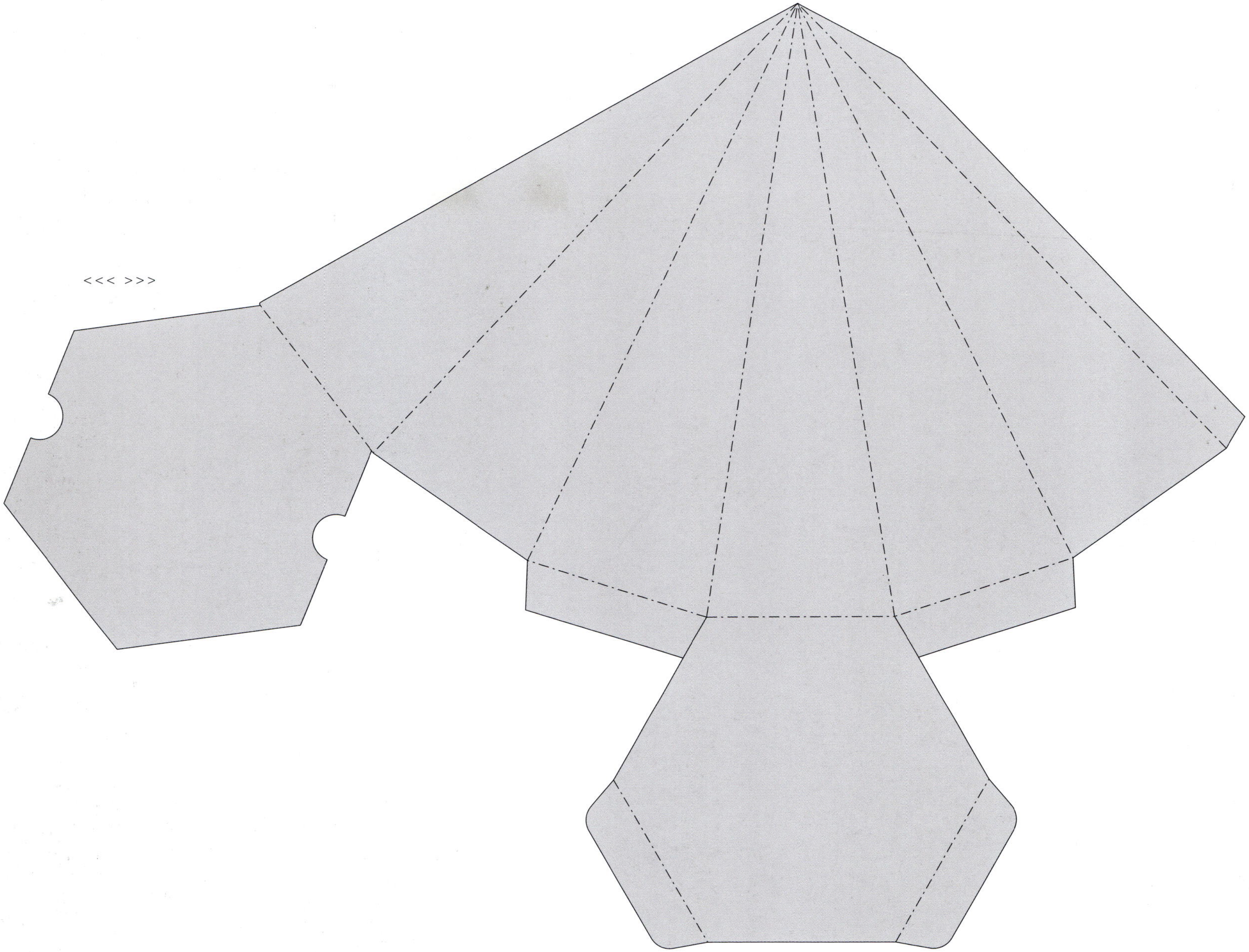

PYRAMIDAL CARTON

The tapered sides and glued tuck flaps create a pyramid-shaped carton, the intermediate slip and lid is optional depending on product requirements. The designer also has the option of increasing or decreasing the number of sides of this carton by altering the side sections and, as such, the shape of the base.

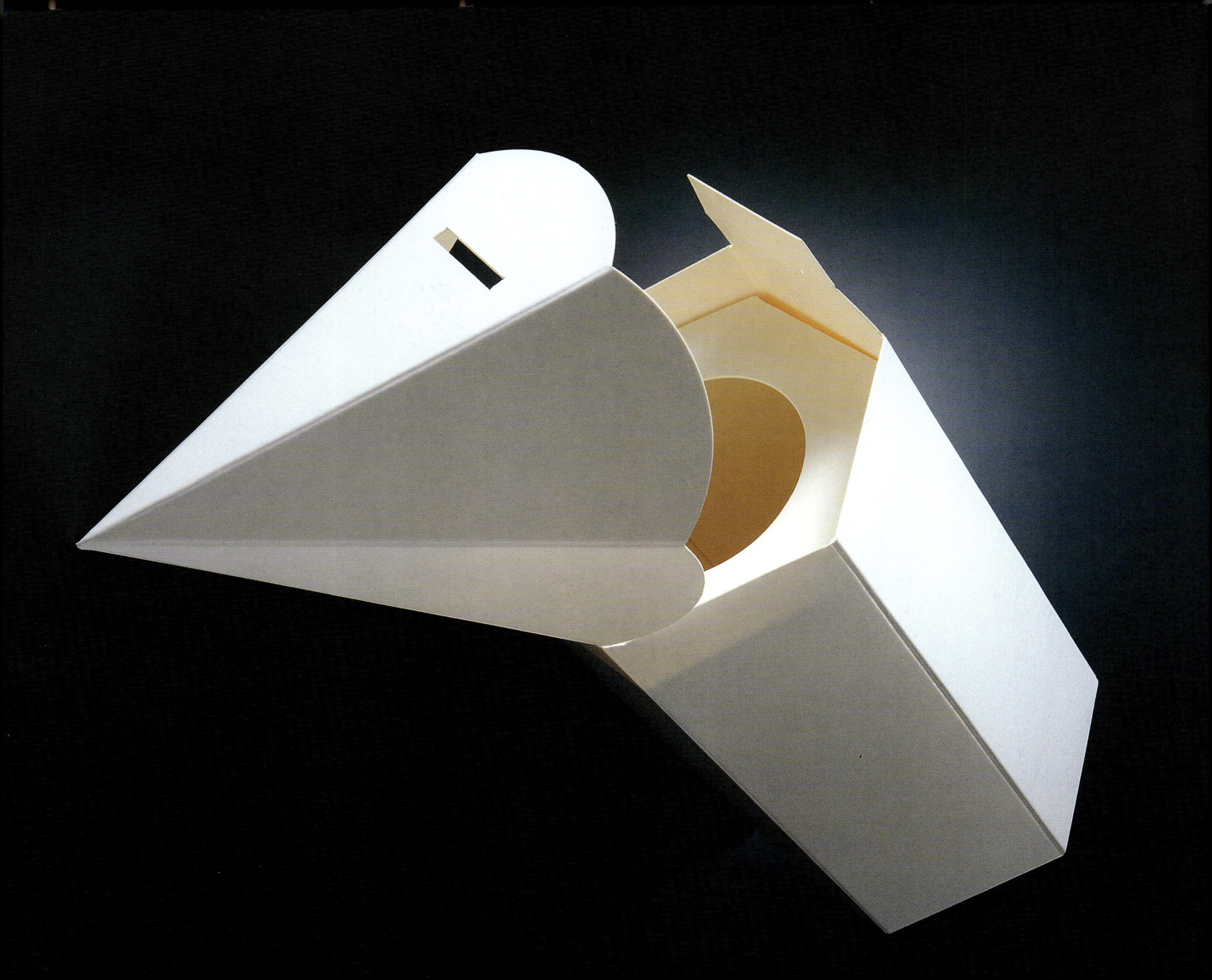

Danapak

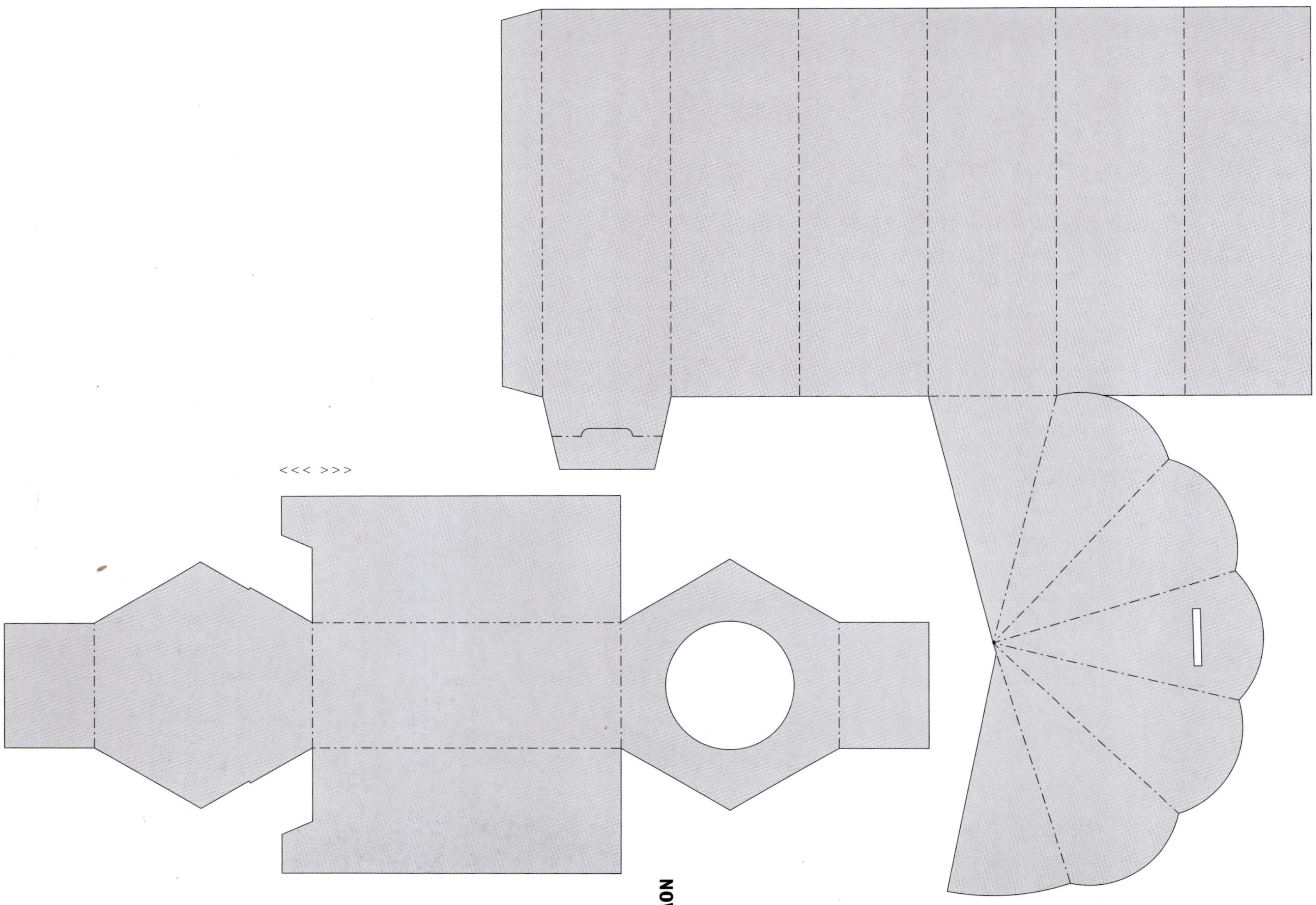

NOVELTY PENCIL-SHAPED PACK

A novelty-shaped carton which may be used to display pencils. The two-part design is suited to promotions or advertising displays rather than large-batch production. The internal hexagonal insert provides rigidity and acts as a useful holder for the contents. This illustrates how novelty point-of-sale cartons can benefit from a design that is very specific to the product contained.

CRACKER-SHAPED CARTON

This is glued along the main tuck flap and stapled together at the ends; with a few ingenious folds, the carton forms a cracker with a storage section in the middle. This design is suitable for novelty or luxury confectionery items, additional use of materials or features could raise the perceived quality of the item contained within.

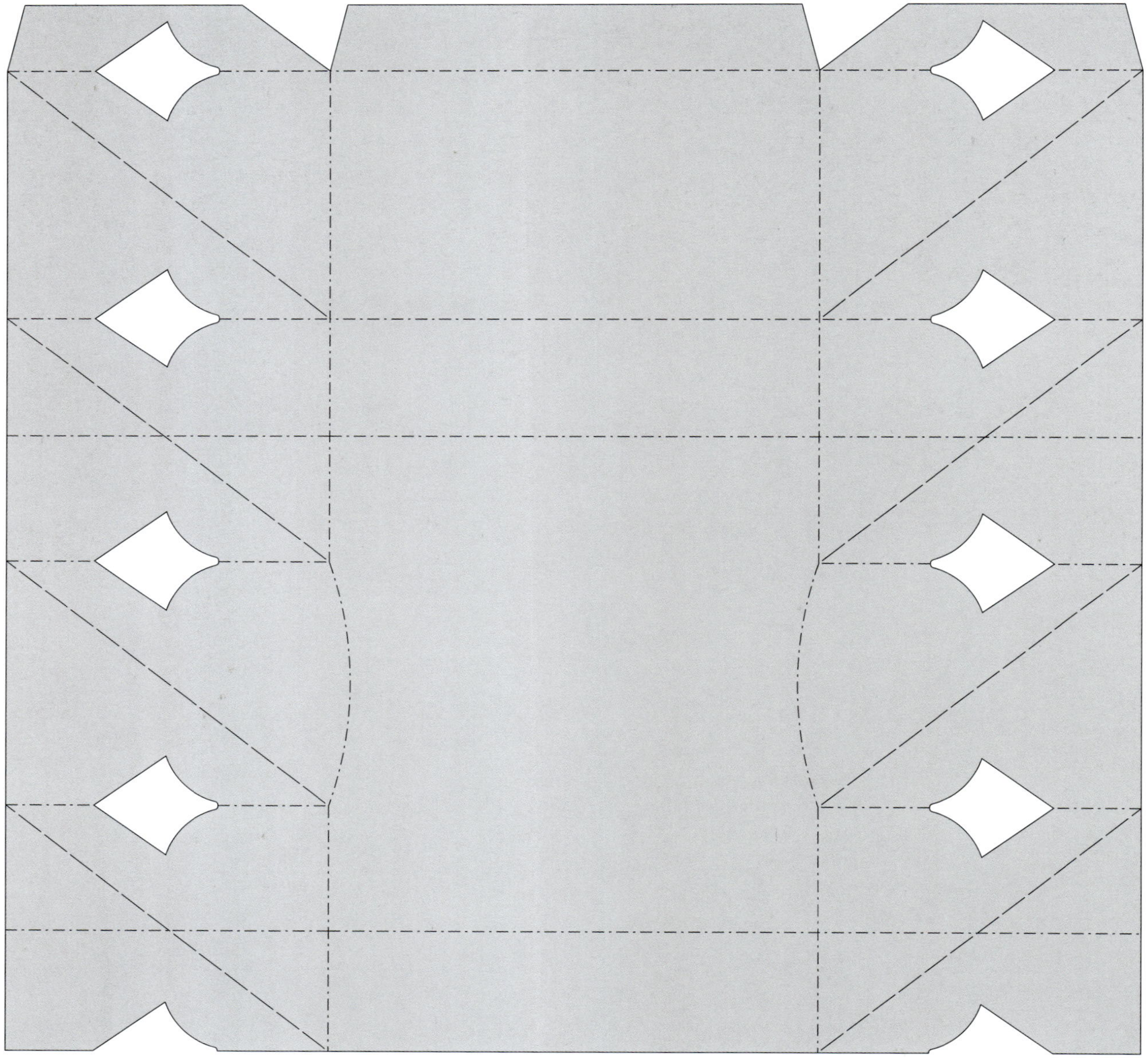

ENVELOPE-BASE CARRY CARTON

This carrier-style carton with contoured and tapered sides is easily flat-packed and erected by pressing the sides, which in turn constructs the press-lock base. Entry is achieved by squeezing the upper sides of the carton, thereby forcing open the aperture at the top. The window display panel is specially cut to follow the contours on the front face of the carton.

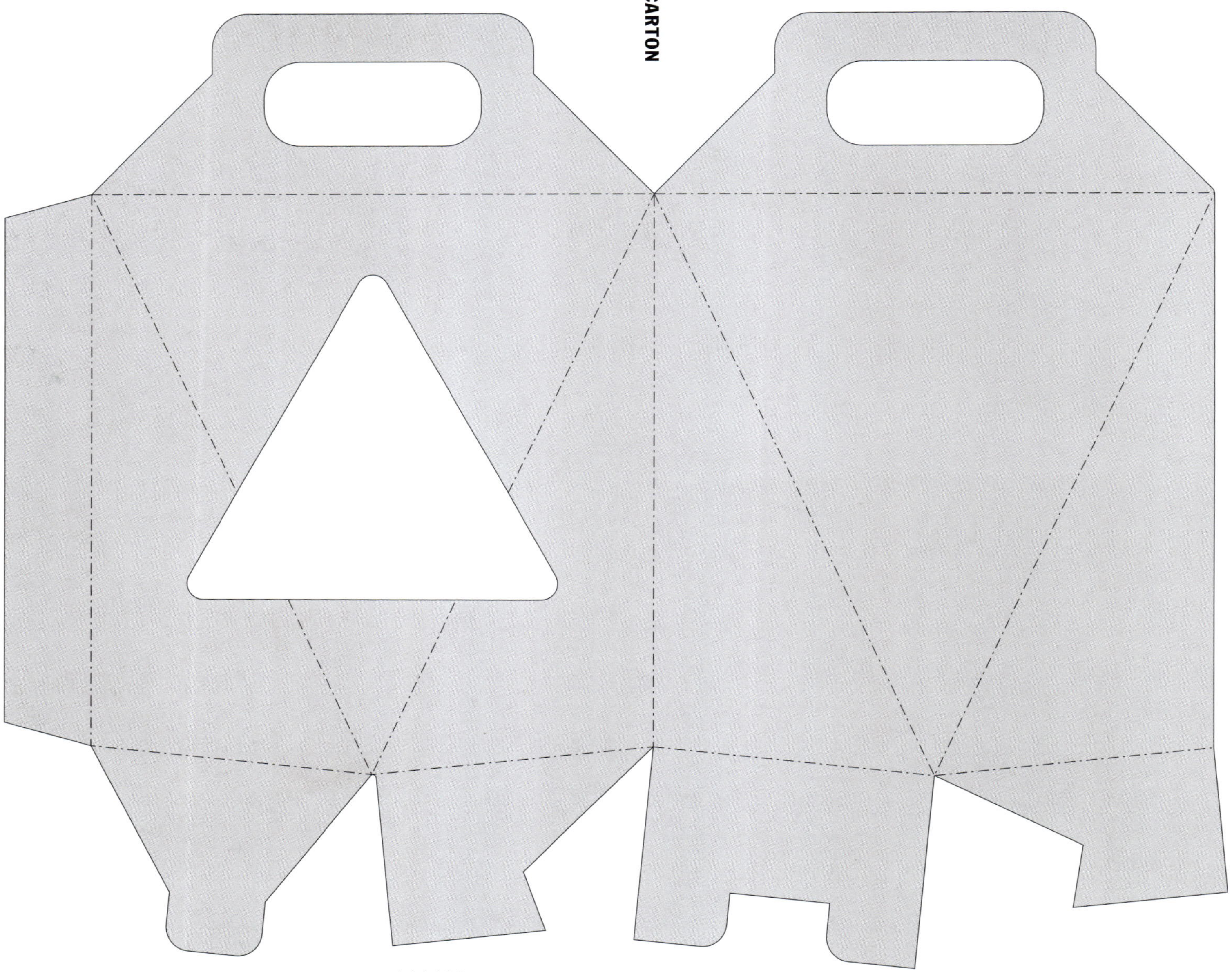

GUSSET-SIDED CARRY CARTON

A cardboard bag with a crash-lock base for use in promotions and for packaging items such as lingerie and clothing. The product determines the required strength of the bag and the strength will depend on the thickness of the card used. This could be used as primary or secondary packaging with the material used greatly affecting the perceived quality of the product or retail outlet that provided the bag.

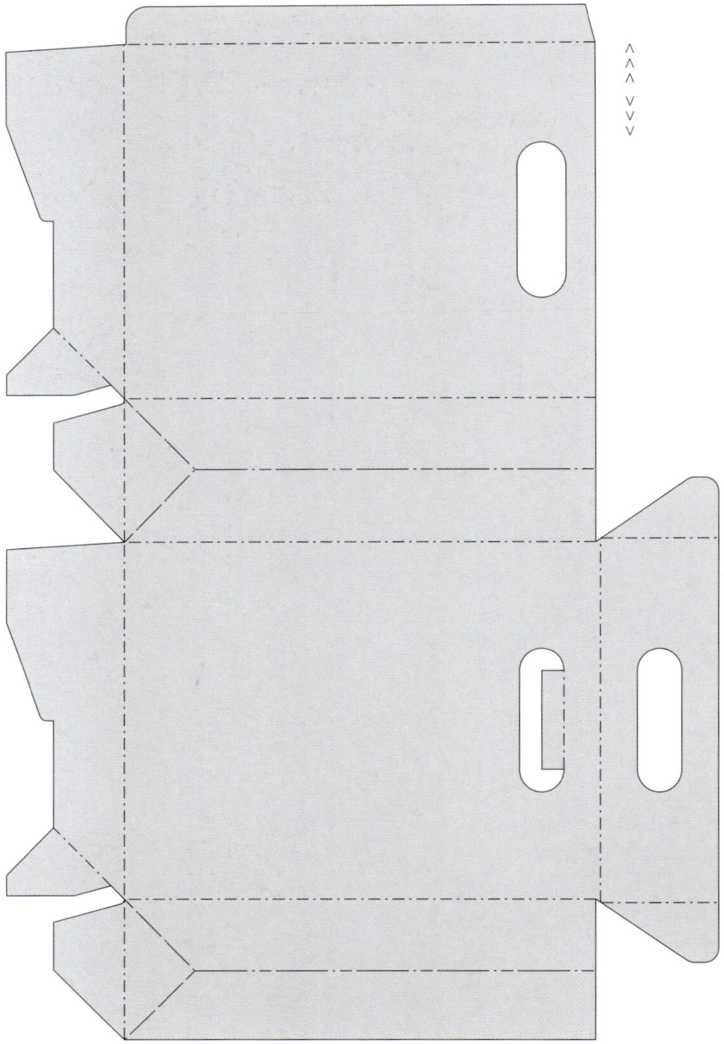

Robor Cartons

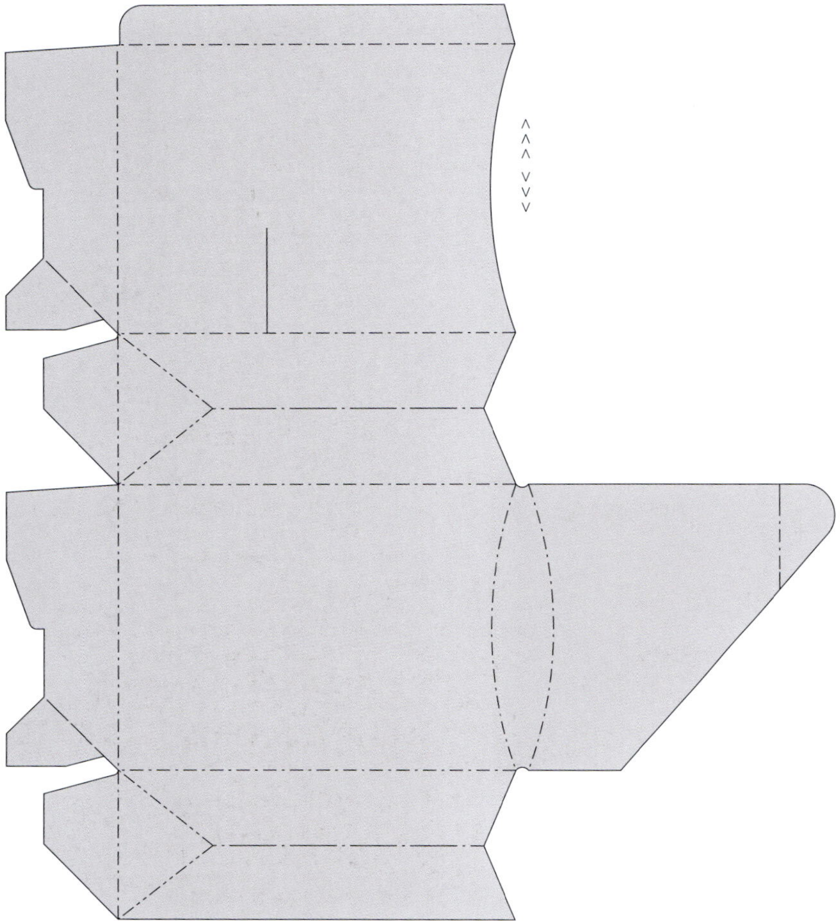

CUSHION-TOP CARRY CARTON

This carton, with a crash-lock base, cushion top, and tuck-flap closure, is designed for gift and luxury items. The use of cord creates a carry-handle and completes this stylish and sturdy carrier. Produced in quality material with the attention to aesthetic detail made a priority, this carton is not merely a transportation pack, but also an attractive package that adds perceived value to the product presented within.

CGM Ltd

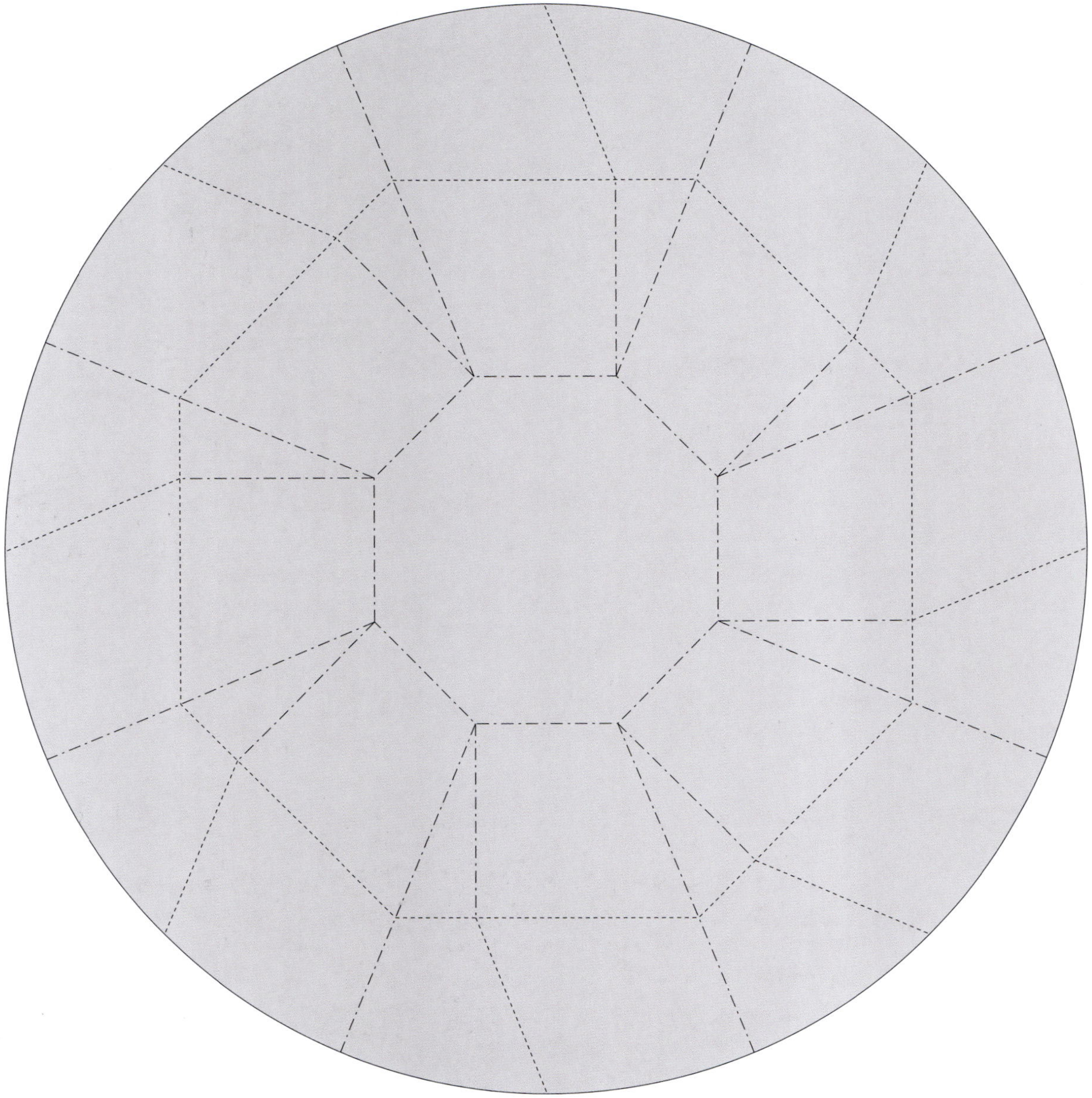

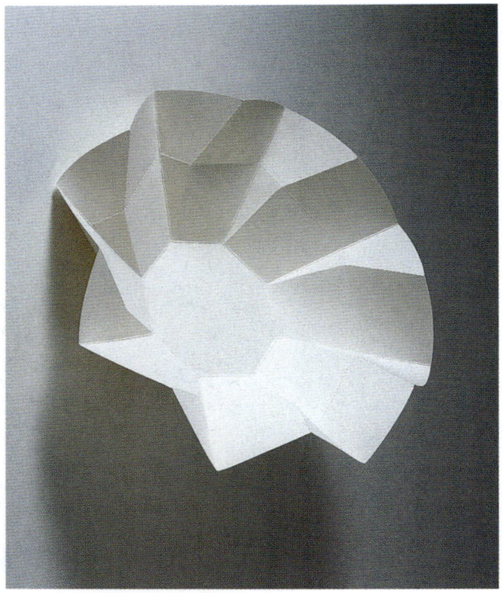

FRILLED CIRCULAR CARTON

An innovative carton derived from the ancient art of origami. This example is octagonal in shape and erects through a series of complex folds and creases to form a rounded box with an attractive frilled closure on the top surface. The complexity of the creasing would demand extra time and skill in production. This is an outstanding piece of paper engineering that creates a box from a single disc of card.

STAR-SHAPED CARTON

An inspirational carton with a star-shaped opening that might provide a decorative container for food, confectionery, or luxury items; it would also be suitable for expensive limited-edition products, or point-of-sale promotions. The use of different colored or textured cards would create interesting variations to this carton. The folding of this complicated piece of origami is illustrated below.

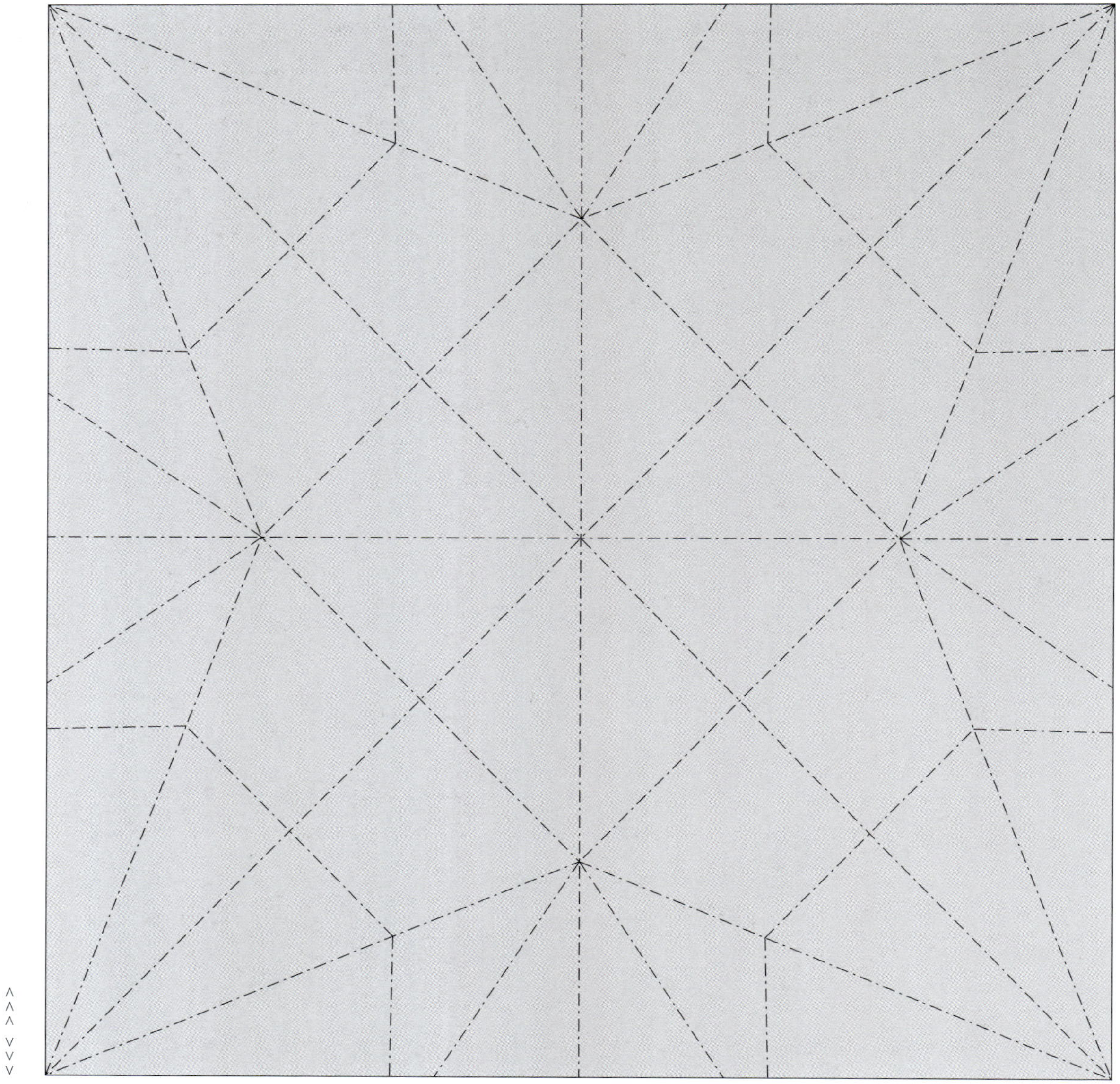

Claire Sheldon, Surrey Institute of Art and Design

DODECAGON CARTON

This carton features a hexagonal lid and base and erects by gluing each of the six tabs. The 12 sides of the carton fold in to form a crystal shape, while the hexagonal lid provides access to the contents. This is a good example of packaging for luxury items such as quality confectionery or other gifts.

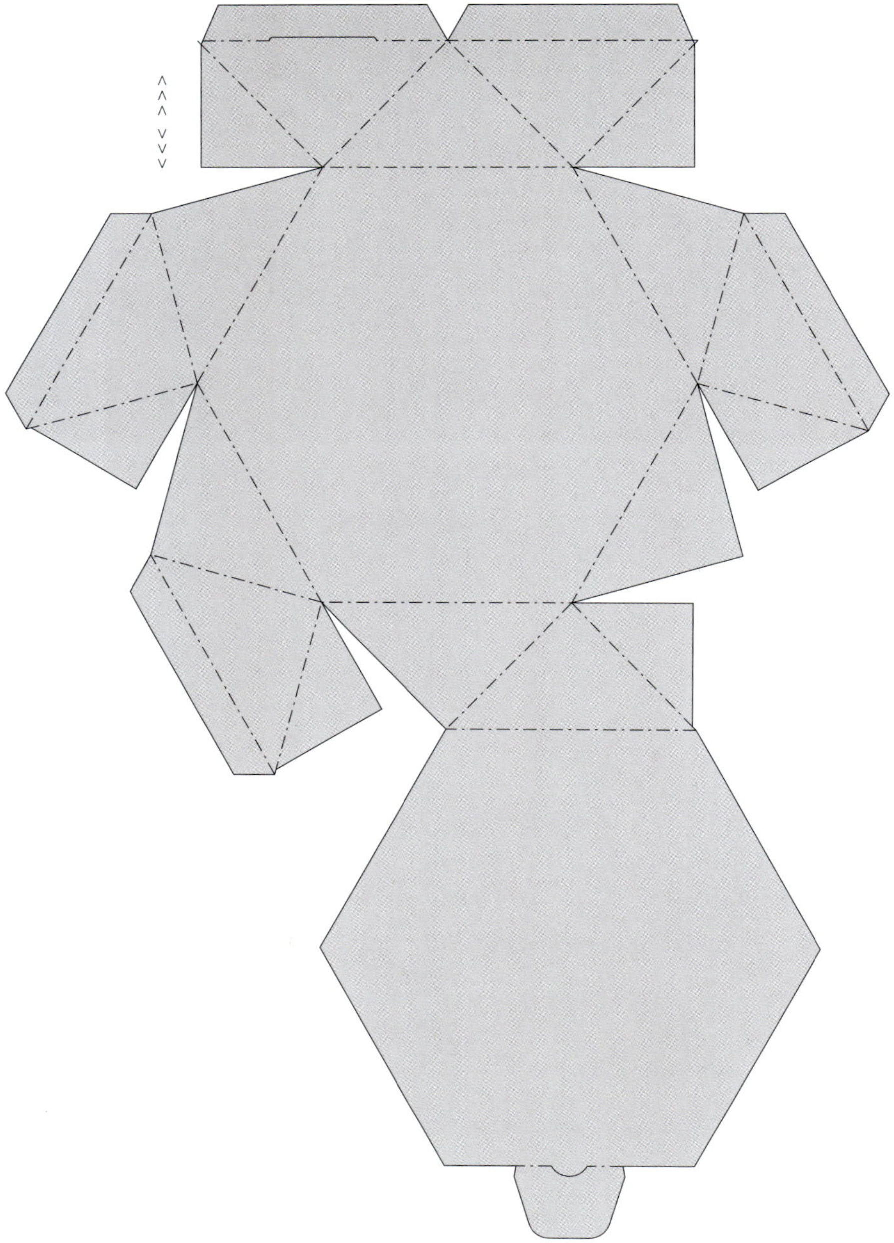

BASE-FLANGED TRAY WITH CURVE-TOPPED LID

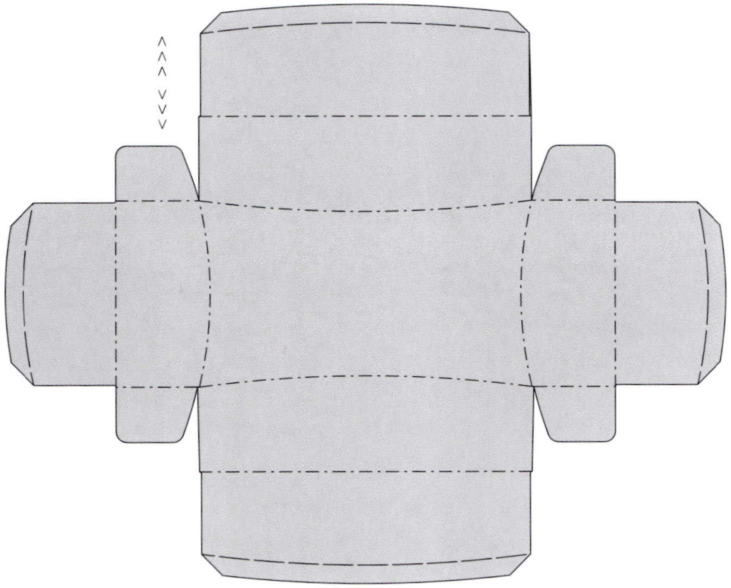

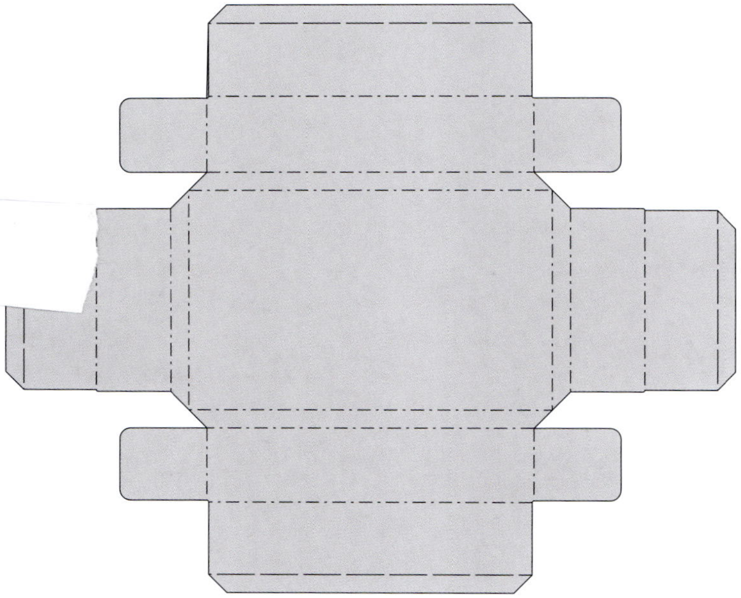

This two-piece carton has a curved, double-walled lid and a flanged base which creates a bevelled lip, giving a degree of luxury or speciality to the package. The flanged base sets the carton apart from others and creates its own space on the shelf. It also provides a platform on which to press down when removing the lid. This might be a particular consideration for the packaging of products for people with special needs.

MULTISIDED TAPERED CARTON

A scored carton with a crash-lock base which can provide three very different shapes. The first is a carton with folded-in sides and a pitched-top lid. The second has a flat top with "ears" on either side which can be folded down, and the third option is the primary design, which has a flat top and reversed tapered triangular sides.

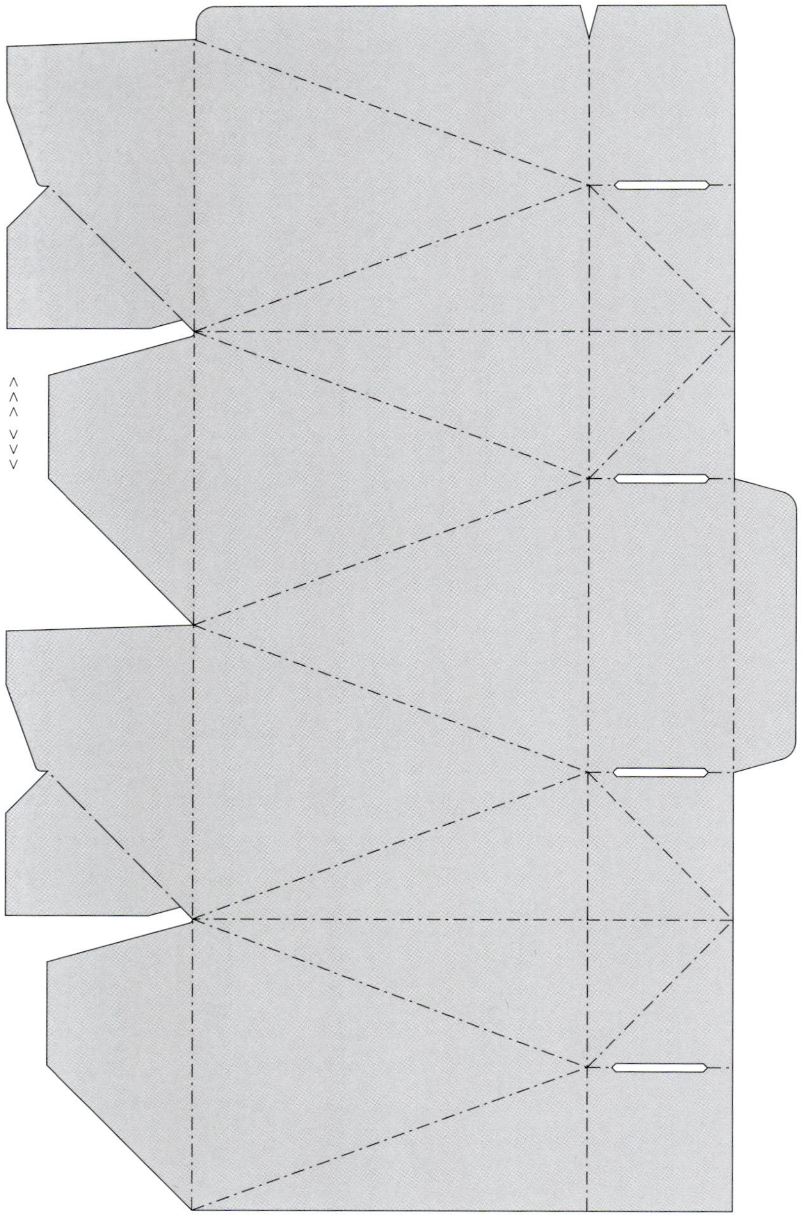

CGM Ltd.

CGM Ltd.

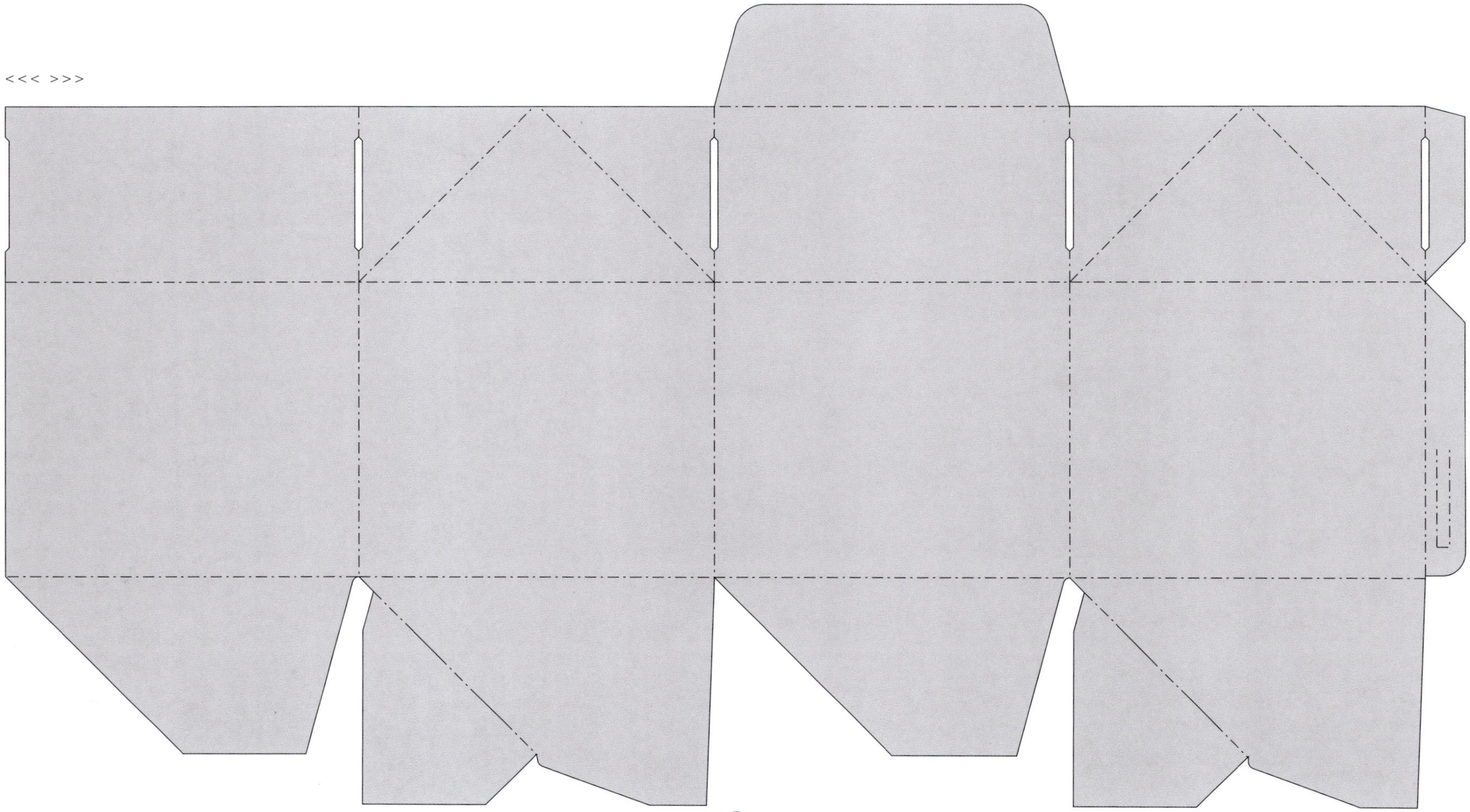

CUBOID CARTON WITH TUCK-IN TOP FLAPS

This simple carton is a cuboid design with a crash-lock base and tuck-in sides and flaps. This is a good example of how the designer can enhance an elementary carton shape with an innovative closure.

WATERTIGHT TAPERED CONTAINER

A cupped container with tapered sides and a larger surface for the top than the bottom, this is ideal as a dehydrated food container with a waterproof-lined inner layer. The design allows water to be added to the contents to provide an instant snack. The design of the folds prevents leakage and requires the use of a multilayered material to obtain the waterproof properties necessary. In this instance solid bleached sulphate board with a polyethylene terathalate (PET) lining has been used.

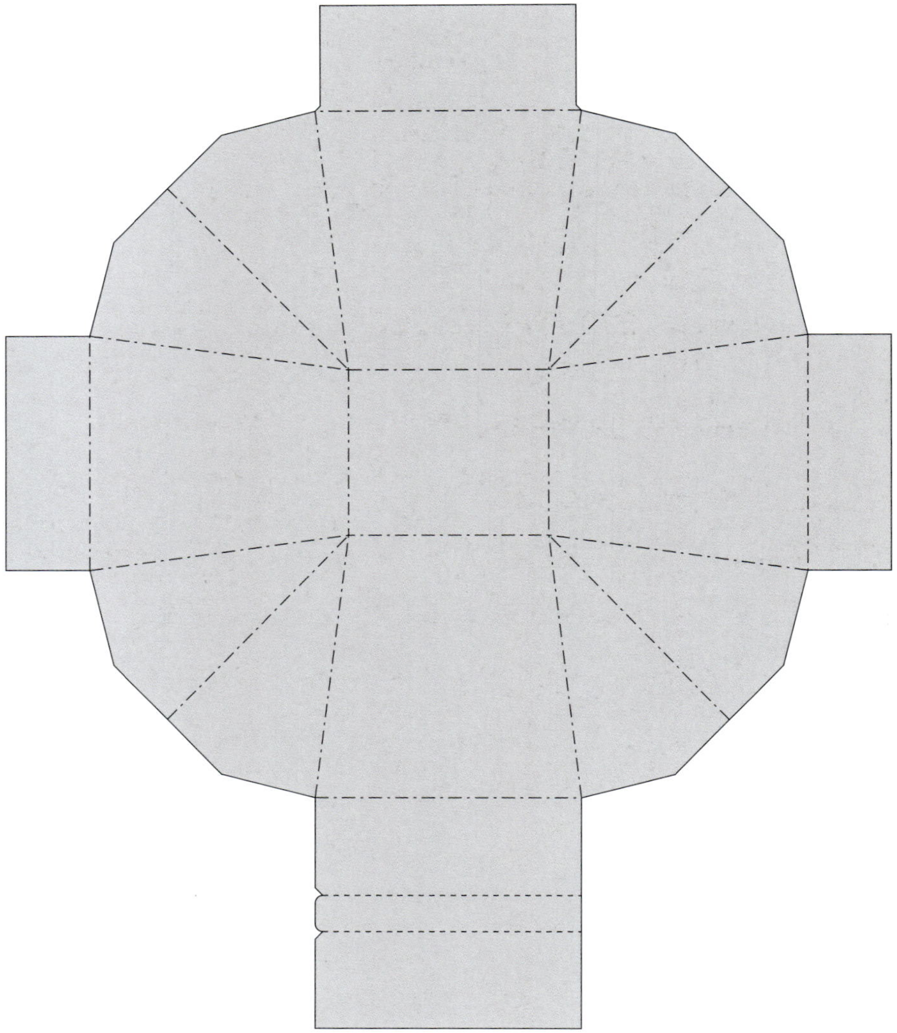

<<< >>>

M.M. Packaging

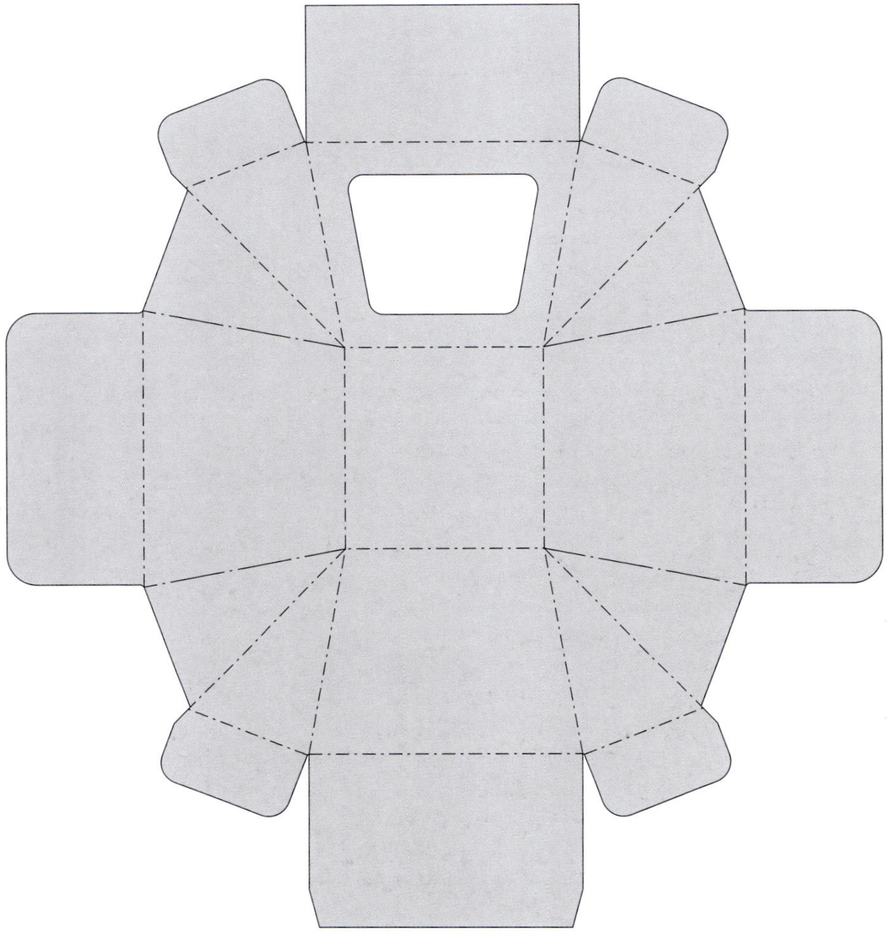

NUGGET CARTON

This is base-glued with tuck-in flaps to form a tapered square carton. The acetate window glued onto the inside of the die-cut shape is an option, all the edges could provide a window to reveal the product within. The score lines can be decreased in number to produce curved sides and corners.

ANGLE-SIDED TRAY WITH TUCK TOP

A tray that creates an image of thickness and rigidity by using tapered hollow sides, this tends to be used as a container for confectionery and gift items. Special locks clip together on the sides of the corners to support the structure. The lid with dust flaps is optional, but if removed this fourth side should be adapted to replicate the remaining sides.

CGM Ltd.

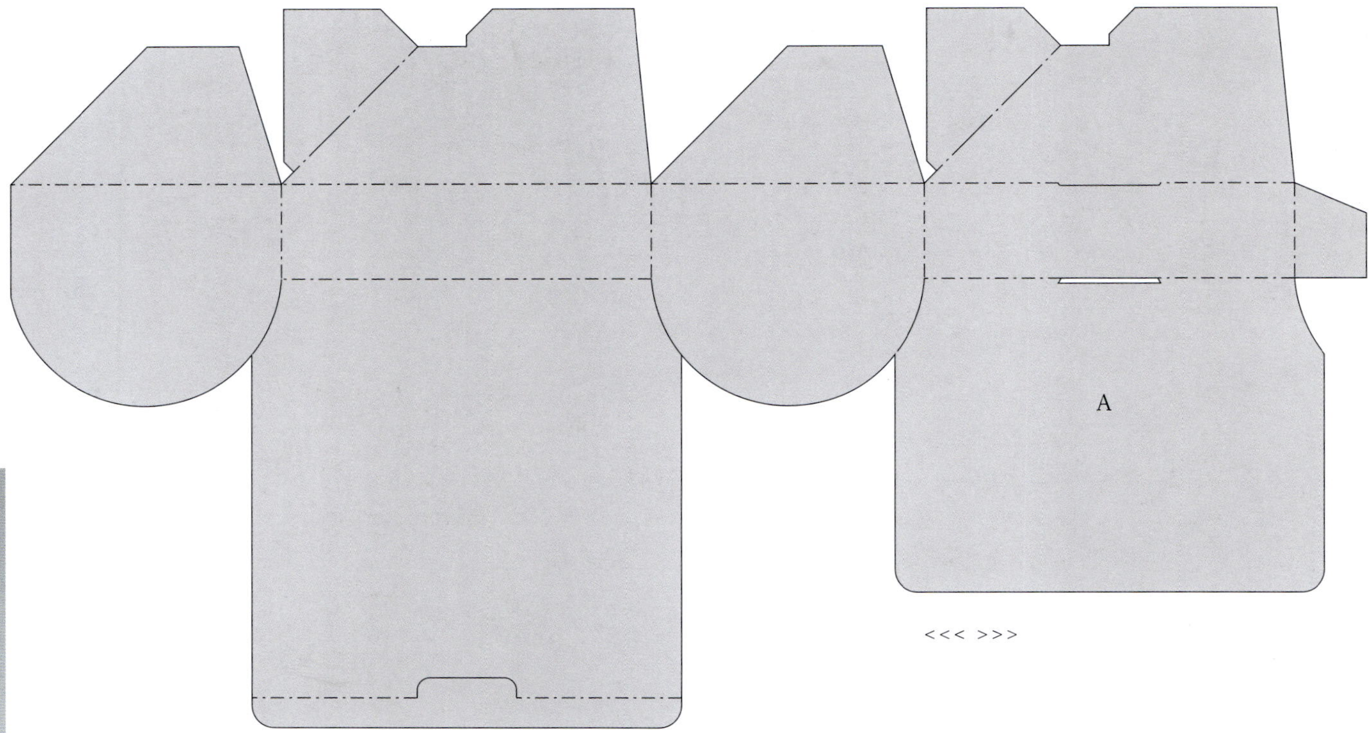

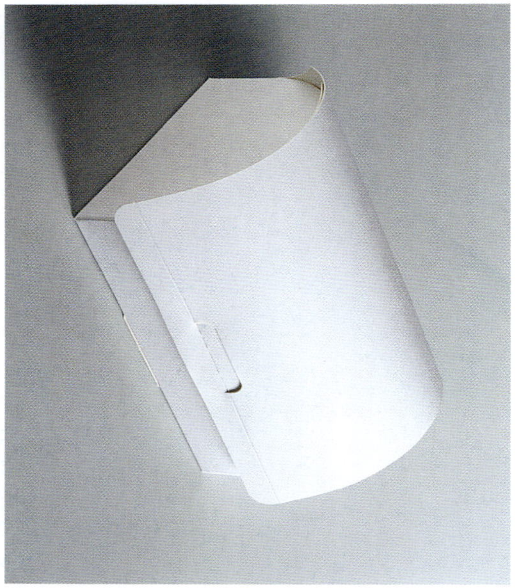

ARC-TOP CARTON

An example of how a standard four-sided carton can be given a more attractive appearance using a curved, semicylindrical lid. Flap A provides greater strength for the lid and protection for the product contained. The closure between the sides and the lid is not very tight, therefore this carton would not be appropriate for products requiring high levels of protection. Possible uses include take-out cartons for decorative food products, whereby the raised lid would provide additional space for the product inside.

PITCHED LID AND TAPERED SIDES

This simple design is a basic container with tapered sides and a narrow base. Access to the product is through the pitched lid at the top of the carton, which in this case demonstrates how a handle can be incorporated into the design as an option to enhance the performance of the package.

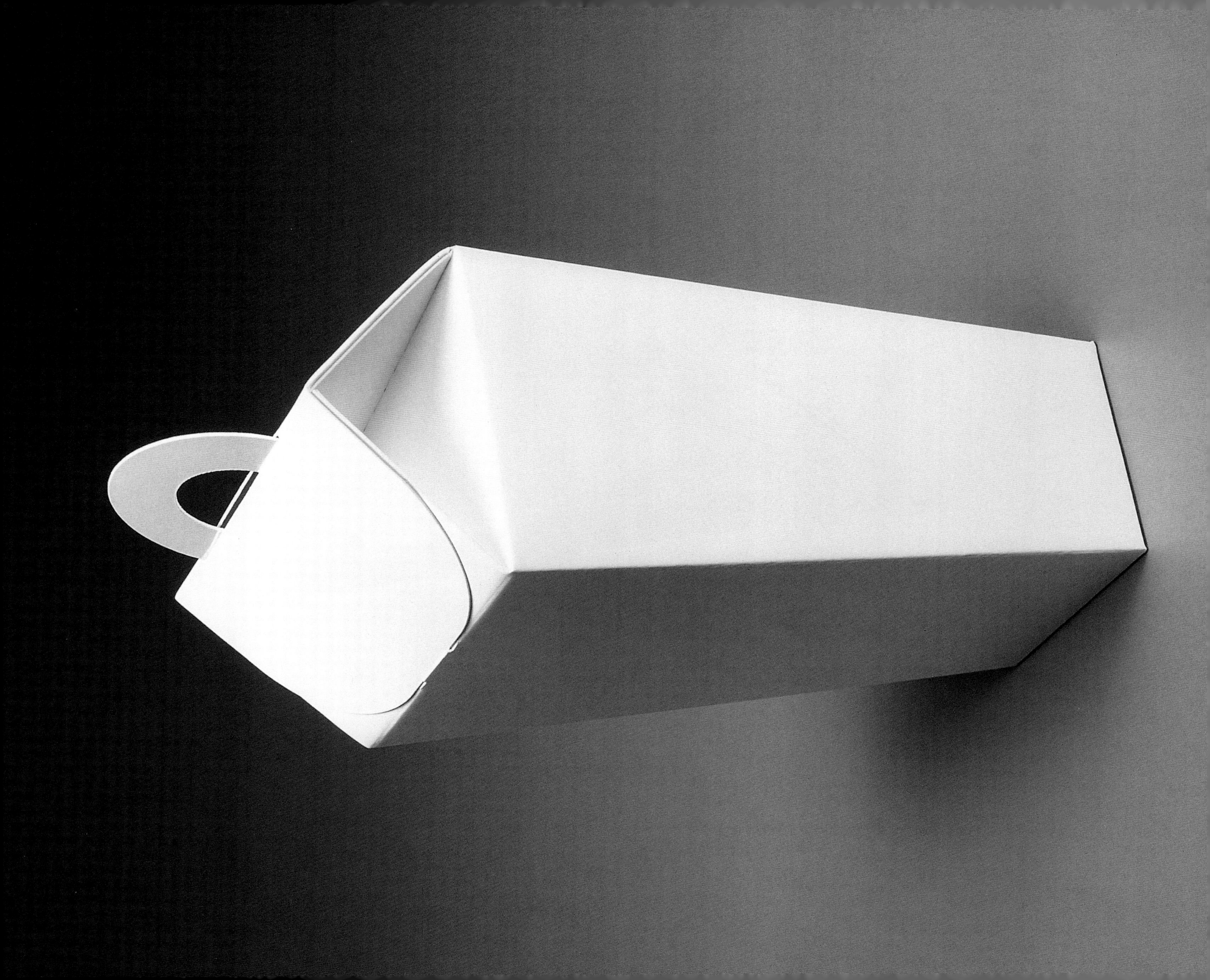

Robor Cartons

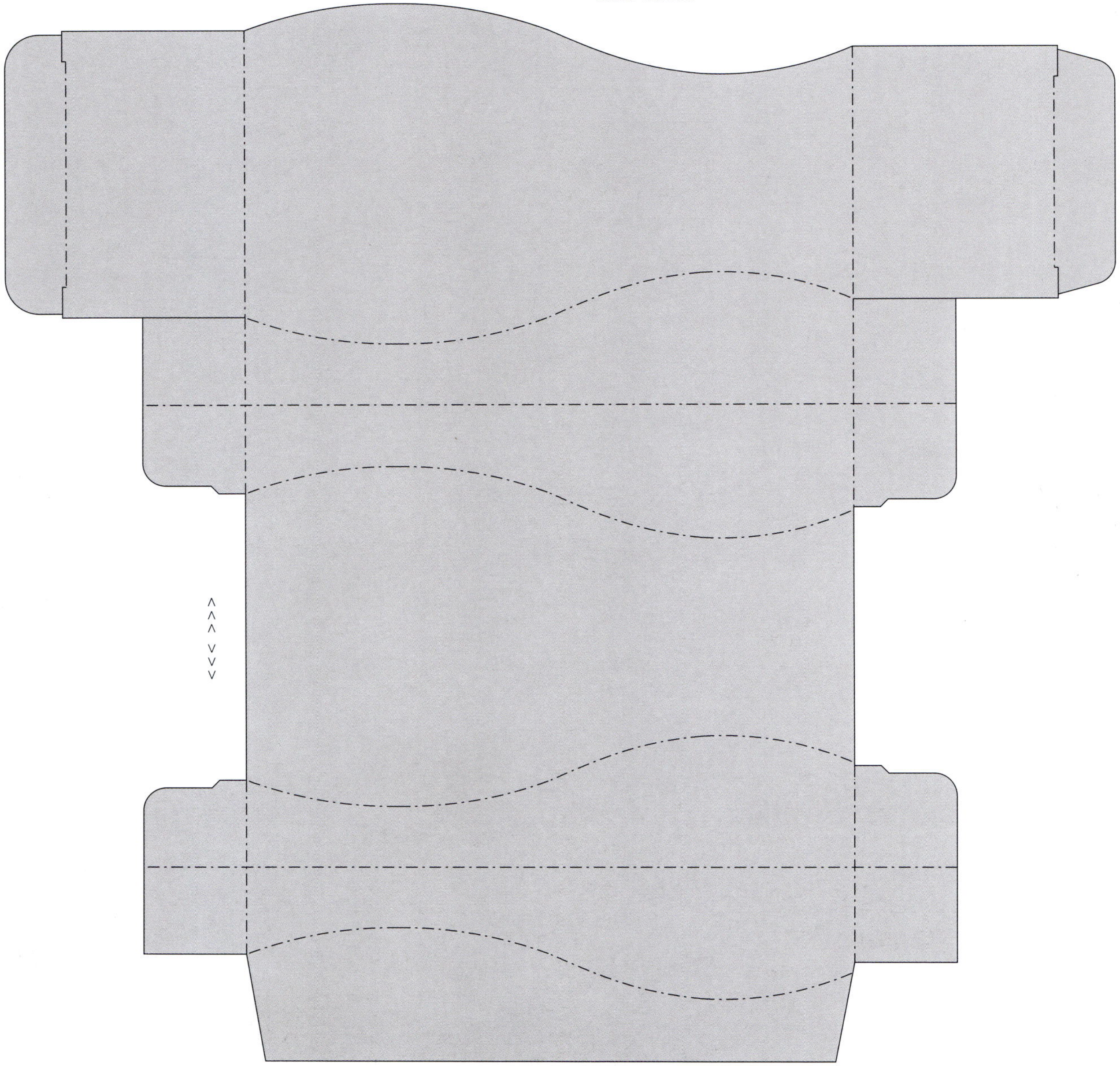

MULTICURVED CARTON

This tuck-end carton makes good use of curved surfaces to create an attractive upright container. The curves and contours provide a visually appealing form that would be an ideal container for fabric, gift items, or confectionery. Glue is required in the construction of this container but it does flat-pack, the container is erected by pressing the sides.

PILLOW PACK

Introducing contours to cartons produces interesting shapes which are visually inspiring. They can also be adapted to accommodate a Euroslot or pull-strips. This carton is good for the storage of soft items such as clothing or a number of small pieces such as confectionery or a jigsaw; it is easily erectable and can be stored in a flat-packed state.

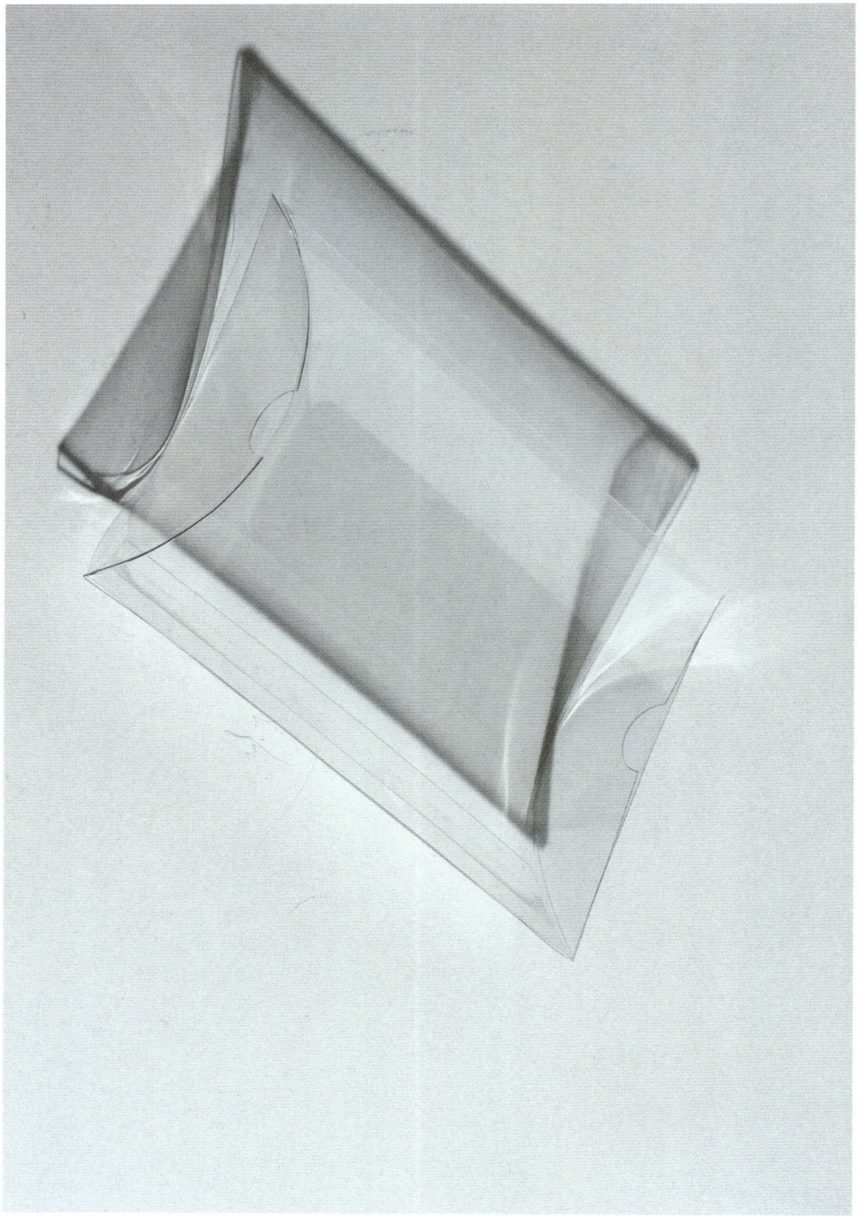

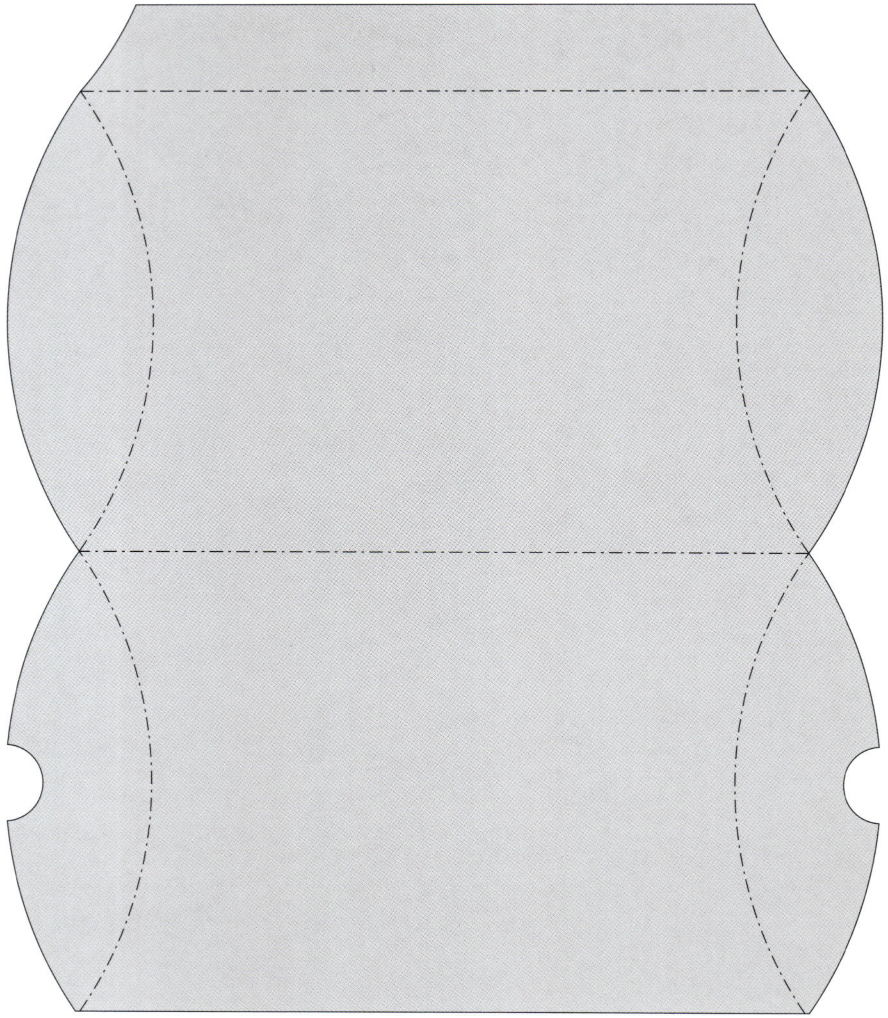

<<< >>>

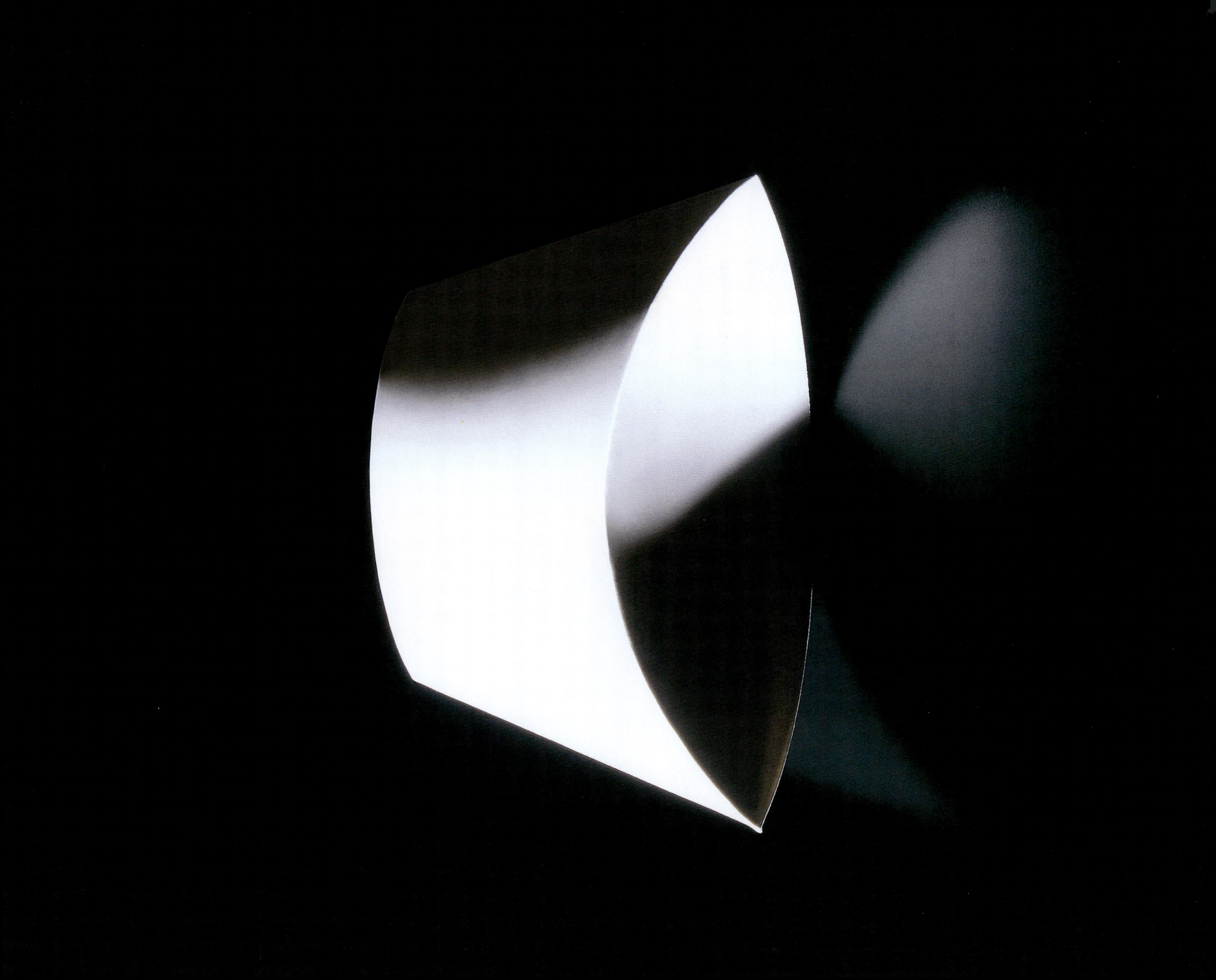

TAPERED PEDESTAL CARTON

A novelty container for presenting and displaying products, this comes in two sections. The four-sided base container has tapered edges, pedestal feet and a clear acetate display lid, shaped in a pyramidal form for stylish presentation of the product contained within. The use of acetate in the lid provides a window, which hinges along one side of the base.

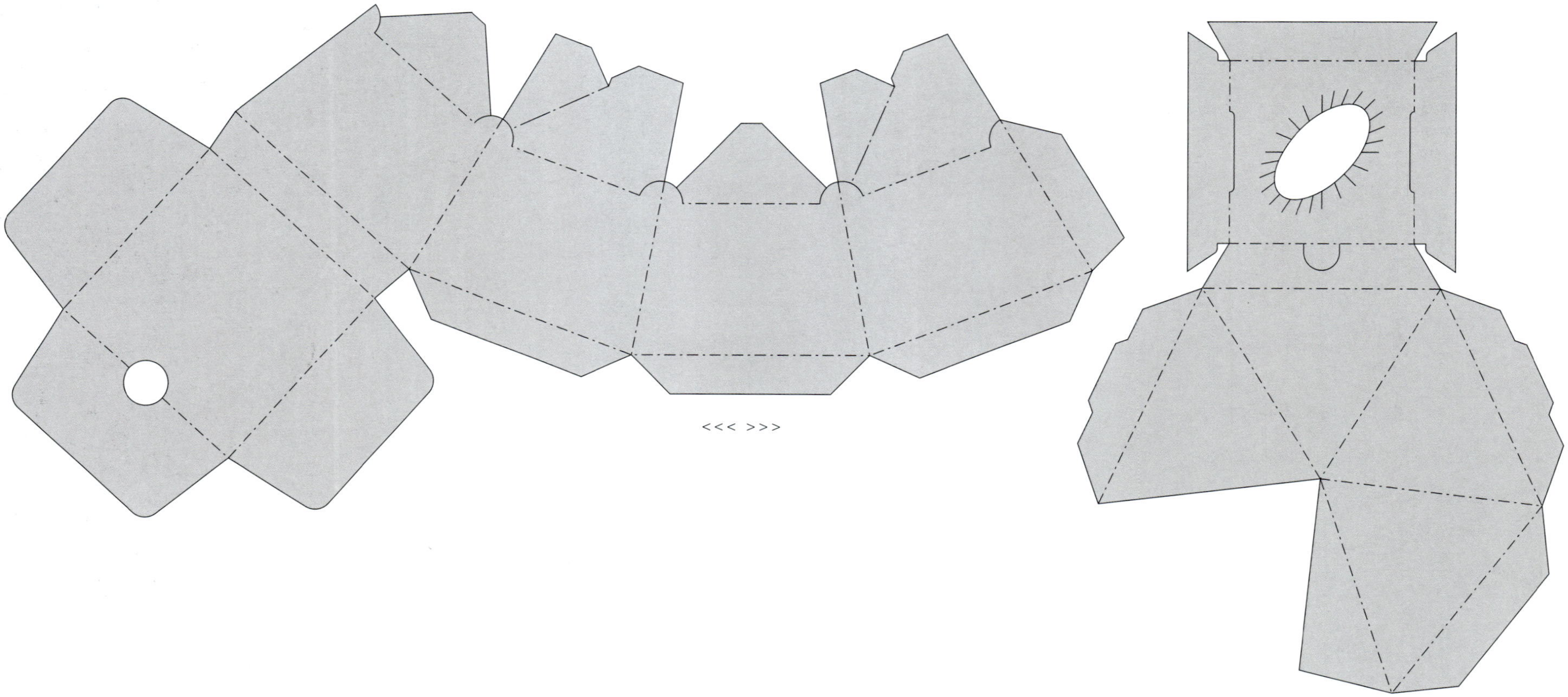

Robor Cartons

Robor Cartons

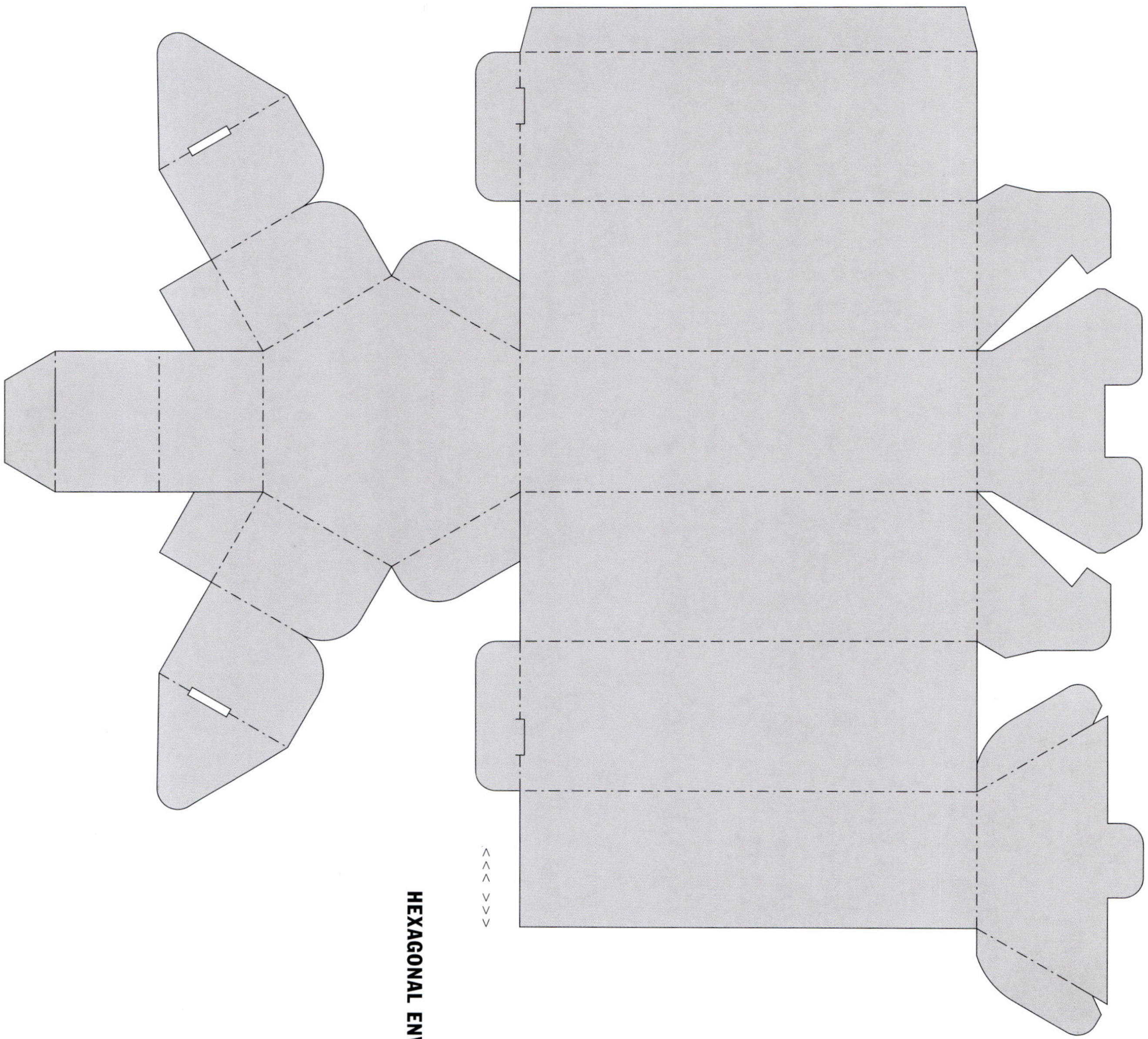

HEXAGONAL ENVELOPE-BASE CARTON WITH TAB-LOCKED LID

This hexagonal carton displays an ingenious method of closure, whereby the walled lid locks in position using tab-locks on the lip of the body. The lid requires no gluing in its construction as a series of folds create a strong and robust structure. An unusually-shaped and sturdy container suitable for a wide range of products that might benefit from having a strong resealable closure system.

CLOVER LEAF CLOSURE CARTON

A slotted disc-top closure carton with crash-lock base. The notched circle segments interlock with the slits to form a clover leaf design. This design is suitable for gift products where the process of opening the package is interesting and revealing. The shapes of the interlocking segments could be varied to provide a number of different shapes when closed.

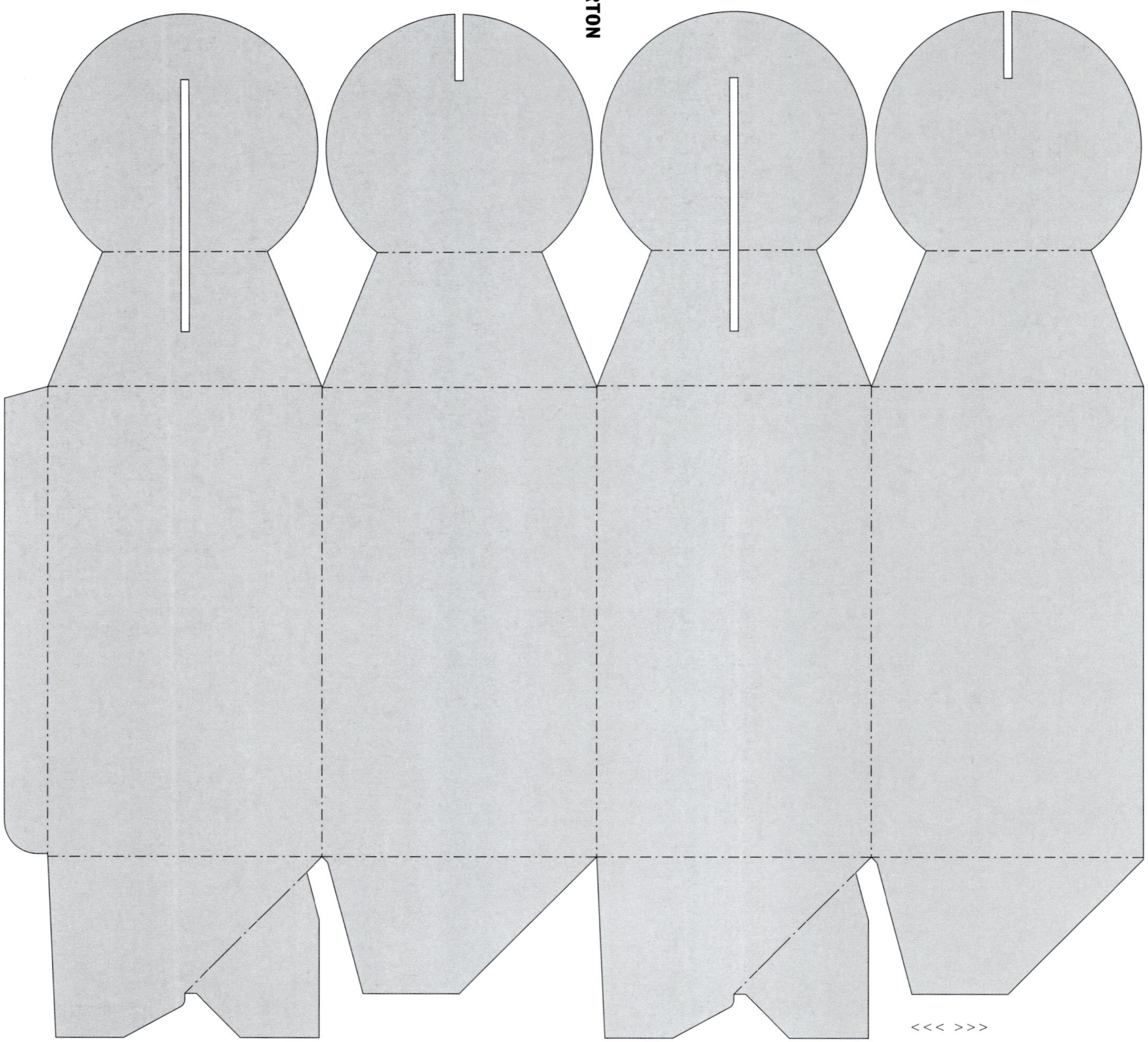

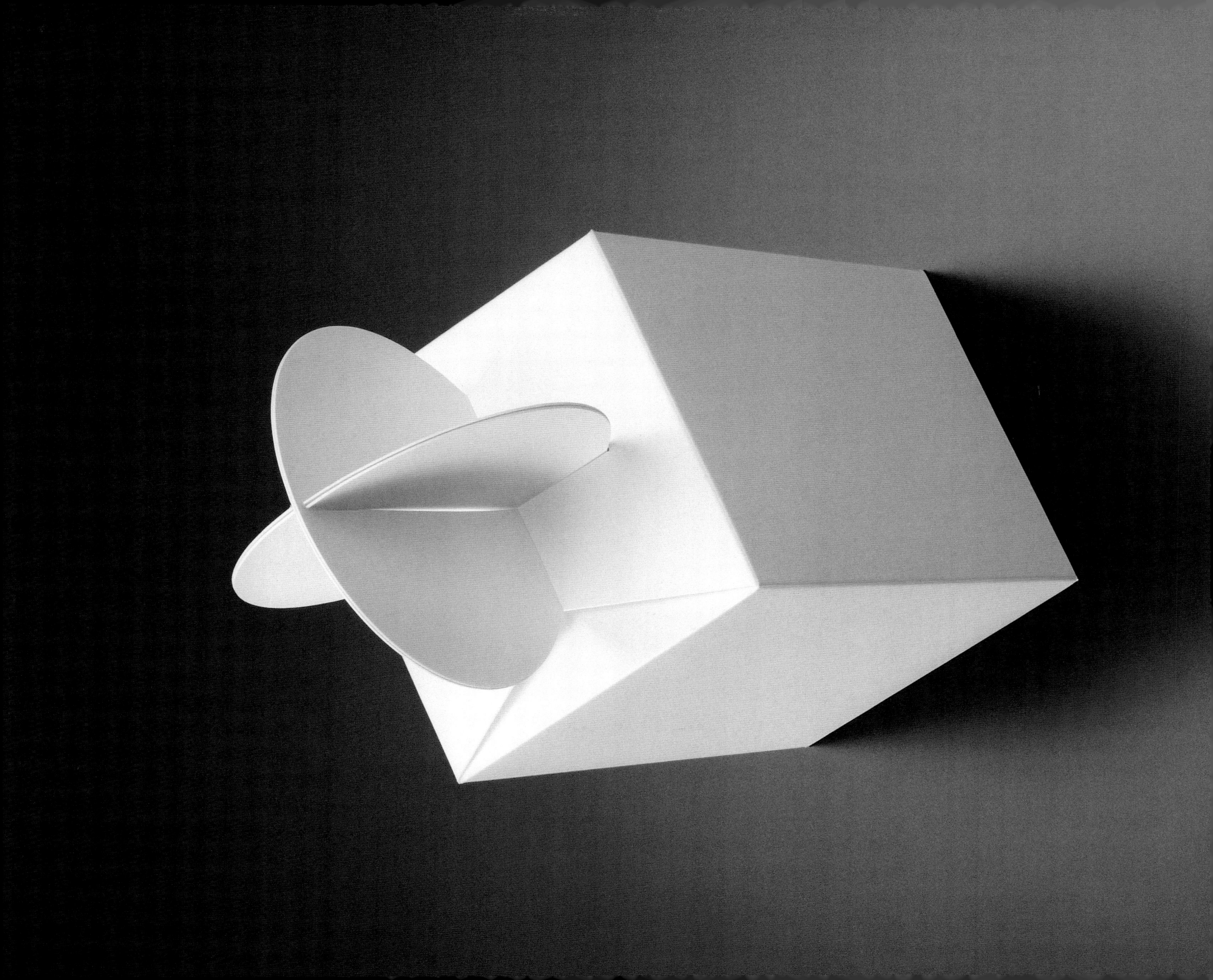

TUCK-TOP HEXAGON

A six-sided carton with flaps which are reverse-tucked, this allows for a square aperture in a hexagonal shape. This might be used in instances where the product requires specific containment within the package, despite the lid being open. The score lines can be softer to reduce the angular appearance of the sides.

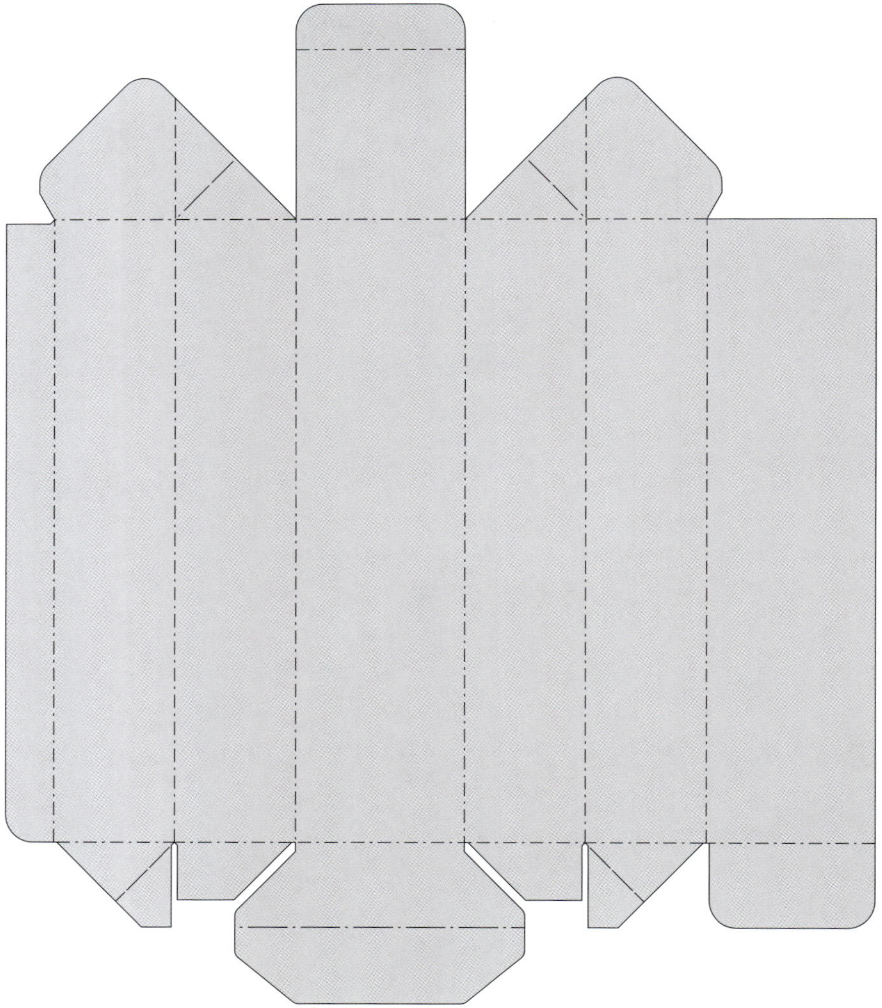

<<< >>>

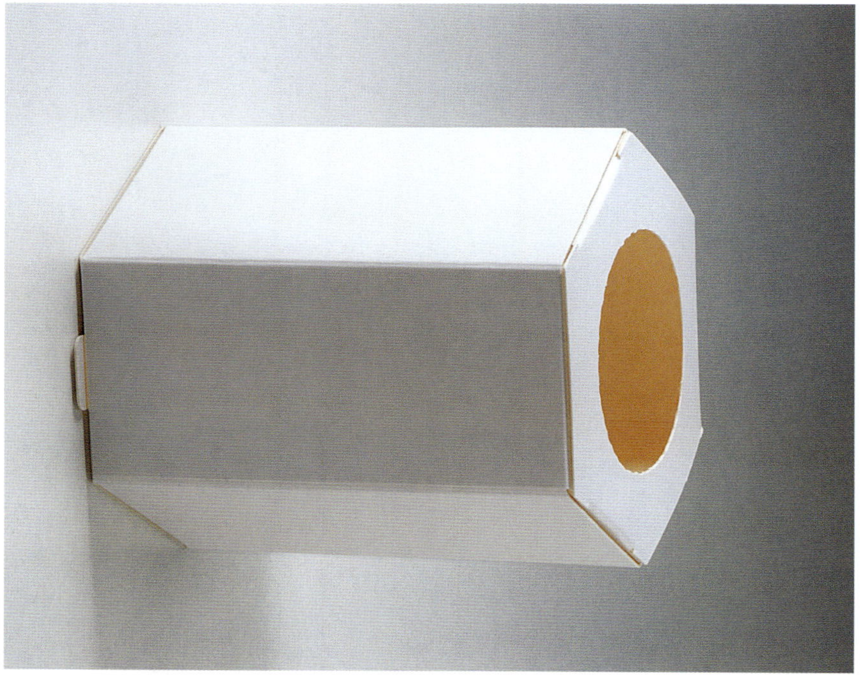

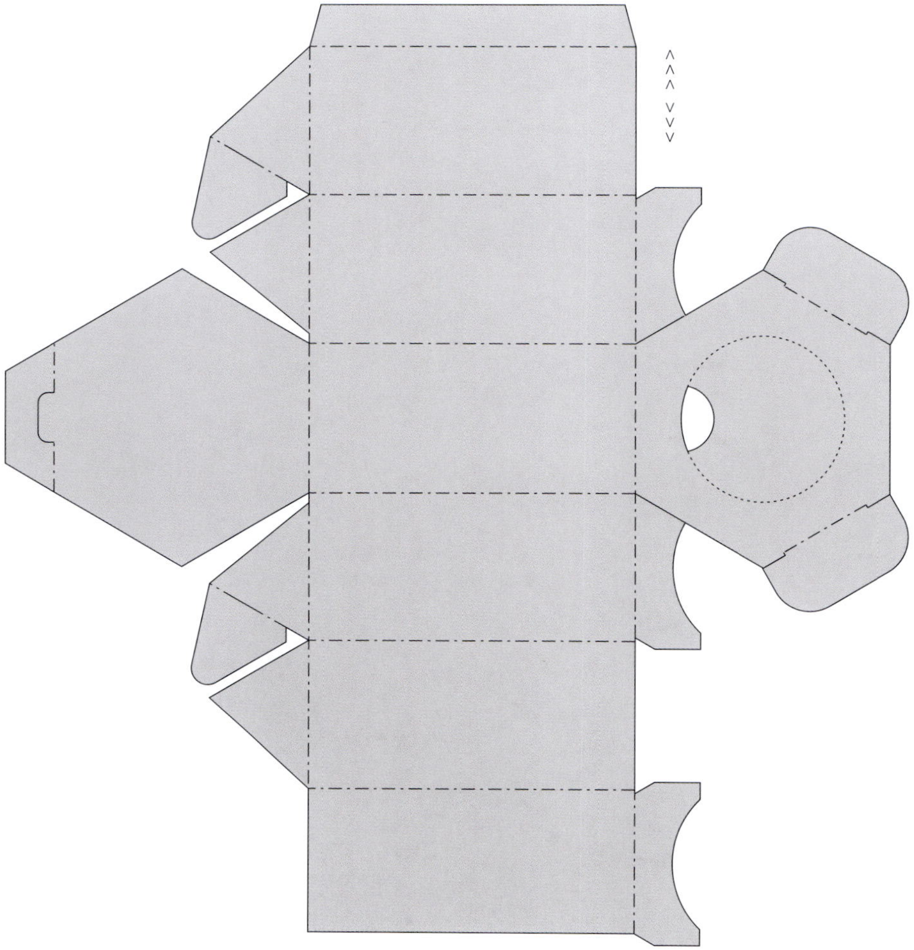

HEXAGONAL TUCK-TOP CARTON

A six-sided container with a glued crash-lock base and push-down bottom flap which is an adaptation of a bellows (gusset) tuck. The top flap straight tucks have slit-lock tucks and the shaped perforation on the top face allows the contents to be accessed. It can be used for bulk display of tall objects such as pens, pencils, or even bottles. With six equal sides there is a good opportunity for the application of graphics.

TWO-PACK INTERLOCKING CARTON

Designed to hold a fragile item, such as a glass bauble, this two-pack carton is constructed from a single piece of card using no adhesive. The complex folds and slots allow a second, identical carton to interlock with the first in order to create a four-pack carton. This type of design reduces the costs of machining and cutting found in other designs of multipacks. The adaptable profiles of the die-cut holes determine the shape that the product will fit into.

<<< >>>

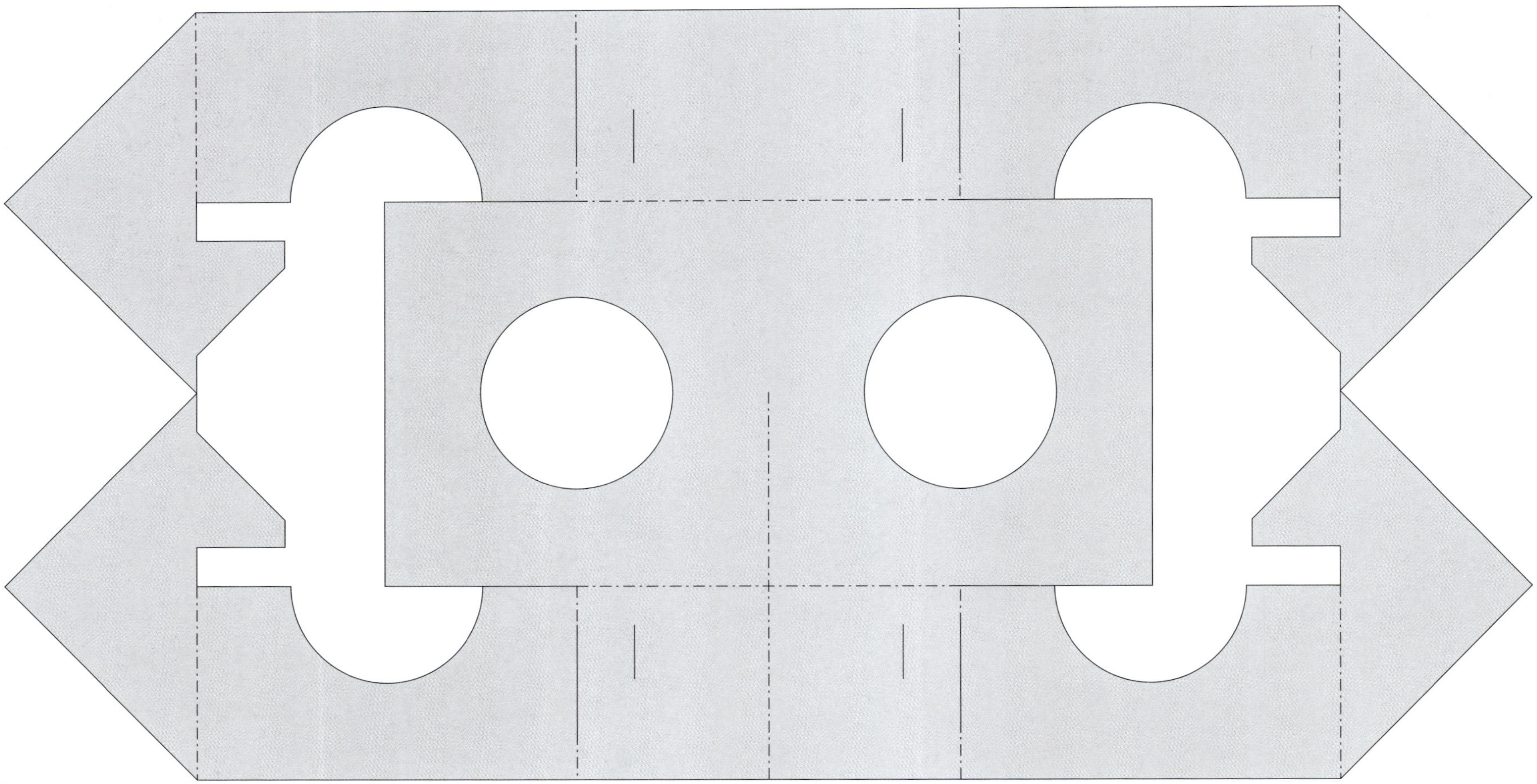

Andrew Robson, Surrey Institute of Art and Design

Danapak

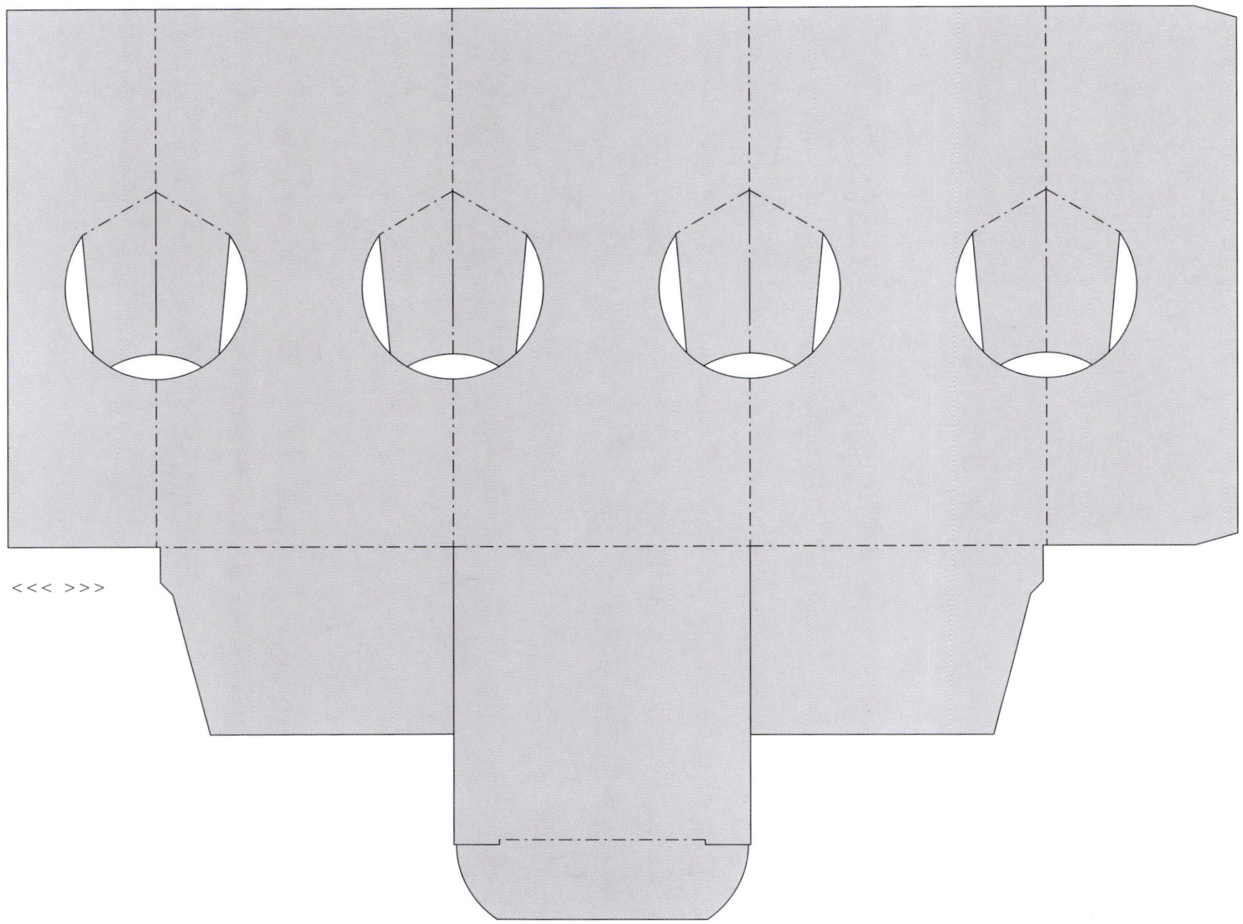

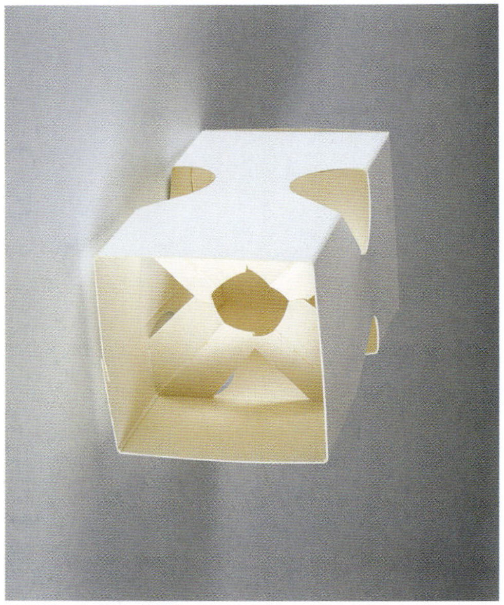

LIGHT BULB CARTON

A one-piece carton which provides storage and safe handling of a fragile item. Cut-out sections within the part-cut circles are creased to fold inward and provide support for the bulb inside. This is an example of a good design feature preventing the need for inserts or additional parts; this principle can be applied to fragile, round, pharmaceutical products. The carton can be manufactured from card, corrugated board, or acetate, but is dependent on the weight of the product and its resistance to the exterior environment.

CONFECTIONERY EGG CARTON

A popular carton for Easter eggs. Constructed from a single sheet, the back of the carton is larger than the front. The front flaps fold inward to be held securely in place by dagger locks; further stability is achieved by fold-in tabs on the side top flaps. The die-cut holes in the front hold the products securely against the back wall of the carton and the dagger locks provide security for the consumer when used in conjunction with a tamper seal that breaks when opened. Display windows can be any shape or size, but the larger the window, the weaker the structure will become.

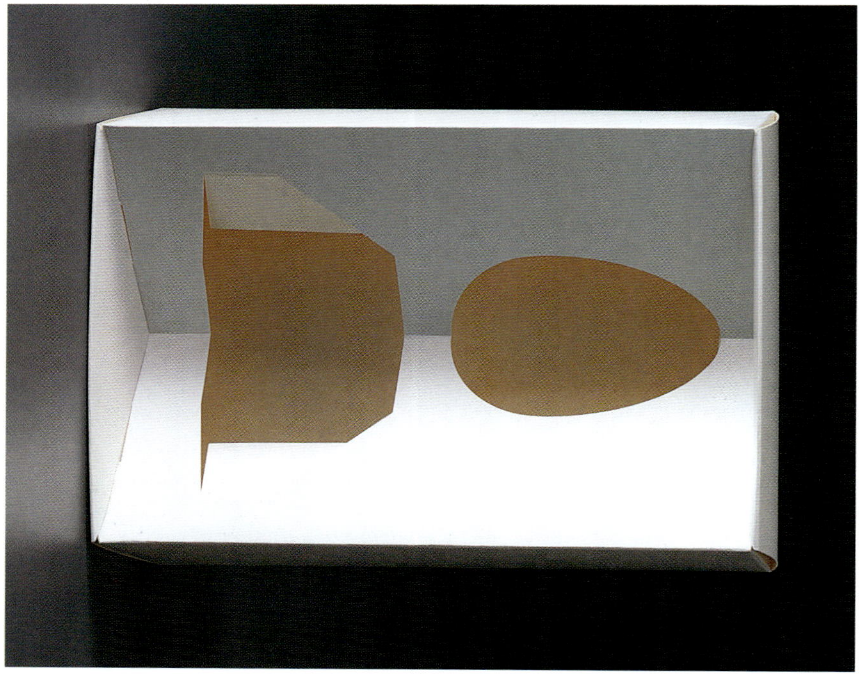

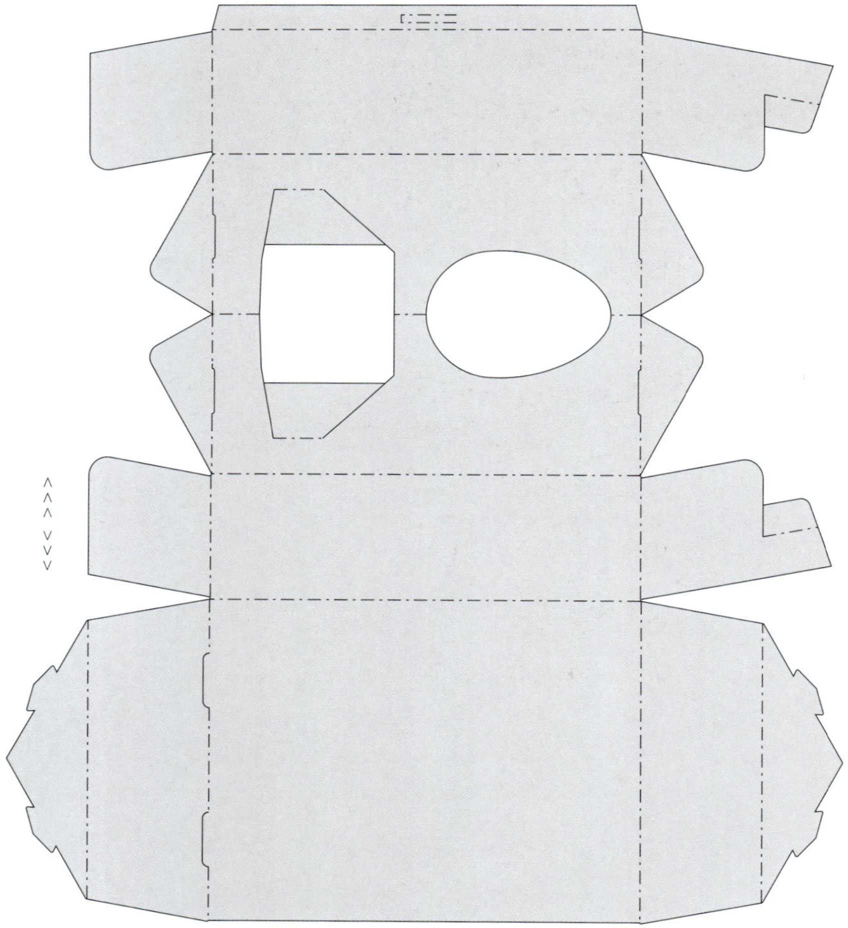

Berkshire Printing Co. Ltd.

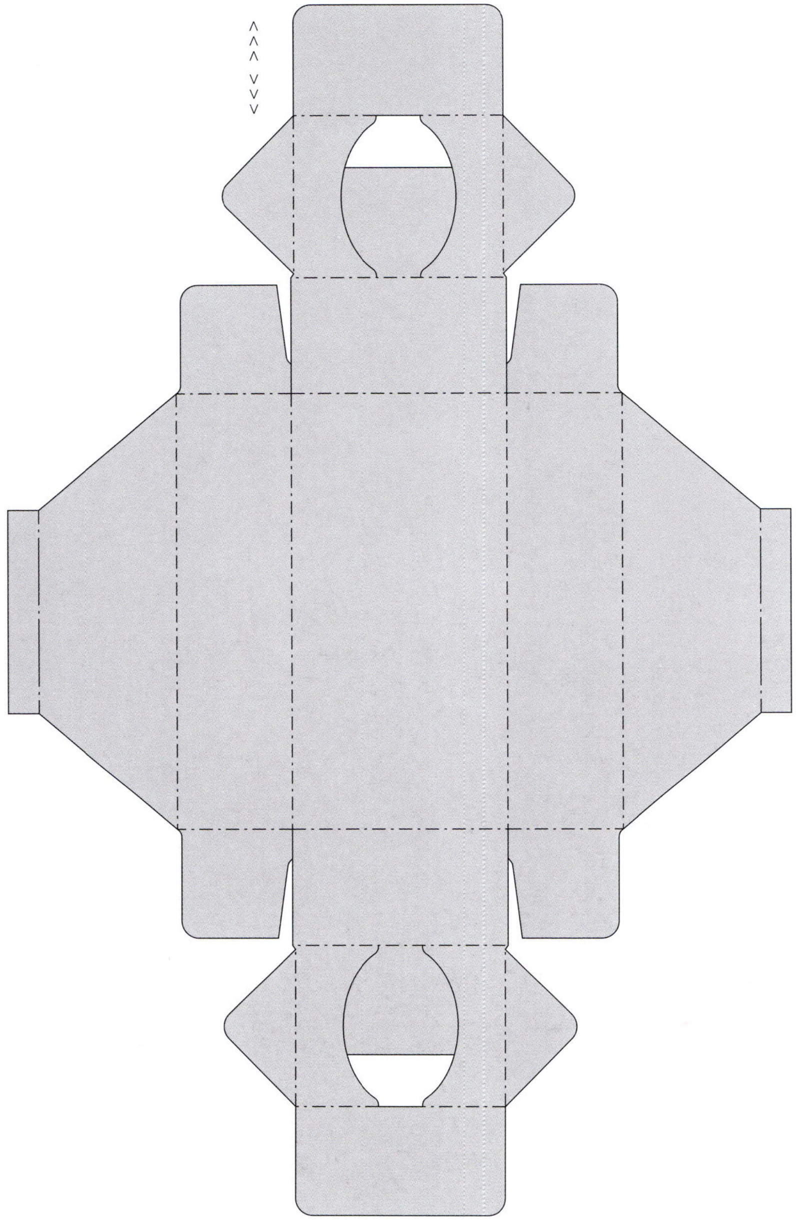

FRAMED-VIEW TRAY

A rectangular carton with a cut-out frame, which enables consumers to see and touch the product contained in the package. Recent examples have encased cosmetics, bottles, and sports equipment. The designer needs to consider the effect of exposure on the product. If, for example, it is sensitive to touch, pressure, or direct light, this package might be unsuitable. An ideal package for focusing attention on the product itself.

CIRCULAR ROLL PRODUCT CASE

A single glued flap made from a single piece of card forms four-sided closure for a roll product. The cut-out flap on the central circular section is used to hold the roll in place simply by being pushed in, preventing the product from rolling out. An ideal point-of-sale unit for rolls where it is difficult to provide consumers with flat areas of graphics.

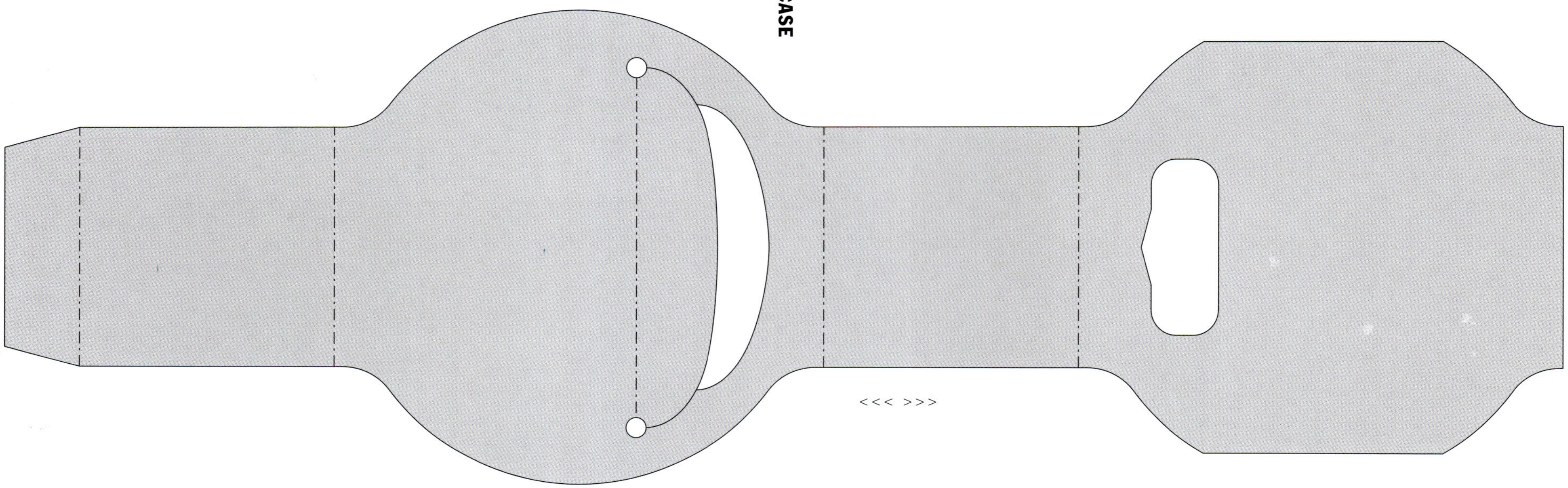

G.C.M. Boxmore

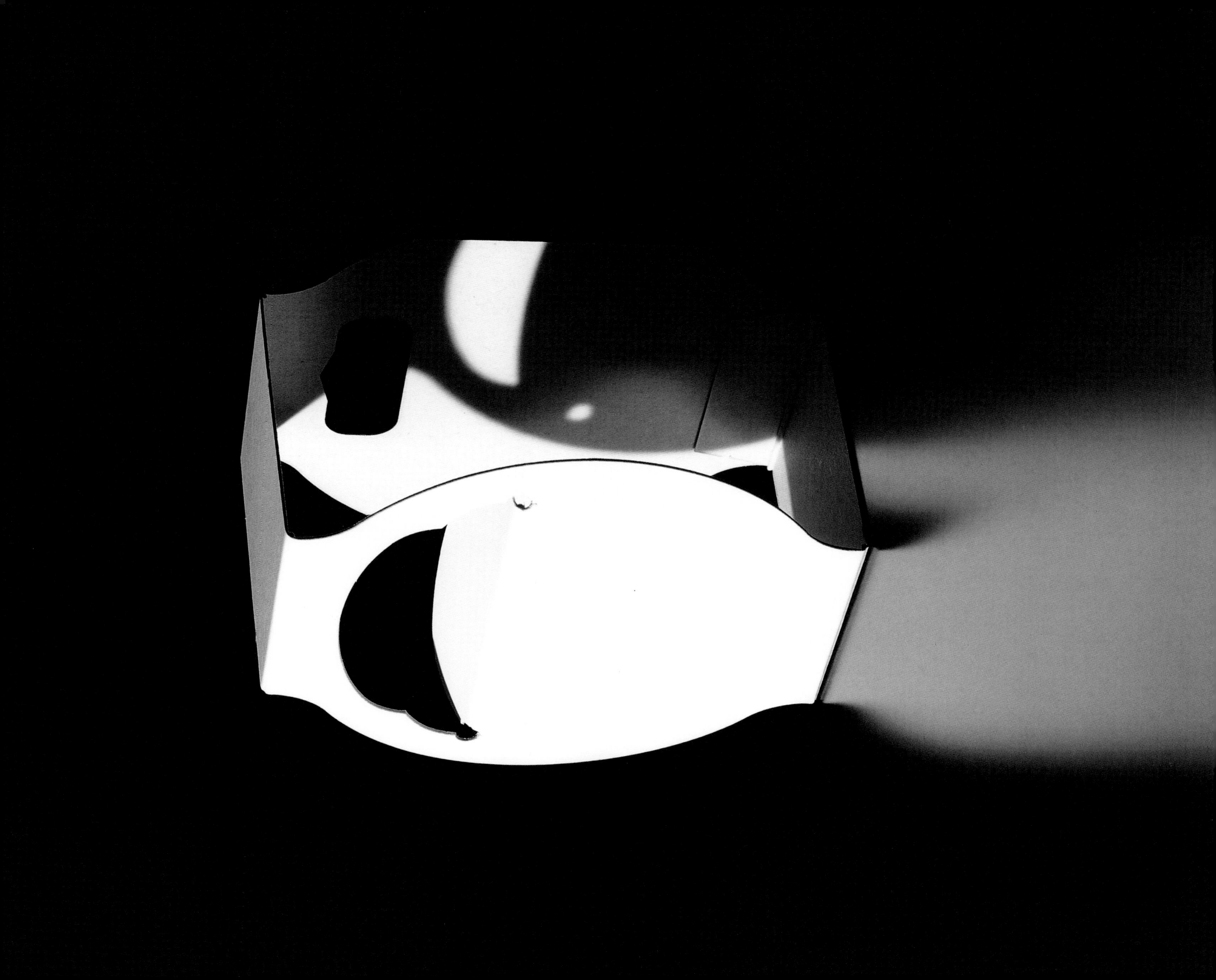

COMPACT DISC WALLET

An ingenious storage sleeve for compact discs which, when opened, offers the disc to the user in a convenient and controlled movement. The folded segments along the score-line hold the compact disc in place as the hinge of the case forces the compact disc out of the package. Other possible uses would be for other flat products needing a pop-up format.

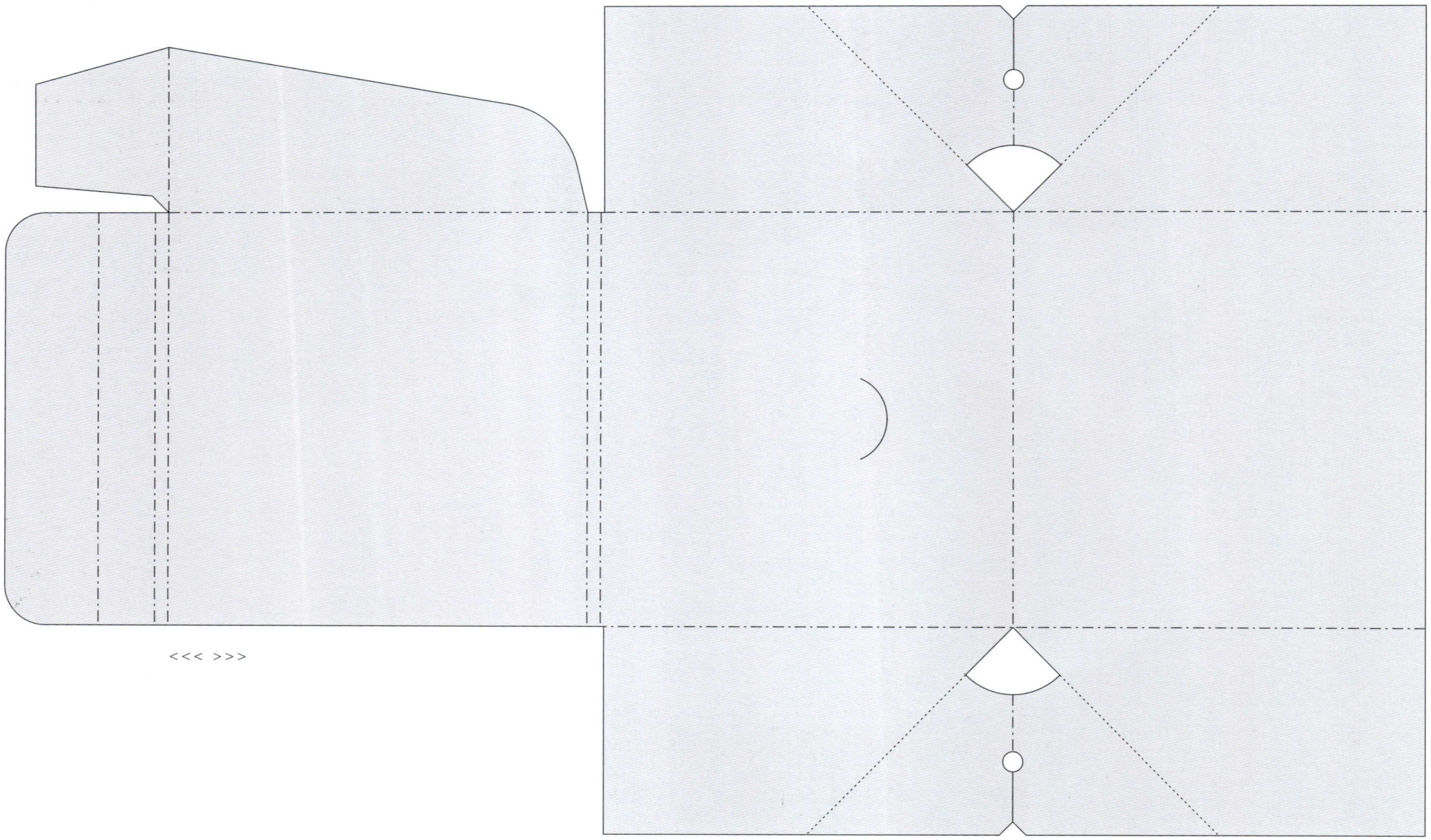

Patent Pending, Danapak

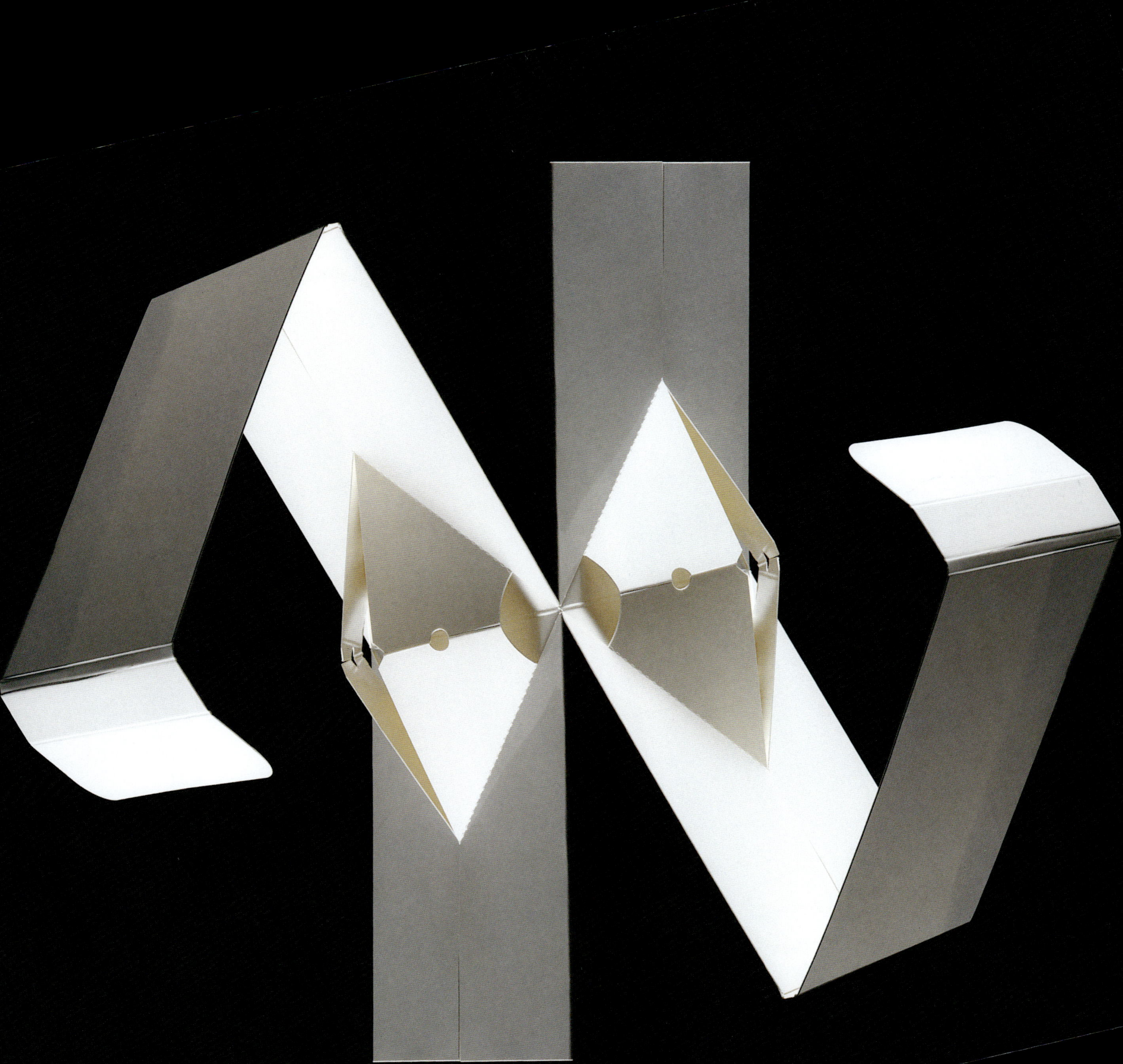

Chris Hobbs, Surrey Institute of Art and Design

WALLET SLEEVE

This simple carton for the storage of flat, fragile objects was originally intended to hold a compact disc, but the principle of the design could be used for a range of different products. The product is contained in a die-cut strap and slot, and is then protected by the rest of the folded package. This provides an attractive sleeve that mounts the product while providing ample space for graphics if required. The type of material used determines the level of protection given, for example corrugated card would provide exceptional protection for a fragile glass item.

DISPLAY CARTON

This skillet carton has an extended lid-flap which opens to reveal a display recess. Die-cut from a single sheet, this tray provides a specific compartment for the product inside. The internal flaps that form the frame also add to the rigidity of the structure as well as helping to protect the product. This design only requires one side of the board to be printed.

Parmjit Bhachu, Surrey Institute of Art and Design

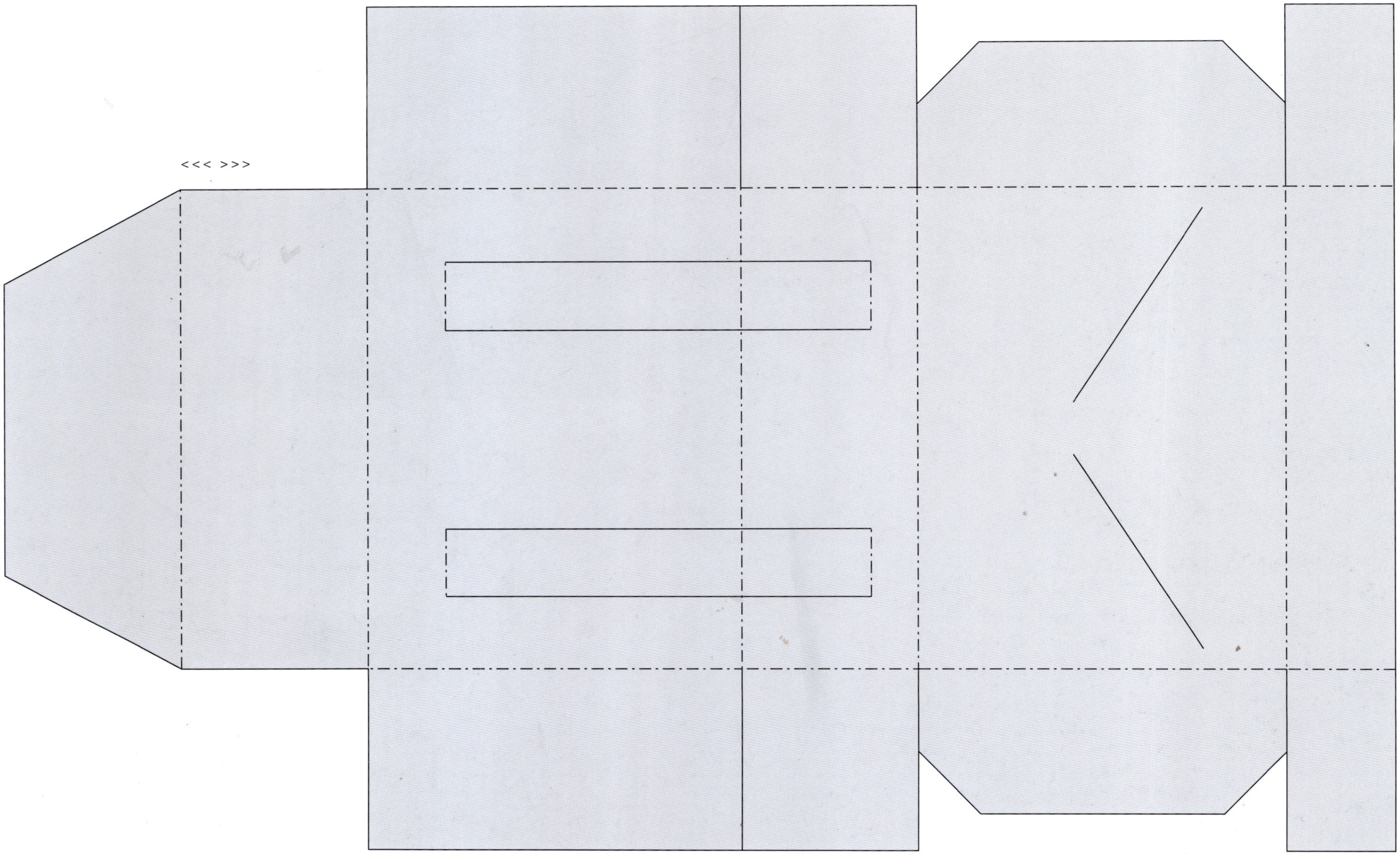

SELF-SUPPORTING CARTON

The design of this carton uses the weight of the product to hold the structure together without needing any form of adhesive. The straps are cut from the same piece of card and used to contain the product while allowing the consumer to touch and feel it. Originally designed for containing soap, this package also allows the customer to smell the product—considerations such as these must be taken into account when designing for a specific product.

ONE-SQUEEZE NOVELTY CARTON

A promotional display carton with novelty appeal. Internally connected with a rubber band, it requires no assembly and erects through squeezing the sides. The elasticity of the rubber band pulls the ends to erect the carton. With use of graphics and minor additional design features this carton could be made to look like an animal or car, for example.

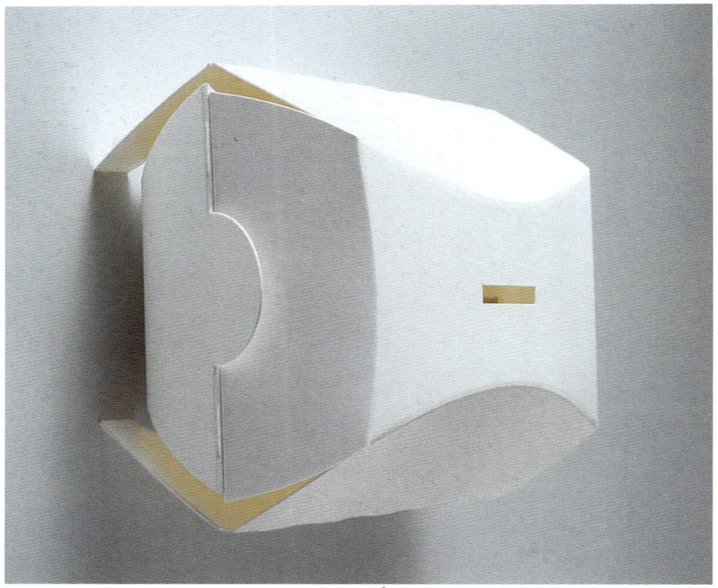

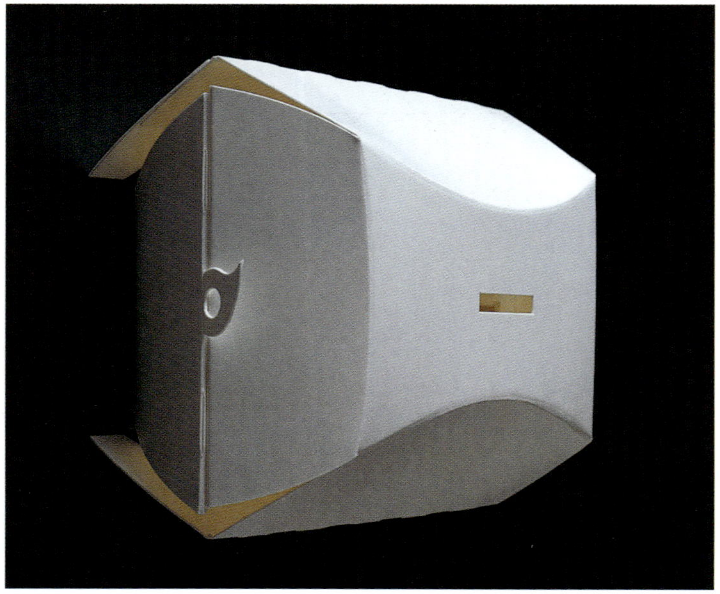

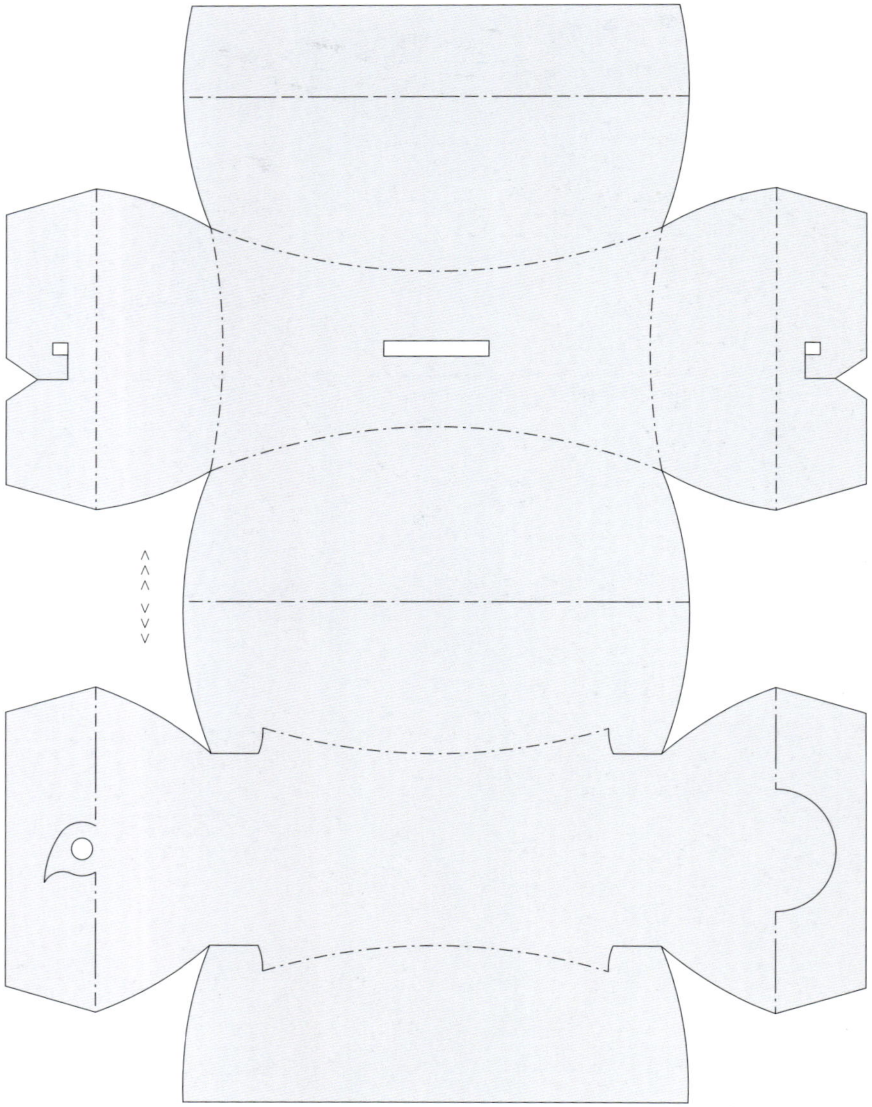

Danapak

"CHARACTER" LID AND STORAGE BOX

If Cardboard Creations

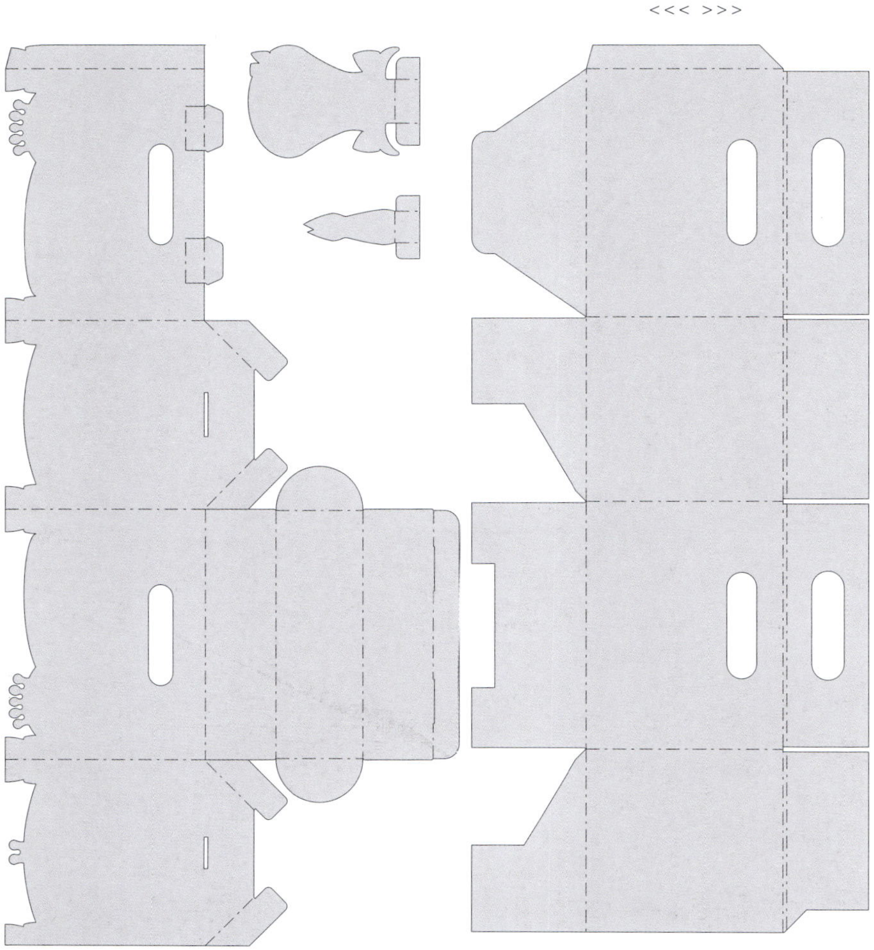

A corrugated cardboard storage container for children's products. This is a good example of how the use of printed graphics and minor additional card pieces can enhance a simple design. The storage box comes in two parts and could vary in size and dimension depending on requirements.

FOUR CORNER GLUED SKILLET-ENDED CARTON

A support base tray for storage of containers or products, this can be used as either primary or secondary packaging. Four sides are used for point of sale while the base is blank and the top die-cut to the required template. It is glued for rigidity with perforated hinge edges and it can be used to hold objects which do not have a flat base such as a pyramidal carton. The depth of the tray could vary depending on the required support of displayed product.

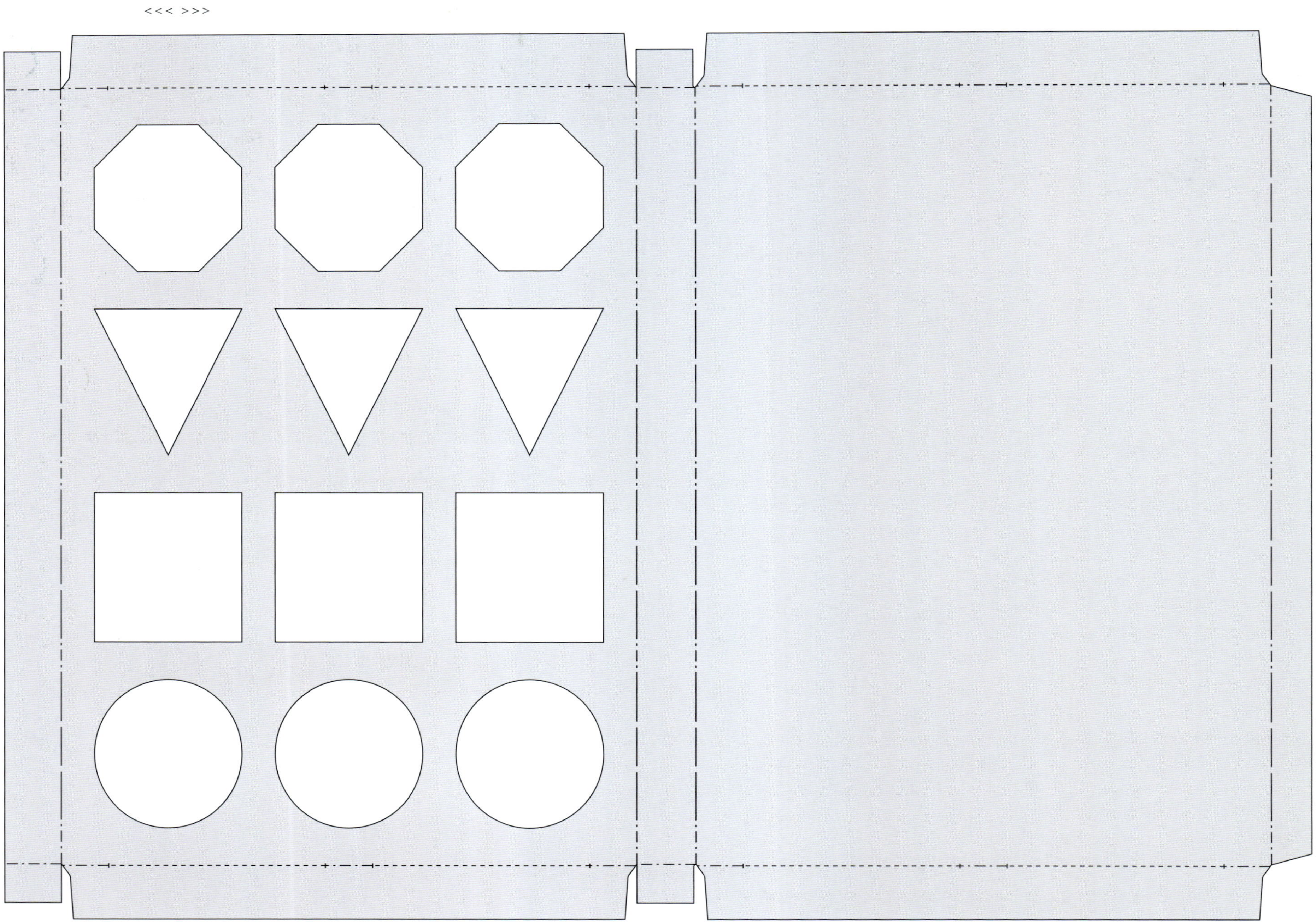

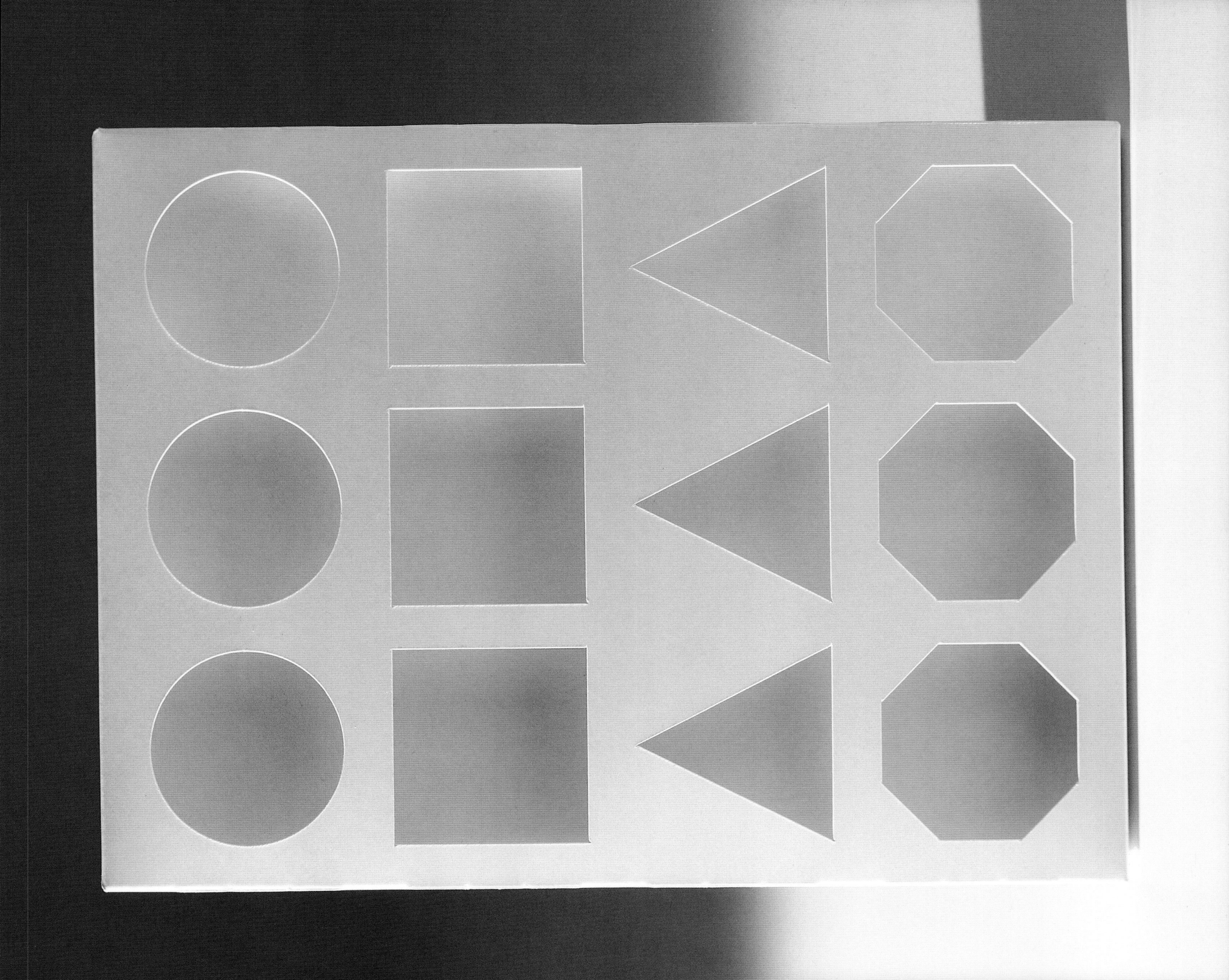

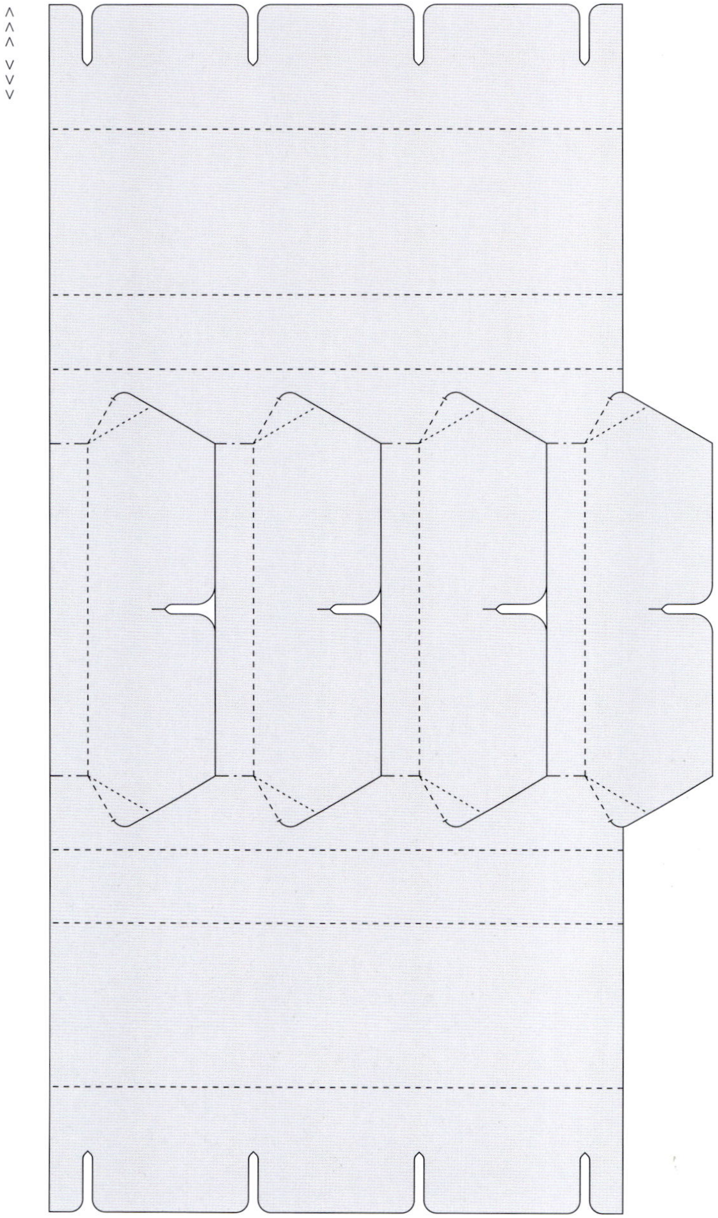

MULTICOMPARTMENT TRAY

A tray carton for the display of groups or ranges of products. The shape of the die-cut spaces may be altered depending on the fit of the product to be displayed. This is an effective way of presenting a whole range of products while also providing point-of-sale information applicable to the whole group. No glue is required in the construction of this multiunit carton.

WEDGED CARTON WITH LID

A two-piece segment-shaped carton with one curved and tapered side and two vertical flat sides. The simple lid-closure fits in place by slotting through a slit on the container and wedging itself in the taper of the rounded side. An appropriate shape for consumable products sold in segments, such as cakes or cheese.

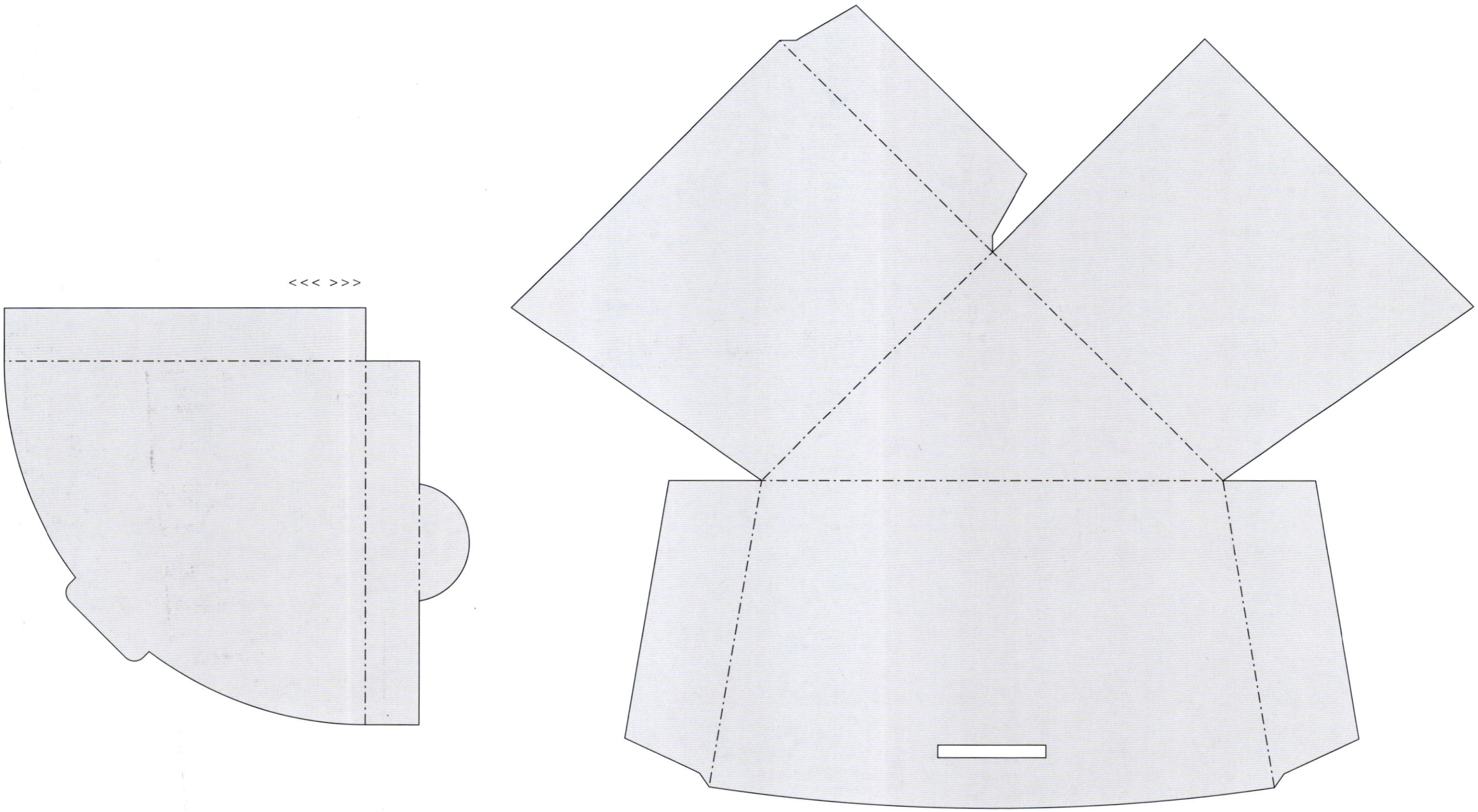

CGM Ltd.

A multiglued container for hygienic storage of food products. The octagonal shape makes this an efficient carton for circular goods. One application of this carton uses a specially designed heat-distribution pad inside the container to distribute the heat equally to the product in a microwave oven. The product is removed from the container and is placed on top of it. When cooked in this way, the action of the distribution pad coupled with the height mean the product is heated efficiently. This is a good example of design aiding the use or preparation of the product.

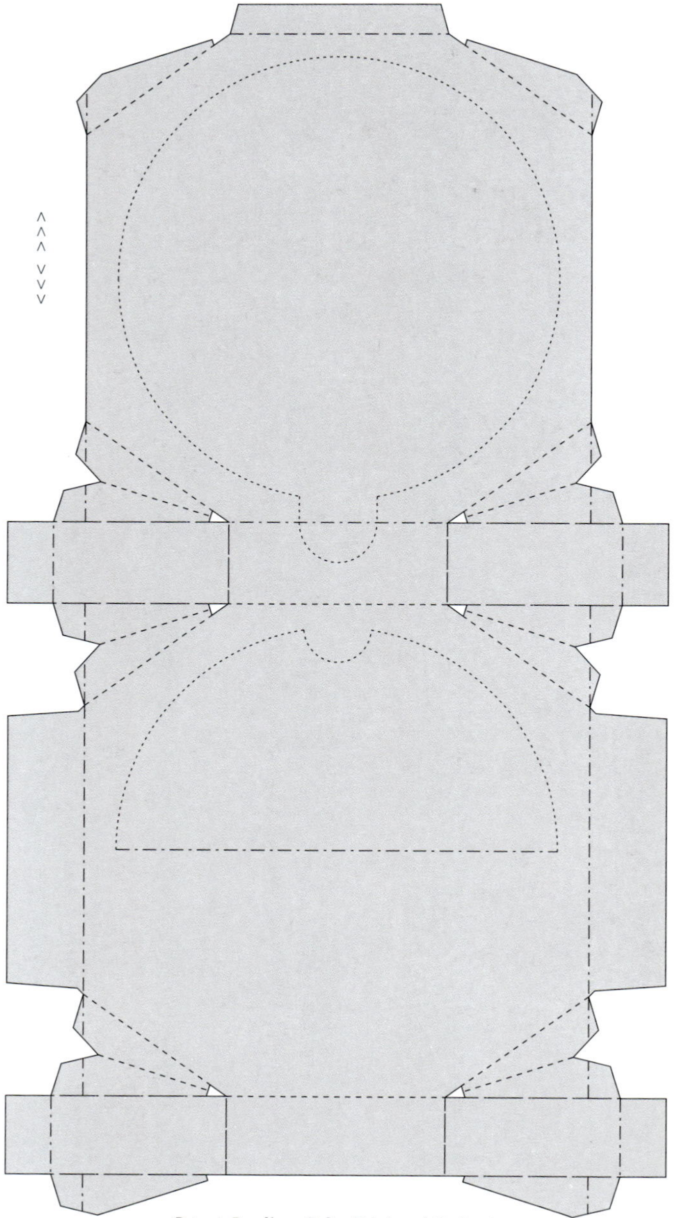

Patent Pending, Delta Print and Packaging

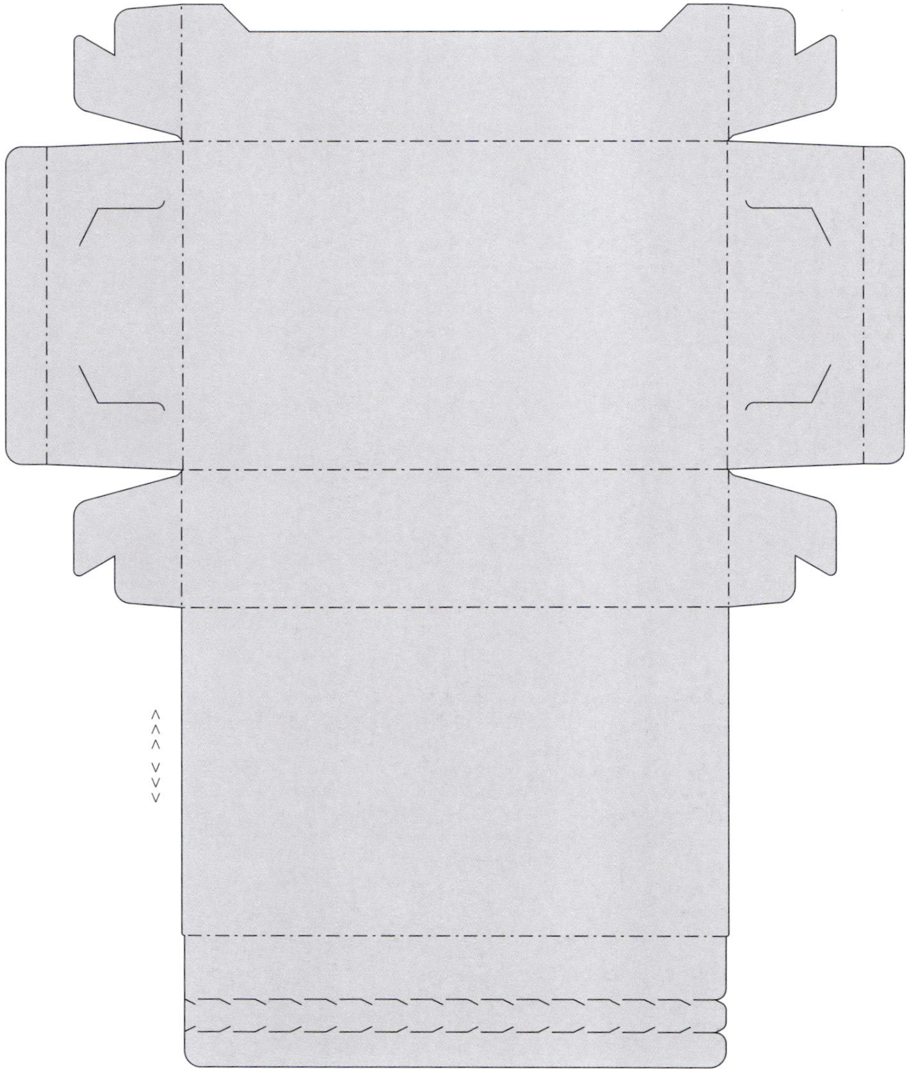

PULL-STRIP CARTON

The pull-strip is commonly used in cartons to prevent tampering as the strip provides an irreversible method of opening. To ensure the pull-strip's accuracy, all the flaps must be glued, or have locking closures, to prevent access to the product via another entrance. Pull-strips can be applied to many different types of carton.

OPEN-ENDED TRAY

This is a simple tray designed to support a group of products on a display shelf. The rear and sides can be larger to include point-of-sale graphics, up to and over the height of the products contained. With glued flaps and score lines the tray can be collapsed easily and quickly for flat-packing.

THREE-SIDED DISPLAY TRAY

A basic carton which provides a point-of-sale accompaniment to awkwardly-shaped food products. This provides partial protection for shrink-wrapped products, but more importantly it provides a visual barrier for the consumer in the form of a graphics platform. A good use of minimal material.

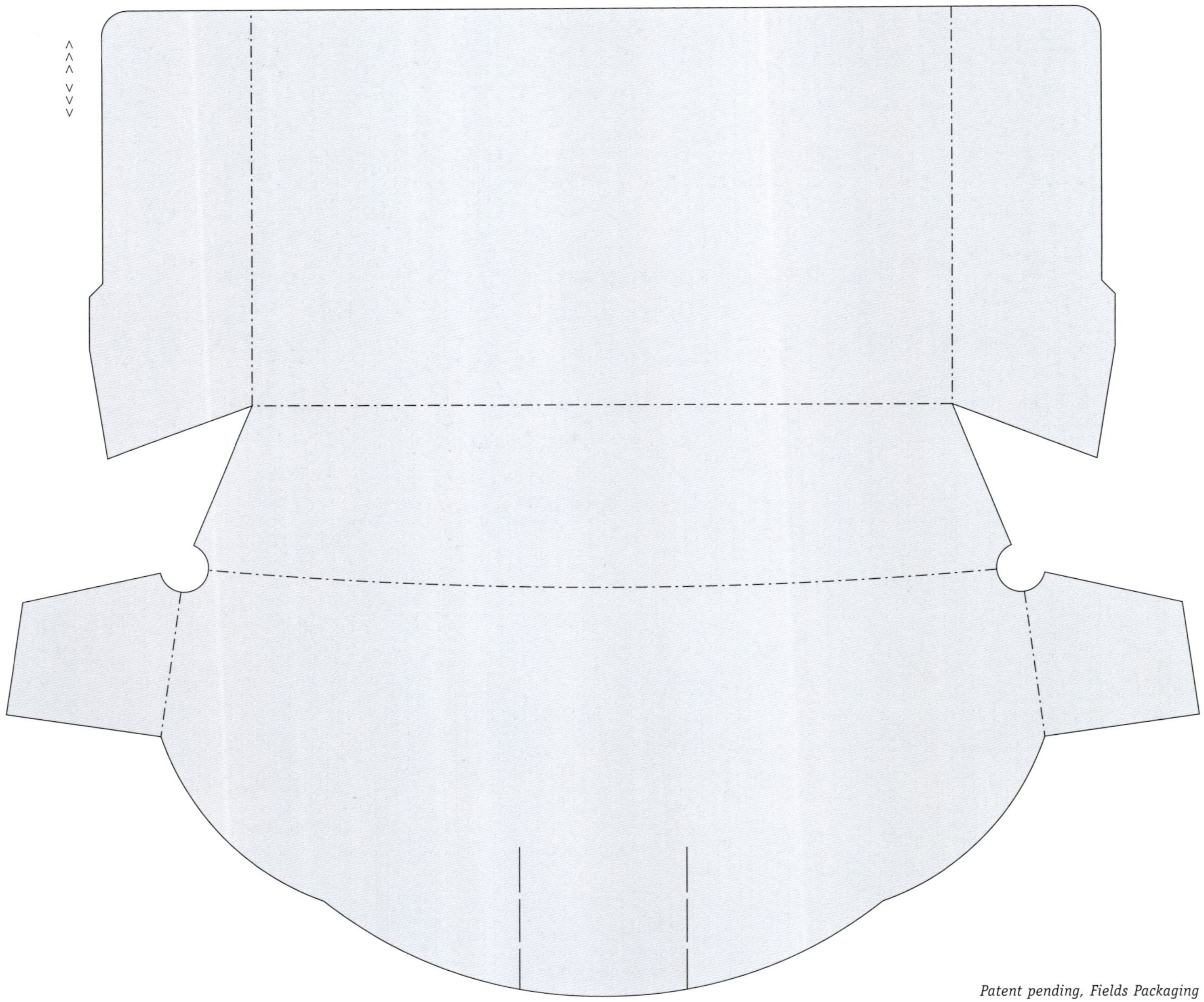

Patent pending, Fields Packaging

FLANGED TAPERED TRAY

A web-cornered glued food tray with flanges typically associated with use in the take-out food market. The flanges on the sides create a lip to hold hot food with without being scalded. The tapered sides prevent food from being trapped in the corners and allow easier serving from the container. The use of special material is essential for this carton to provide water-resistance and rigidity when full of hot liquid. Cartons with tapered sides are excellent for saving space in transportation and storage as they can be stacked, which might be an important consideration in some applications.

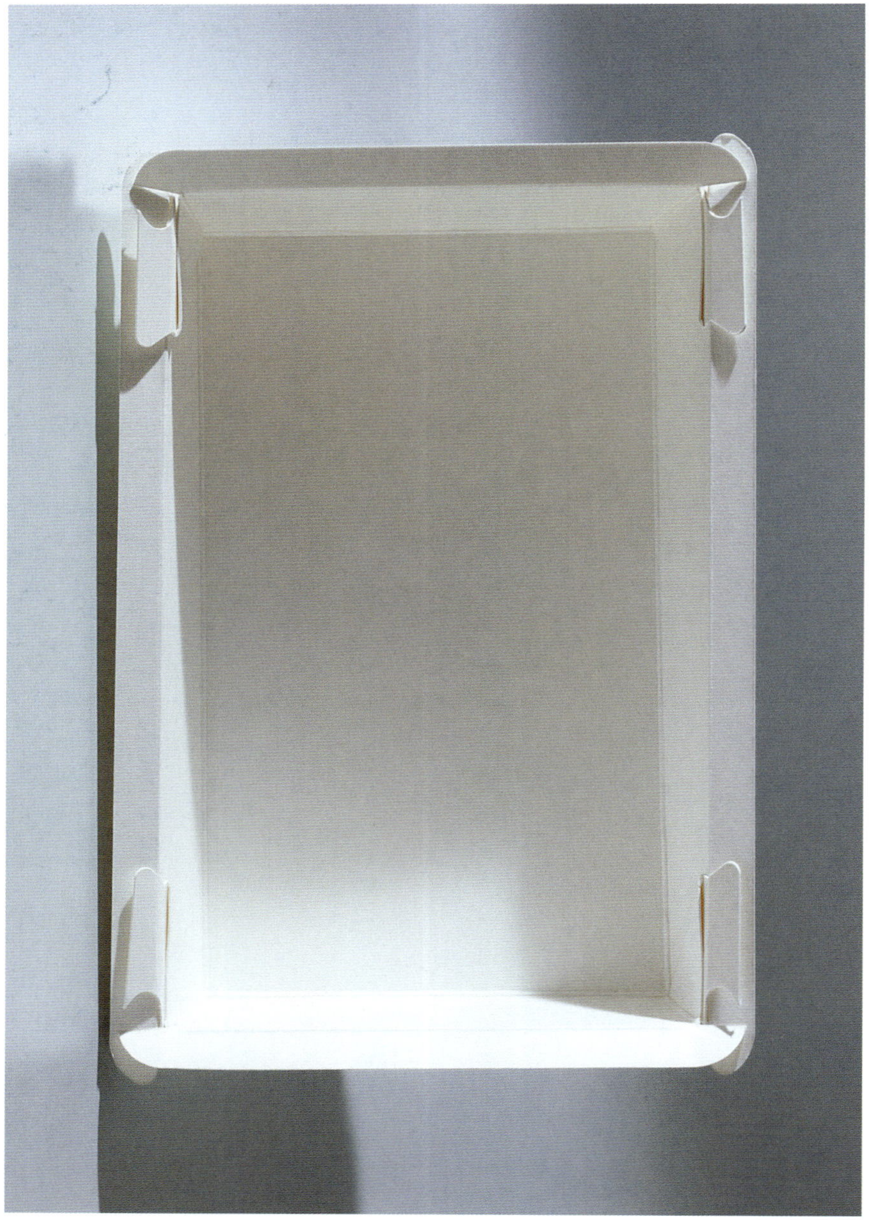

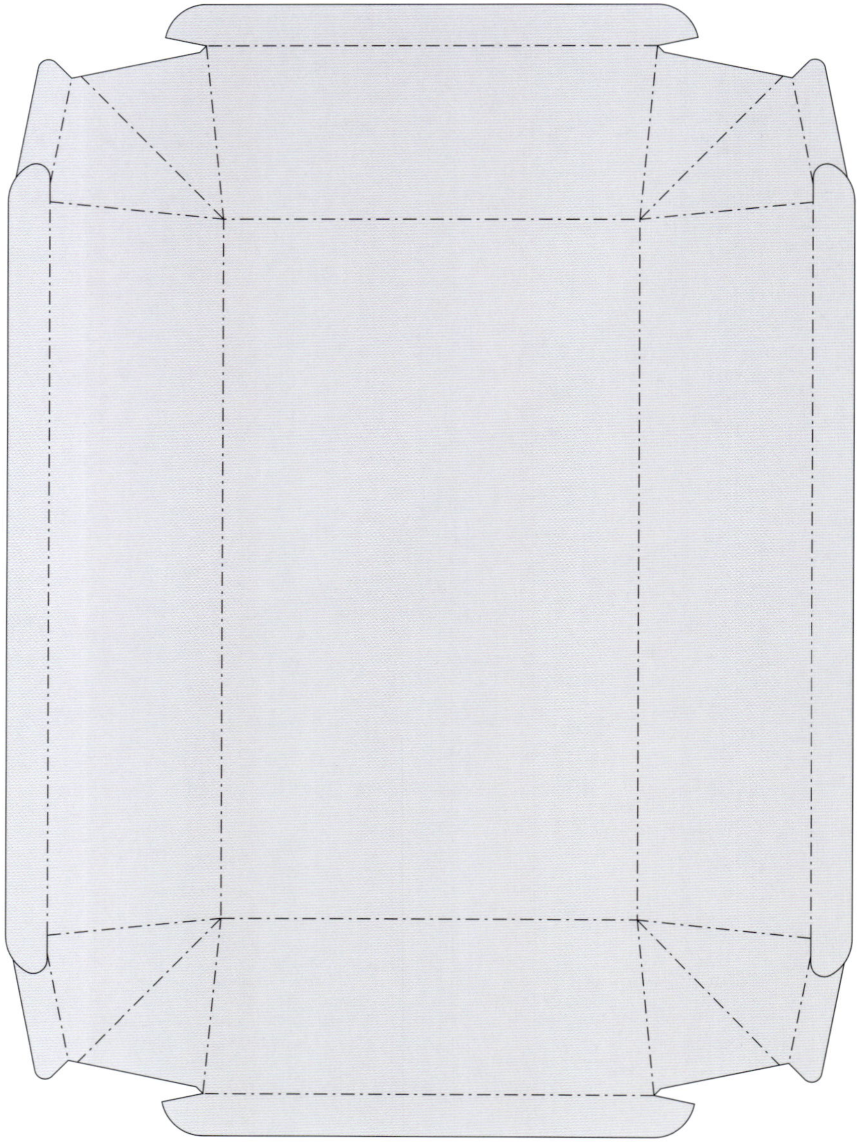

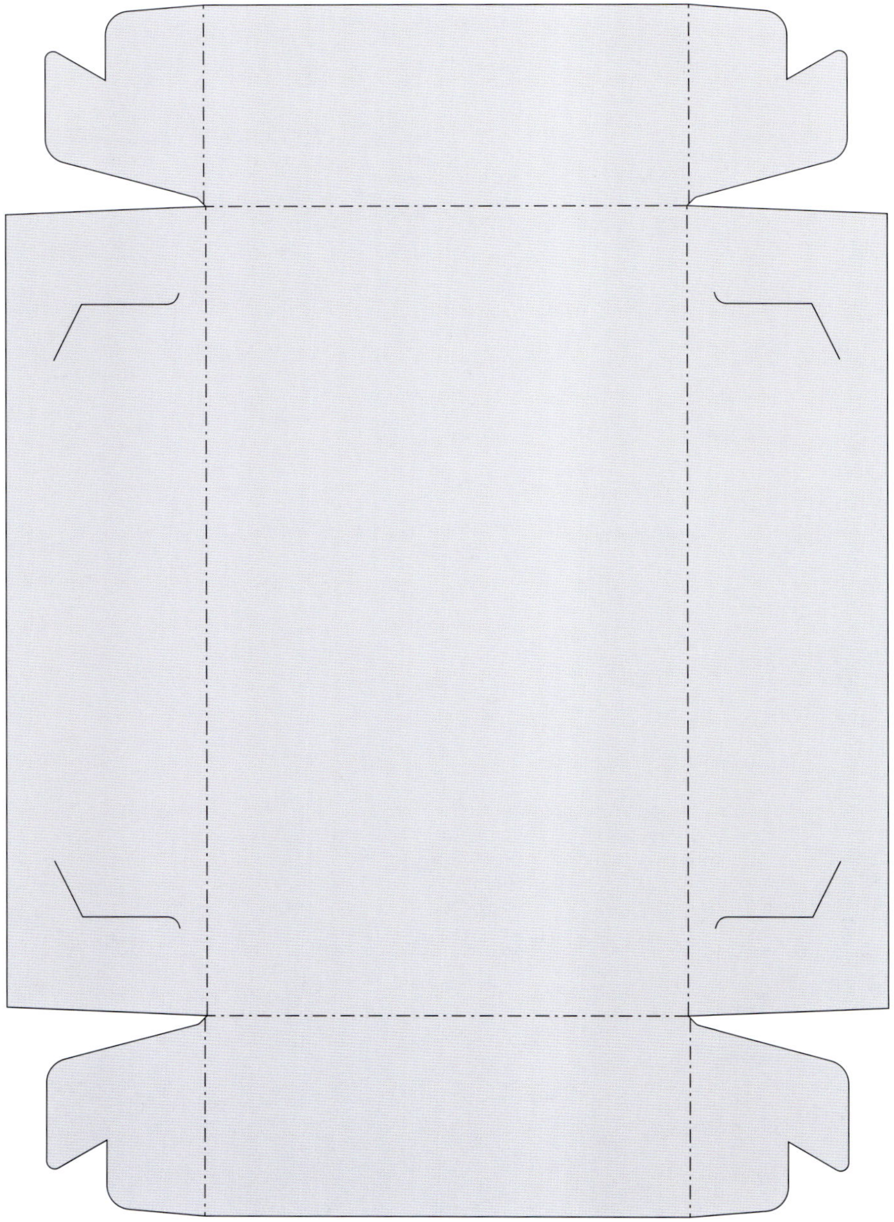

CORNER-LOCKED TRAY

A machine-erected four-sided tray with corner locks used for general purposes. It does not need any glue in its construction, but uses the tongue and slot on the corners to create the sides. The tray can be formed into a container with a lid by using one tray as the base and a slightly larger one for the lid.

IN-FOLD/OUT-FOLD FOUR CORNER TRAY

A simple easy-erecting carton used for the storage of many items which are later to be shrink-wrapped while in the carton. The sides also act as an adequately rigid barrier for the contained product when removed from the shrink-wrap. The sides of the tray are strong when pushed outward because they are designed to fold inward—there are many different variations to the way this is achieved. The creases allow the tray to be flat-packed and stacked; the tray self-erects just by lifting the sides.

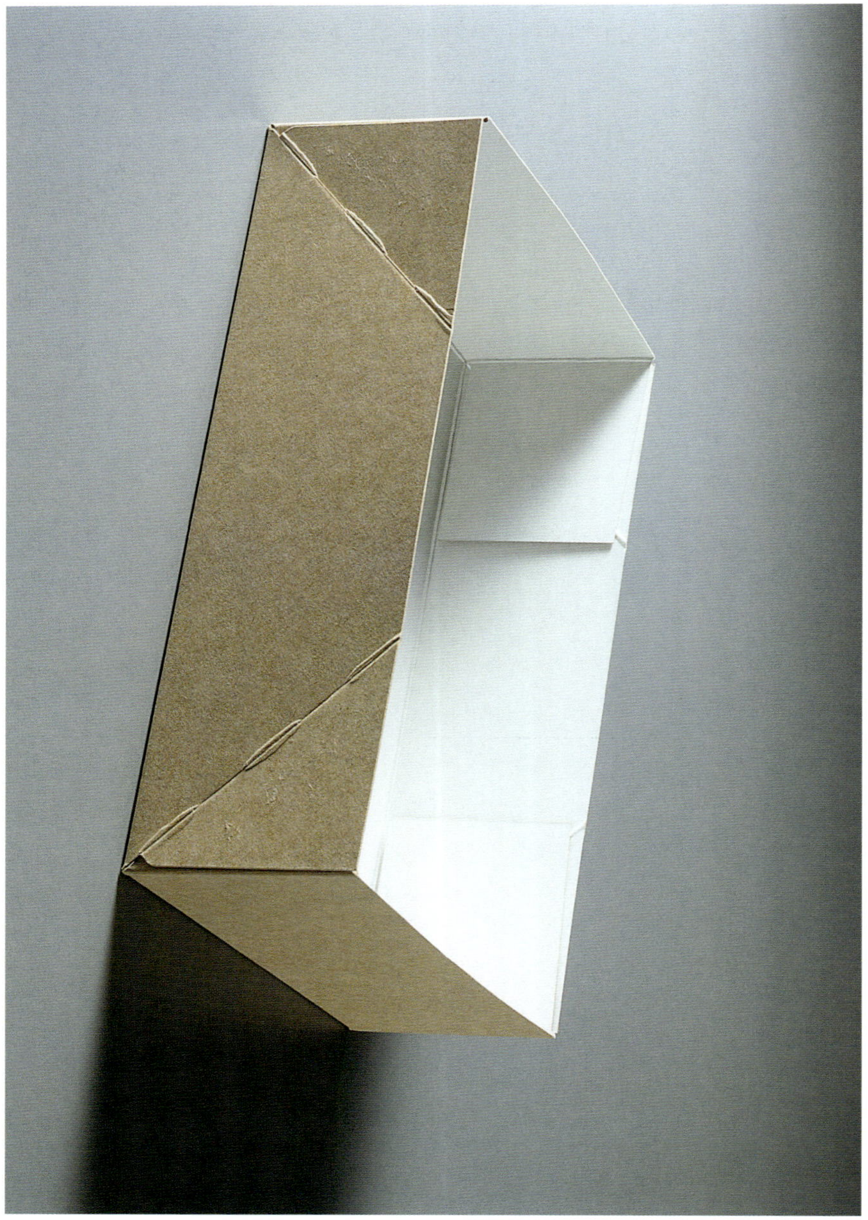

<<< >>>

ONE-PIECE FOUR-CORNER GLUED LID AND BASE

A carton with in-fold hinge walls which incorporate the lid along one side. This simple carton with a combined lid demonstrates another type of tuck/closure pattern. When the carton is opened fully it forms an extended tray with side walls. As a closed carton, it provides a fast and effective container mechanism suitable for the fast food industry—grease-proof, or moisture-resistant card would have to be used for this particular application.

TRAY WITH CRUCIFIX FLAPS TO MEET

A corrugated carton used for transporting product units safely, as such, it traditionally functions as a secondary package. It could be used for primary packaging but the thickness of carton board would limit the size and scale of the container. Corrugated board would normally be used for its shock-absorbency and strength. This carton is typically sealed with adhesive tape/hot melt adhesive/staples.

FOLD-OVER TRAY

This tray has fold-over glued sides and fold-over ends. To increase the tray strength, a thicker or more rigid material might not always be the best solution; by doubling over the sides and ends all the faces can be strengthened in a different way. This type of design allows graphics to appear on the inside of the tray while only printing one side of the card. The tray could be glued, or T-shaped ends could be added ensure that the flaps do not open out. However, generally the weight of the product carried ensures this does not happen.

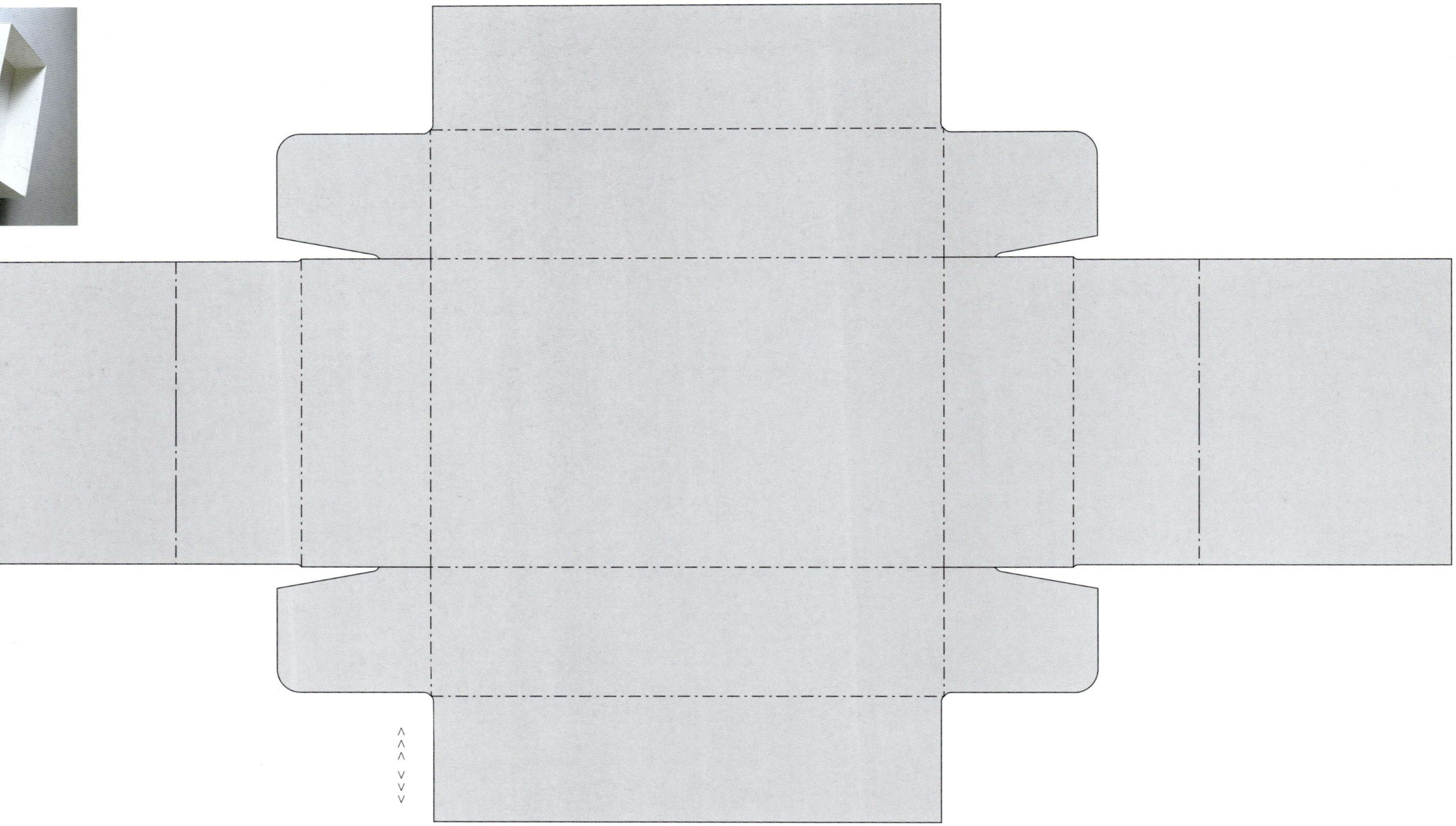

SIDE-ENTRY HEXAGONAL CARTON

This is a simple, yet effective, variation on the regular design of rectangular containers. The hexagonal shape makes the carton more eye-catching and provides a more interesting shape on the shelf. The orientation of this carton could be either vertical or horizontal. In the vertical state, it might be effective for containing bottles for special-edition products. The length or number of sides are easily changeable in this carton.

FOUR FULL FLAPS SKILLET

A particularly effective carton for heavy items, as the base and the lid are double thickness, due to their overlap. This uses B-flute cardboard which is the most popular material for providing the greatest shock-absorbency and crush resistance. This is primarily used for secondary packaging, although design alterations could broaden its applications.

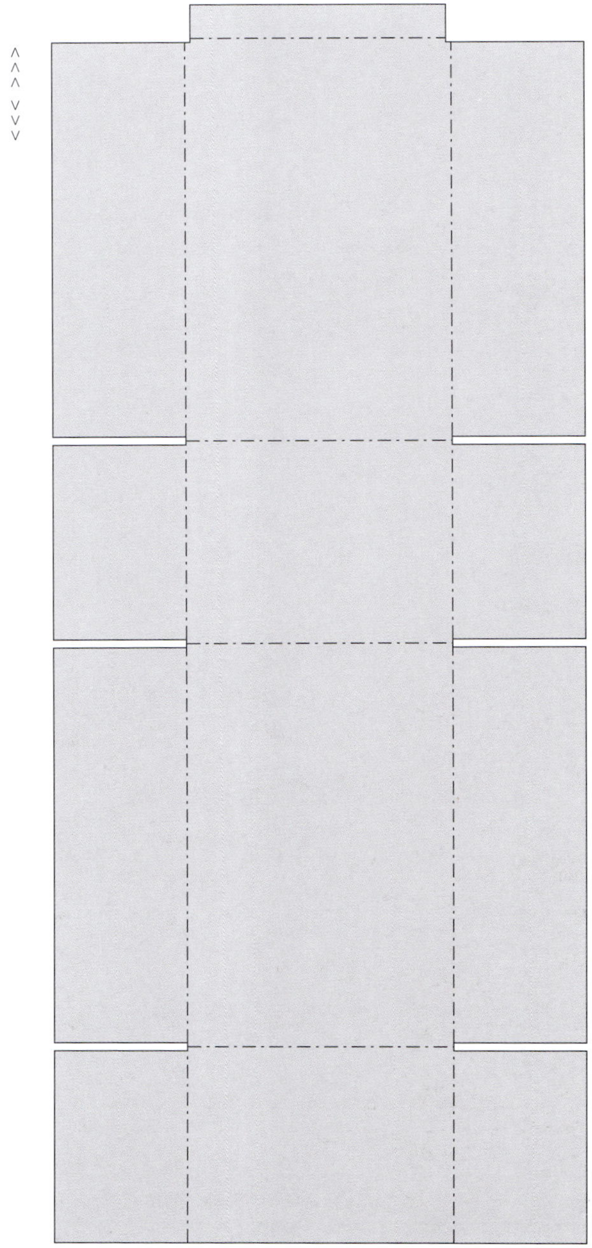

FOUR OVERLAP-FLAPS SKILLET

A traditional transit carton normally used as secondary packaging. Due to its simplicity and structure, it tends not to be used as a point-of-sale package. Excellent strength properties mean that this container is used for packaging anything from small cartridge products to television sets. Its proportions are easily manipulated to required dimensions and scale.

This two-part container has four flaps to meet on both the base and the lid, and the lid is full length. It can be manufactured with shorter sides to ease removal or placement, although full-length sides give better protection against side impact. B-flute corrugation provides shock absorbency, while the outer face provides a surface which can be printed on using line art or letterpress printing methods. More complicated designs can be preprinted using flexography or rotogravure methods which use kraft, white clay-coated kraft, or clay-coated bleached liner materials.

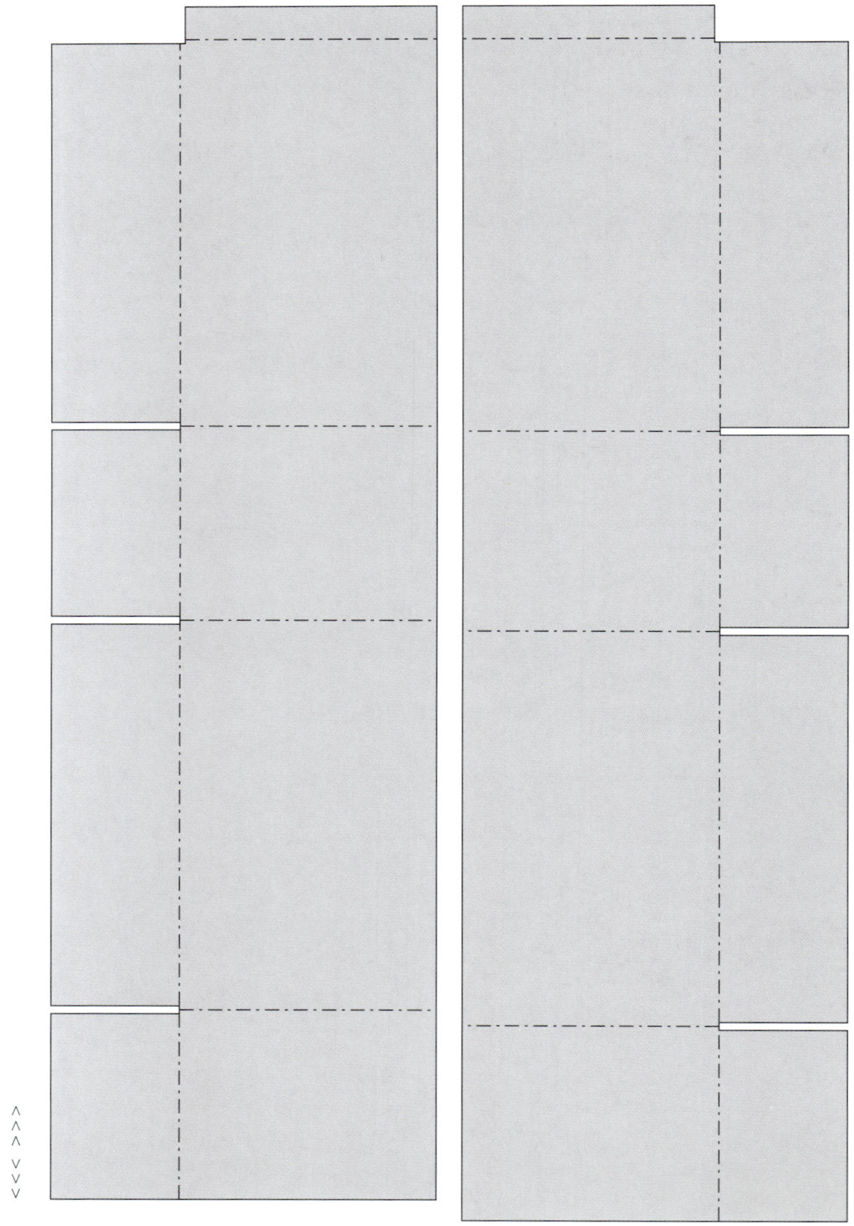

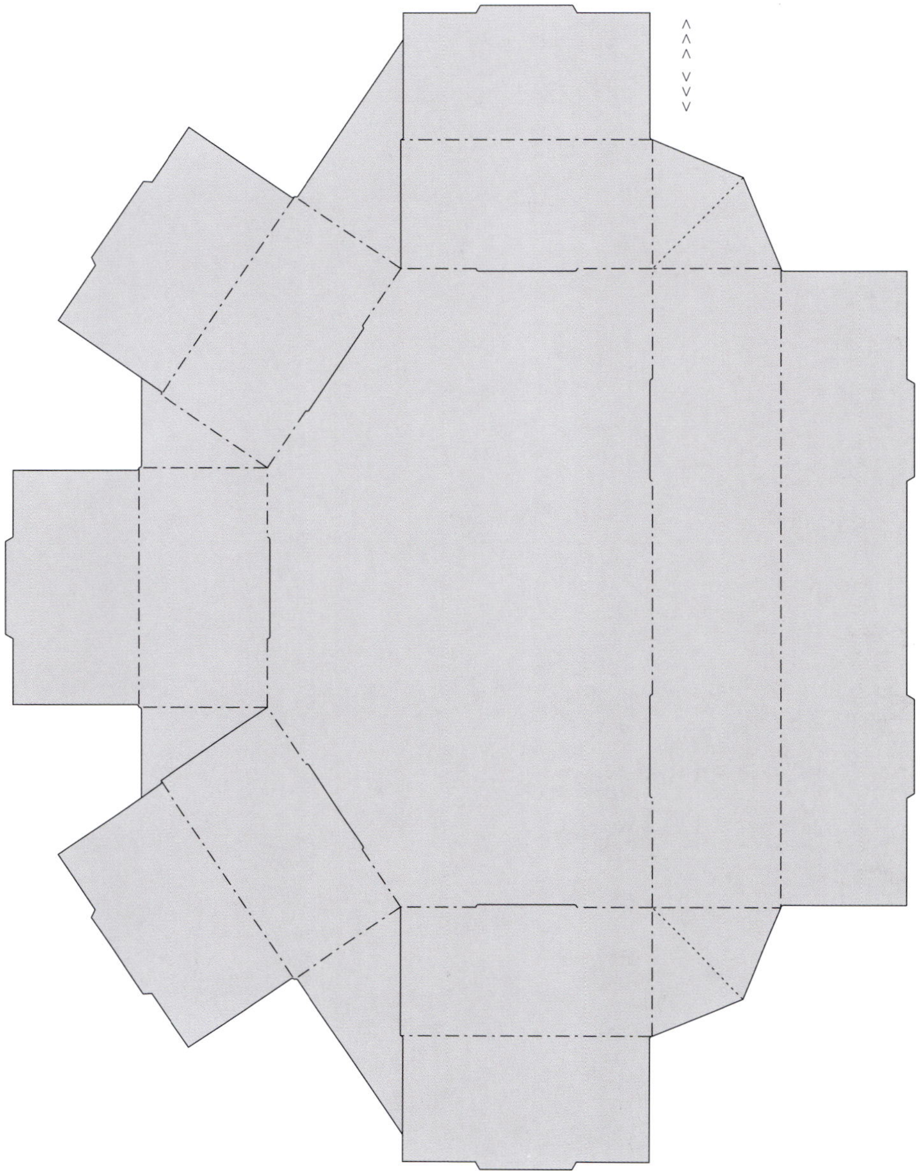

An example of a tray that uses folds, rather than glue, to form its structure and strength. The tuck-flaps fit through recesses in the base formed by the pedestal feet. This is a particularly useful design for trays that require construction by hand in large batch quantities.

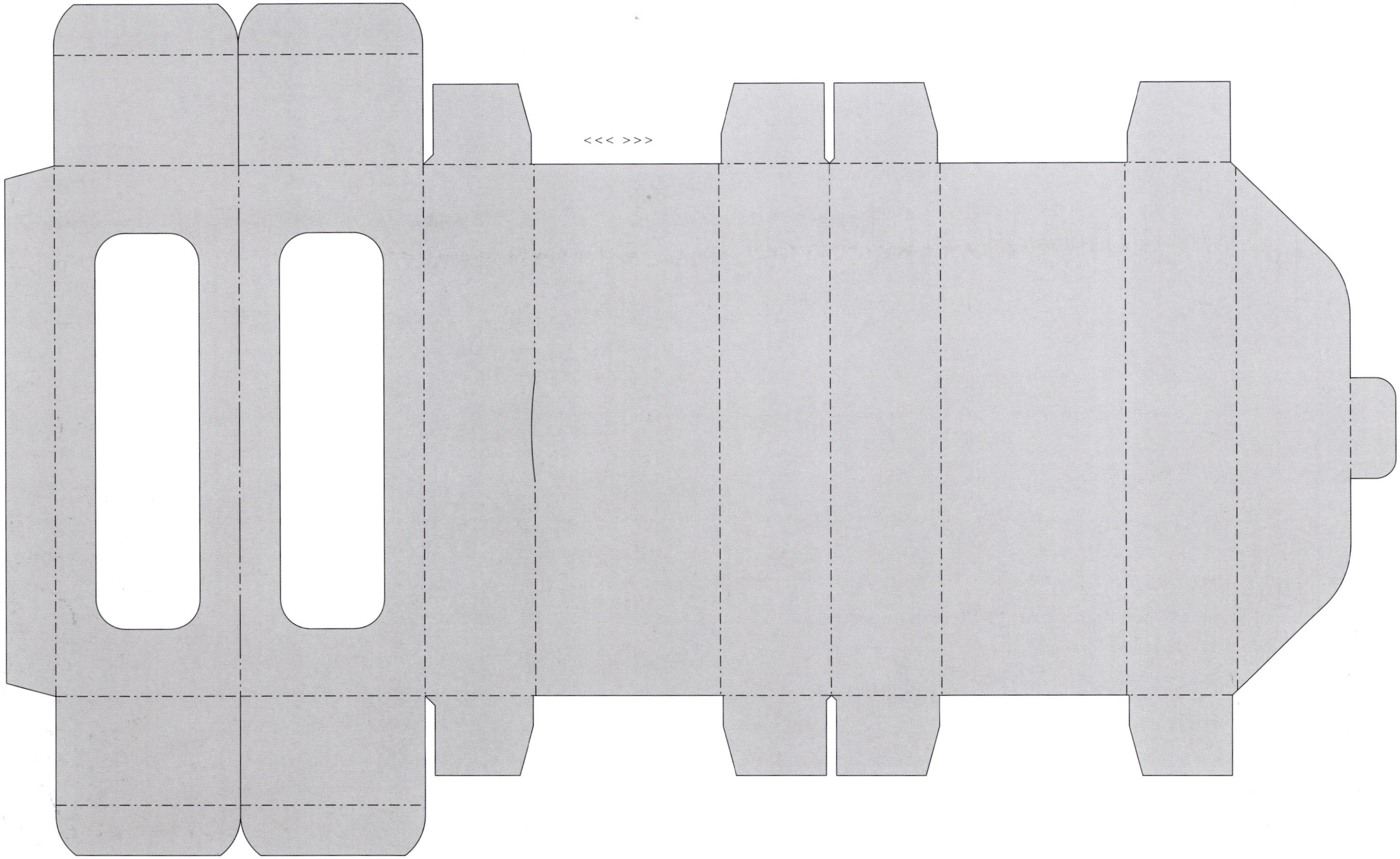

HINGED DOUBLE CLOSET CARTON

As a sealed, closed container this carton initially appears to have a single compartment, but when opened up it displays two equal—yet separate—chambers. Made from a single piece of card, this carton is an ideal package for grouped or twinned items. The compartments are hinged together along their length and have display windows on their front faces. These form an attractive display panel when the package is opened out. The design is ideal for gift items, where the product is concealed inside the package and a level of unwrapping is required.

This is an effective transit carton, delivering goods safely to the point of retail. The carton may then be transformed into a point-of-sale unit by sliding the sides down to form a freestanding display of the product inside. The addition of graphics would provide information and aesthetic appeal, minor design alterations could allow many types of product to be contained in this way.

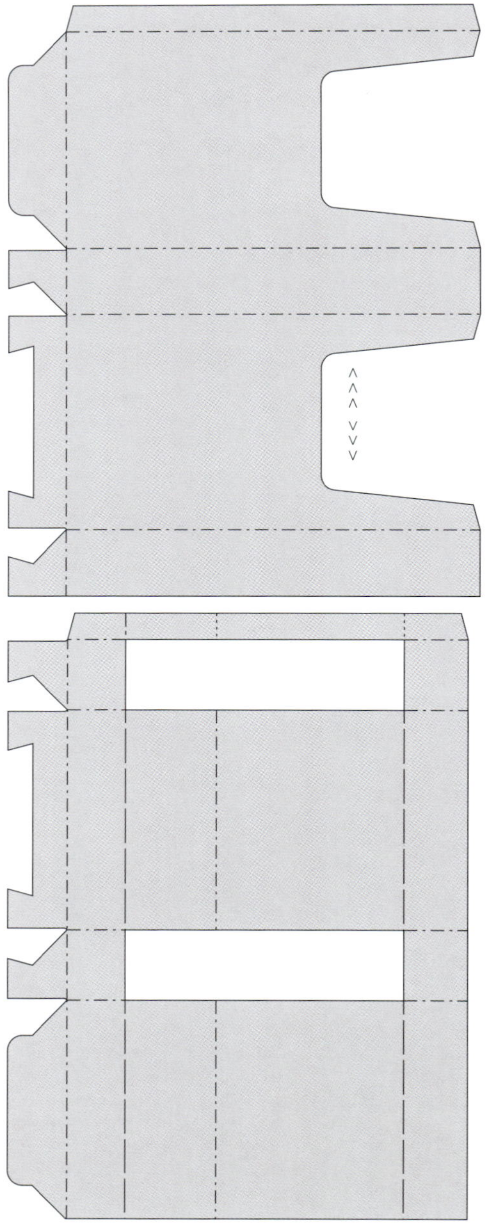

P.I. Design International.

OUTER SKILLET CARTON WITH CENTRAL PERFORATIONS

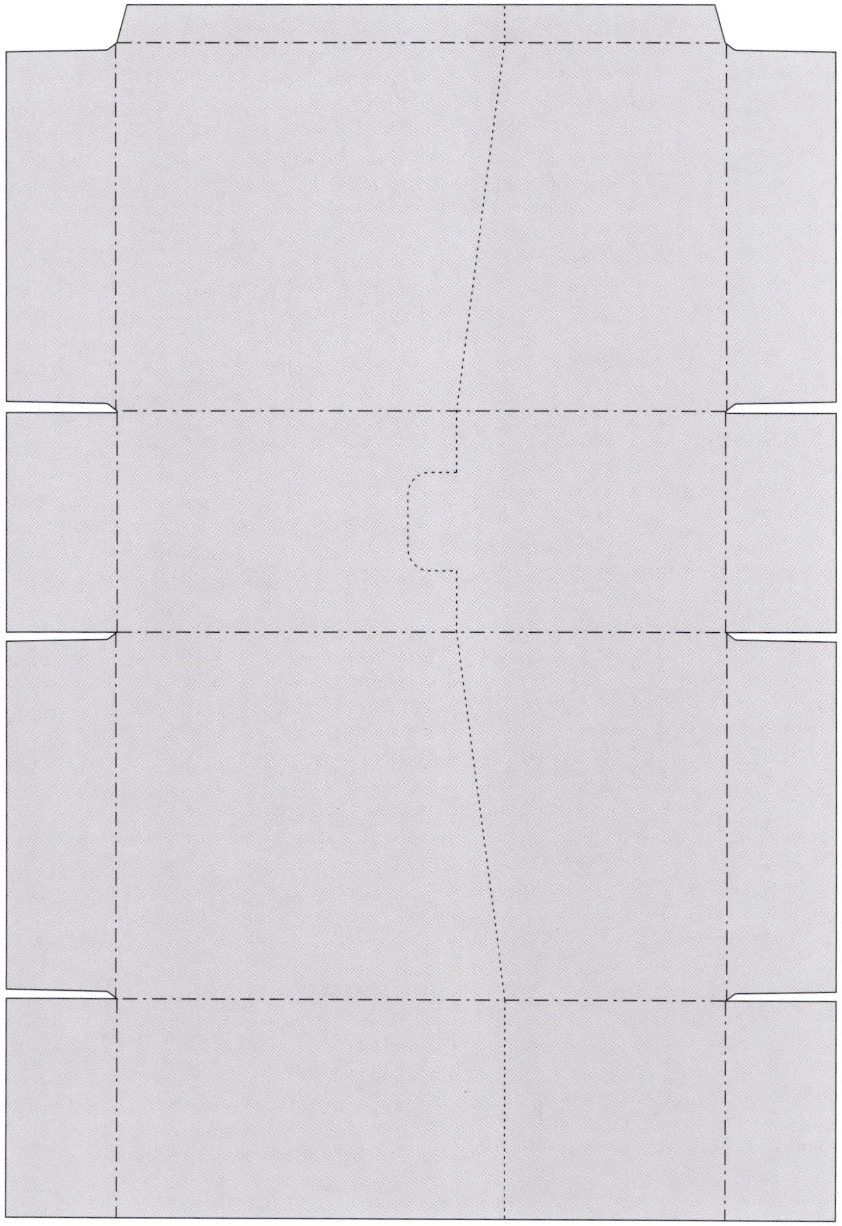

A traditional transit carton used as secondary packaging that converts to a useful point-of-sale container. This is particularly suited for use on filling lines where automated taping machines seal the filled cartons as they pass along the production line. The flaps meet in the center of the top and base of the carton, allowing an internal flat area which is especially suited to freestanding products. Applied perforations along a designated angle around the girth of the container allow the base of the container to perform as a point-of-sale support after transit.

CARTON WITH INTERNAL COMPARTMENTS

This is an example of how a simple design from a flat sheet can separate the items inside the carton. This reverse-tucked example has three compartments, but this could be increased or decreased with little effort, depending on specific requirements. It is ideal for products that need separation for a specific reason; to avoid contamination from neighboring products such as soaps or to provide protection for fragile items such as light bulbs. Windows could be easily added to the front of each compartment to provide the consumer with a view of the products.

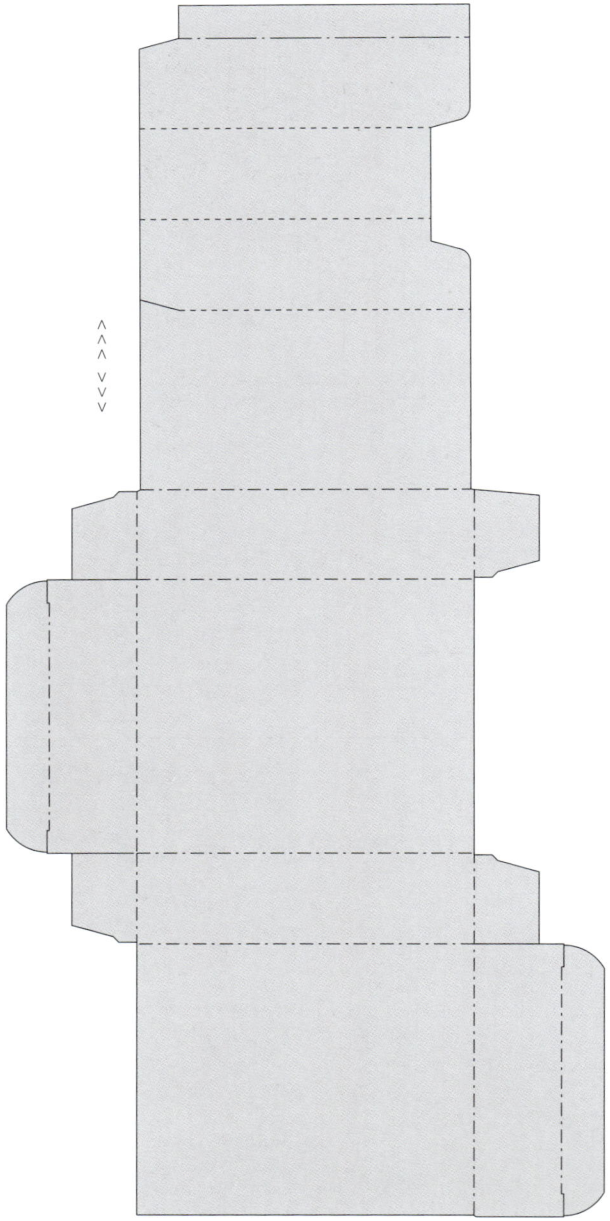

FLIP-TOP CARTON WITH TEAR-AWAY ENTRY

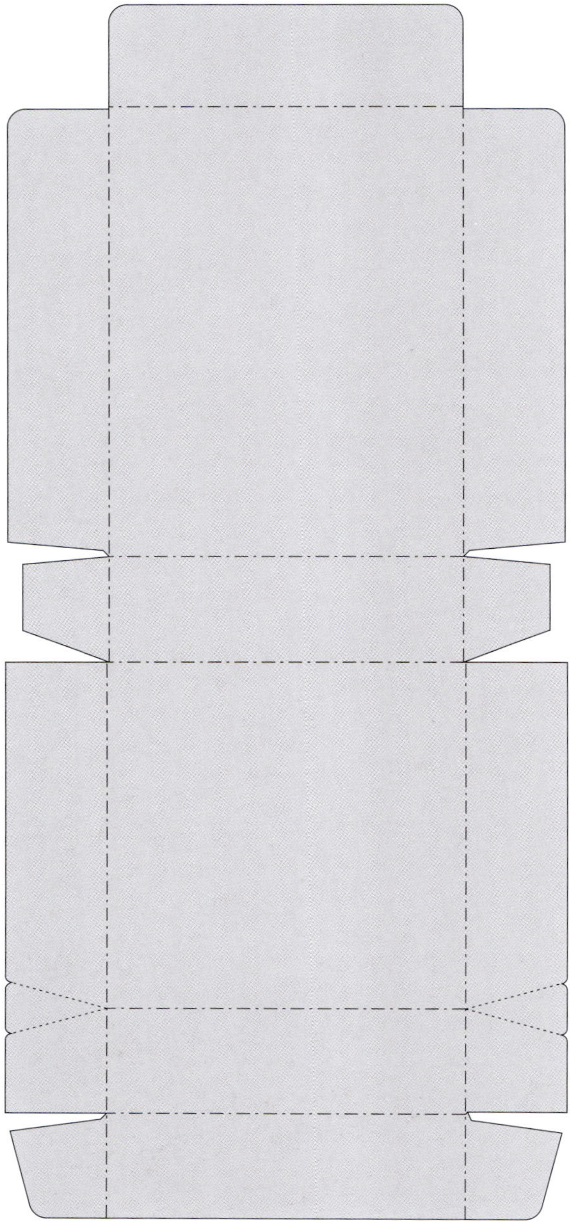

This is a carton which allows access to the product by tearing away the end panel along designated perforation lines. Commonly used for small confectionery; other possible uses are for clothes which would need to be pulled from the carton. This type of opening is good for products that require a level of security or tamper-proofing. Attention should be paid to the amount of glue used around the entry, so as not to hamper access.

POUR SPOUT CONTAINER

This is an example of how a simple closure can determine or transform the purpose of a carton. By scoring along the correct axis, the consumer is able to open the spout and gain access to the product. This container is most commonly used for detergents and breakfast cereals. The designer has the freedom to alter the dimensions of the container considerably, providing the dimensions of the aperture remain large enough for the product to pour out freely.

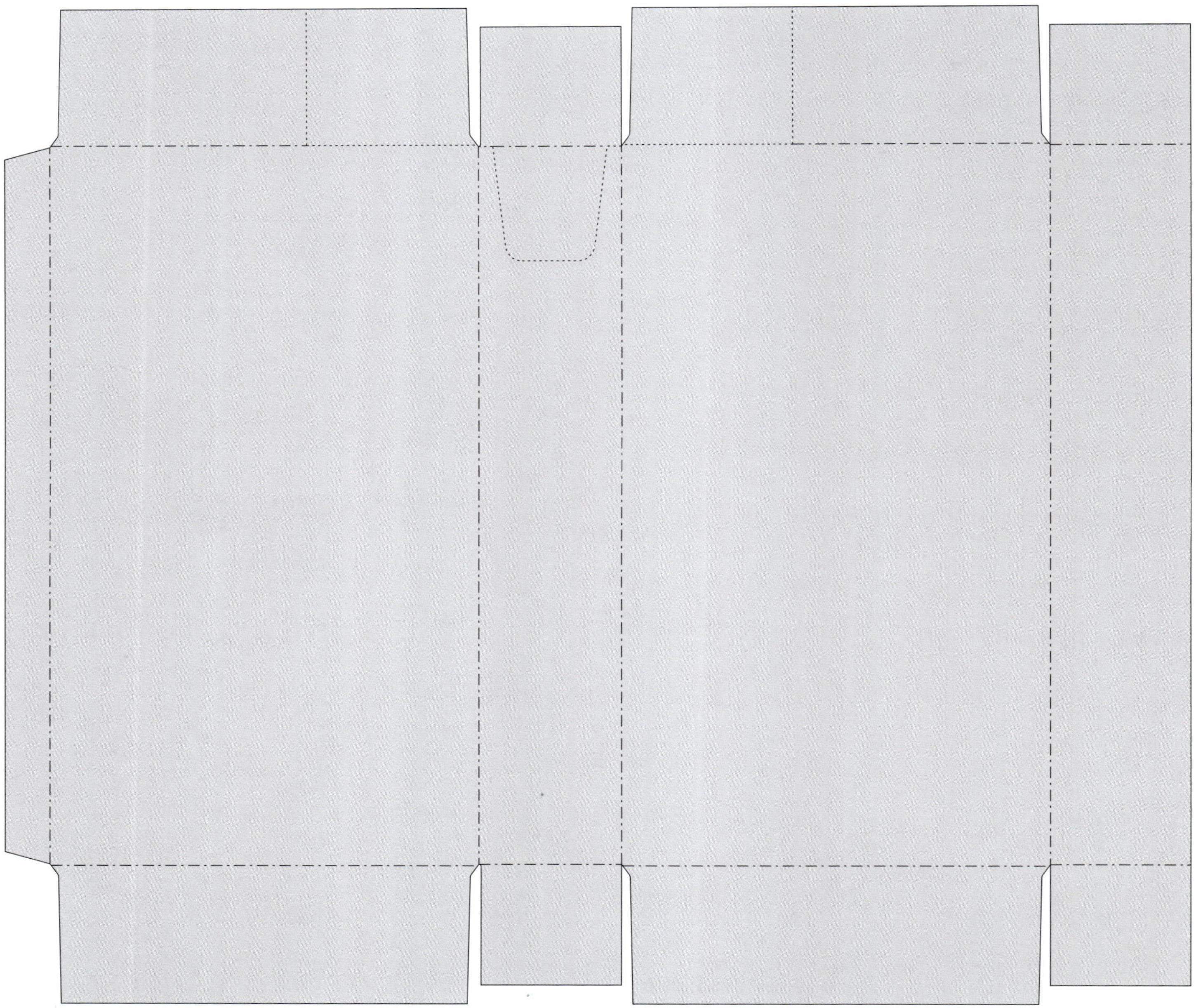

This carton shows how perforations can be used to provide access to the product, rather than having any openings at the ends. The perforation could be of any shape that would most usefully provide access to the product; complicated shapes are inconvenient because they could be difficult to remove. Attention must be paid to the strength of the carton without the perforated panel since some rigidity may be lost once this section is removed, although the sides have a full overlap to provide added strength. As this kind of opening is irreversible, it is not convenient for perishable goods, but is practical for nonperishables such as paper tissues.

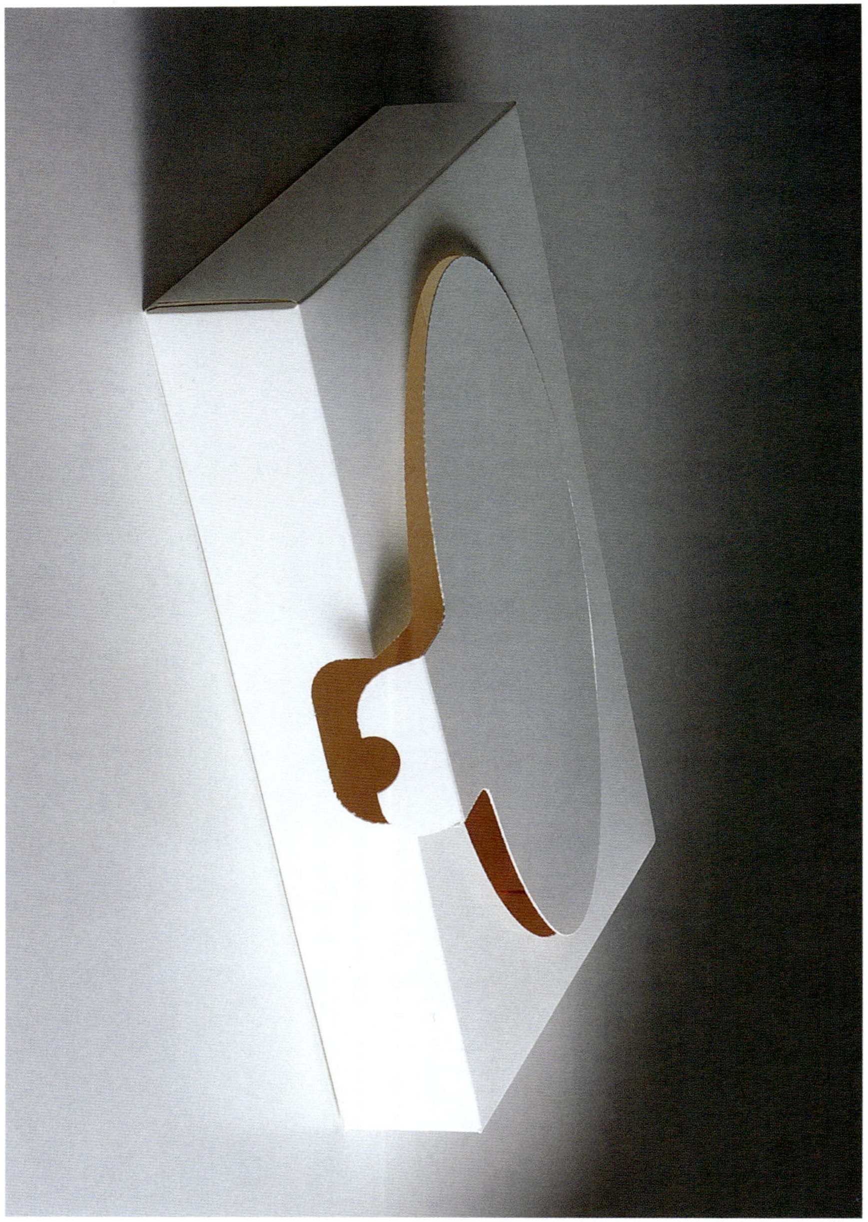

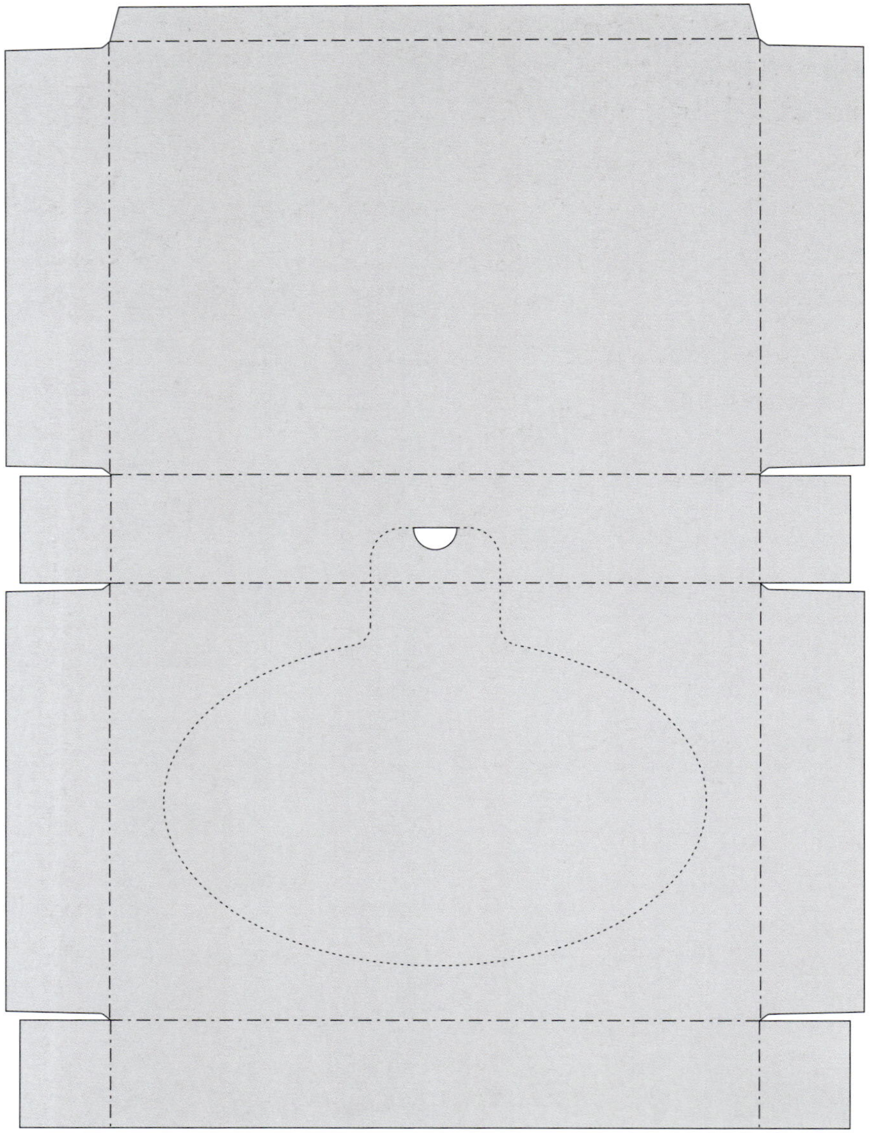

<<< >>>

P.I. Design International

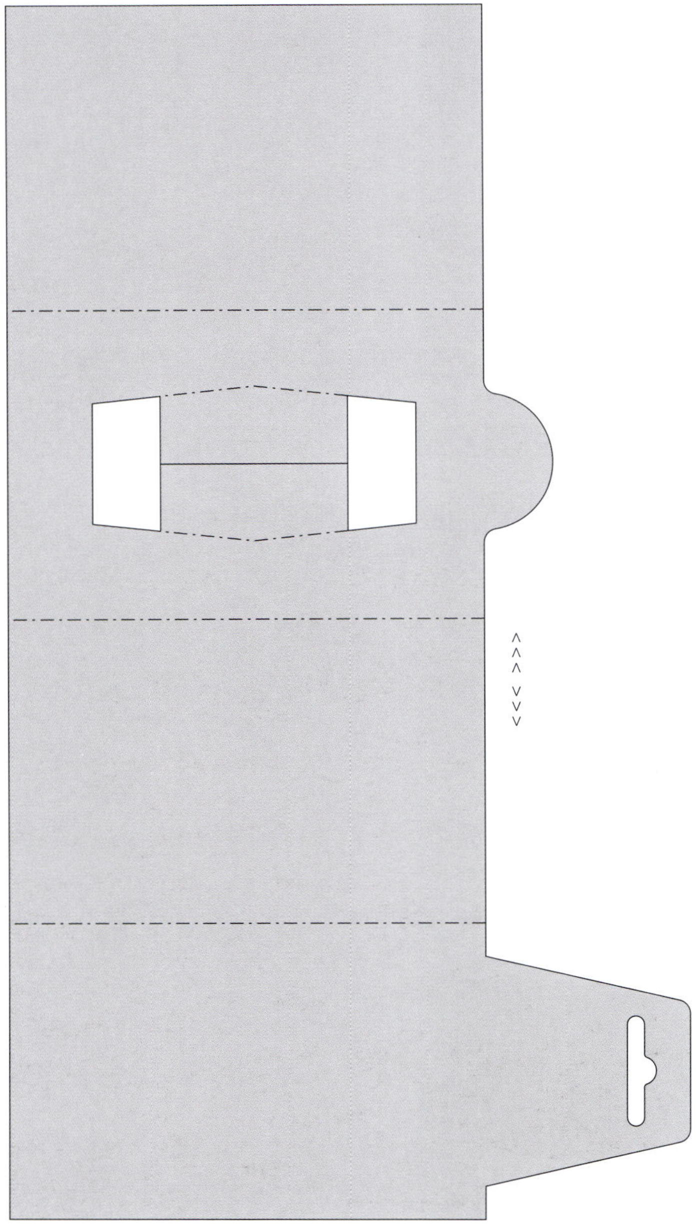

FABRIC DISPLAY CARTON

This carton allows the consumer to touch and feel the product it encases, while providing a point-of-sale display unit. Designed to display fabrics such as chamois leather, it uses a minimal amount of card to frame the product rather than contain it. The frame has a window through which the consumer can touch and feel a sample of the fabric, the rest of which hangs below. This demonstrates how a carton need not fully contain and protect a product to enhance the value and perceived quality of the display.

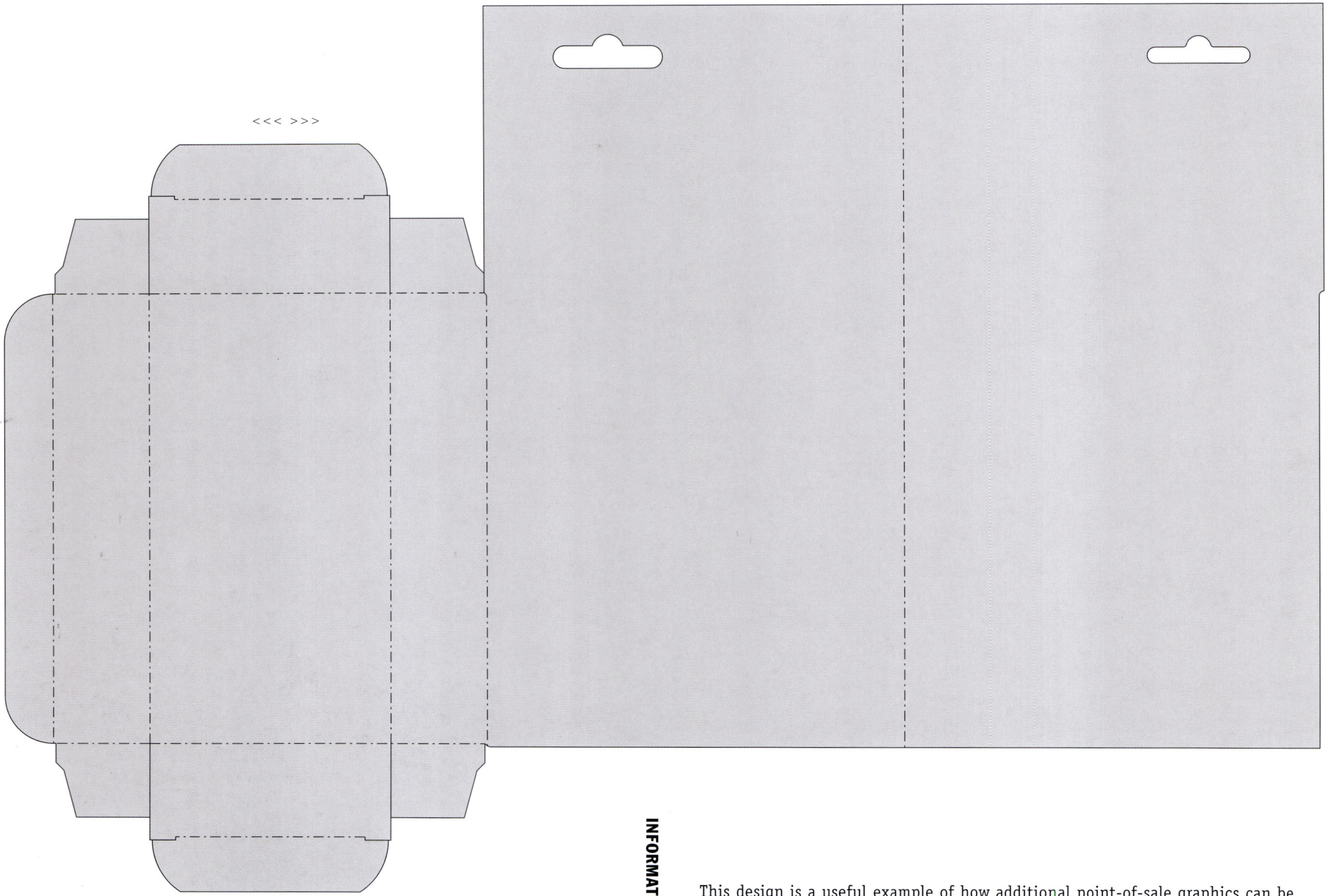

INFORMATION DISPLAY CARTON

This design is a useful example of how additional point-of-sale graphics can be given space on a small product package. The oversized rear panel can be doubled over for extra rigidity, while allowing printing on just one side of the flat sheet before folding. Textual information relating to the use of the product might appear on the reverse of the panel, thereby preventing the need for an instruction leaflet inside the carton. The Euroslot is optional but attention must be paid to its position, this should be above the center of gravity or the package would fail to hang straight.

PODIUM MOUNT

A simple, innovative design for displaying long objects such as pens or tools. This design, cut from a single sheet of card, makes good use of material with very little waste from the die-cutting process. The profile of the pedestal is cut from the flat sheet before being folded to create a 3D supporting structure for the product. The shape of the backing card or the pedestal could change to suit requirements.

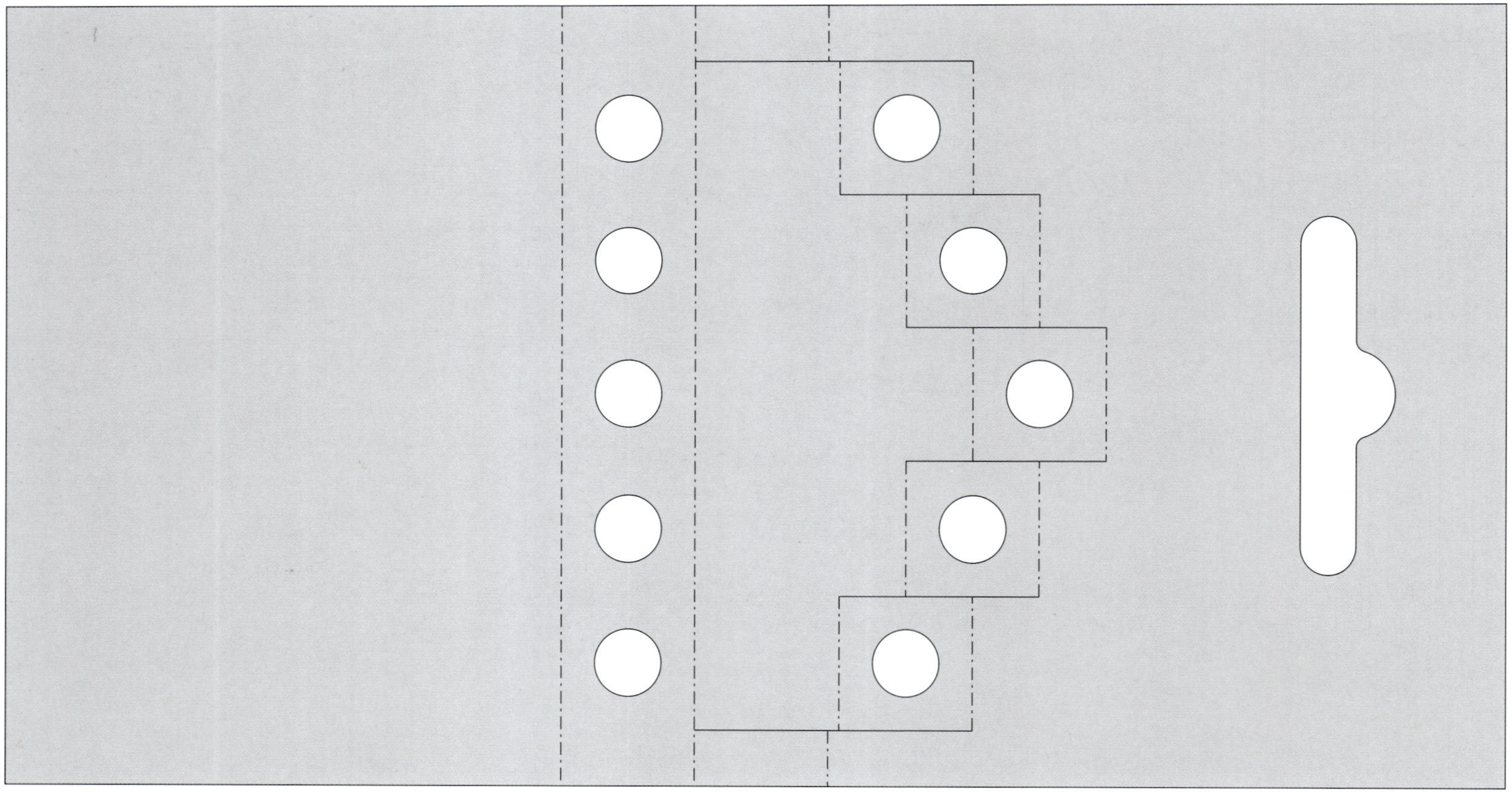

<<< >>>

Mathew Chin, Surrey Institute of Art and Design

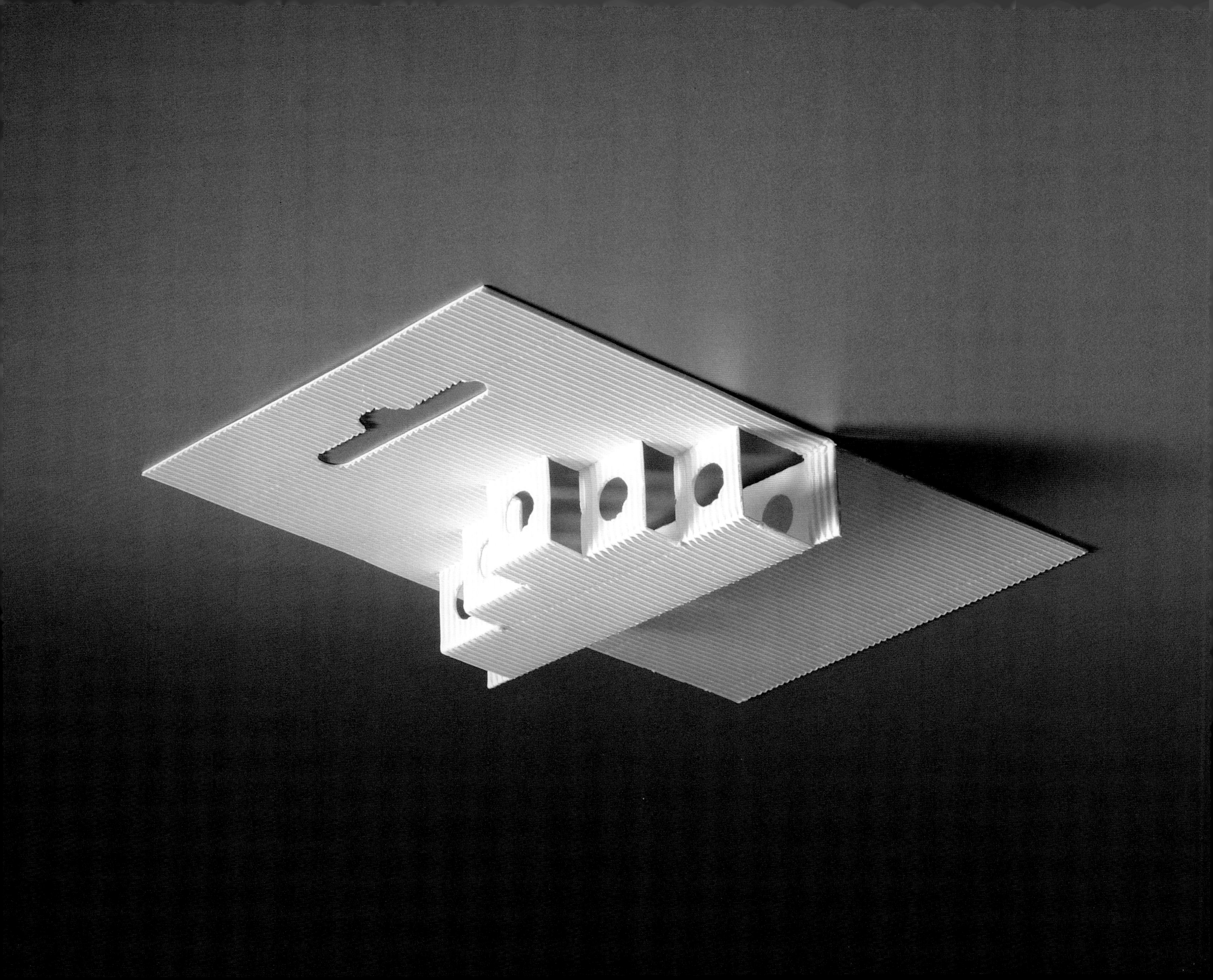

P.I. Design International

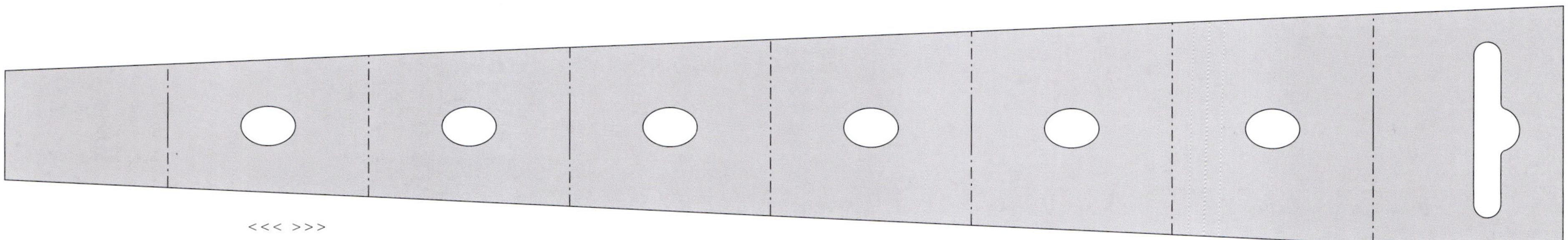

CONCERTINA-SHAPED CARD

This simple display pack can be manufactured from any sheet material. The pack effectively protects the product while allowing the consumer to see and touch it, it is particularly suitable for long narrow items such as pens, cutlery, or needles. This was originally designed to pack a new cardboard pen and therefore continue the environmentally-conscious theme of the product. One important consideration is that the holes be fractionally larger than the product's width, as the card will be at an angle and the holes consequently will be smaller.

SPIRAL BINDING DISPLAY

An inspirationally simple and attractive design to hold any kind of long object, this would be suitable for novelty or point-of-sale packaging. The backing card could be of almost any shape and contain point-of-sale graphics and a handle or Euroslot could also be added. The product is placed on the backing card before a strip of similar material is wrapped around in a spiral through the die-cut slits. The spiral effect provides an attractive method of containment, while the consumer is able to touch and visualize the product. No glue is required in the construction of this packaging.

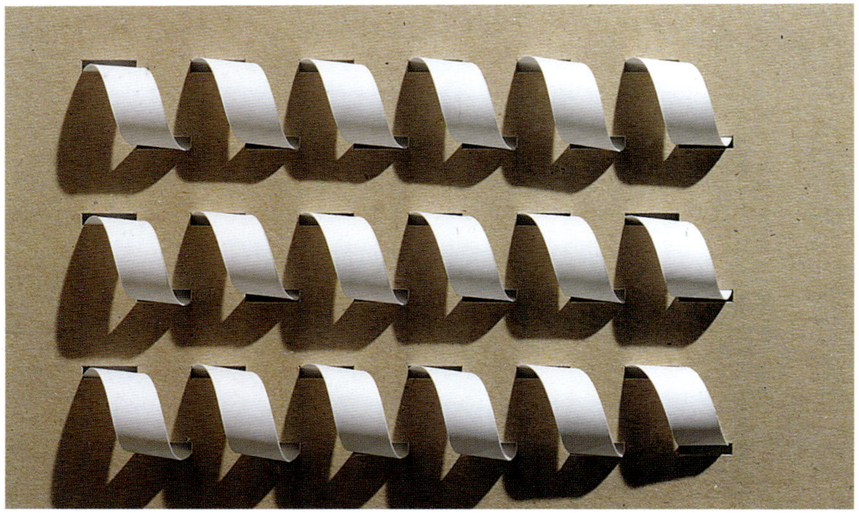

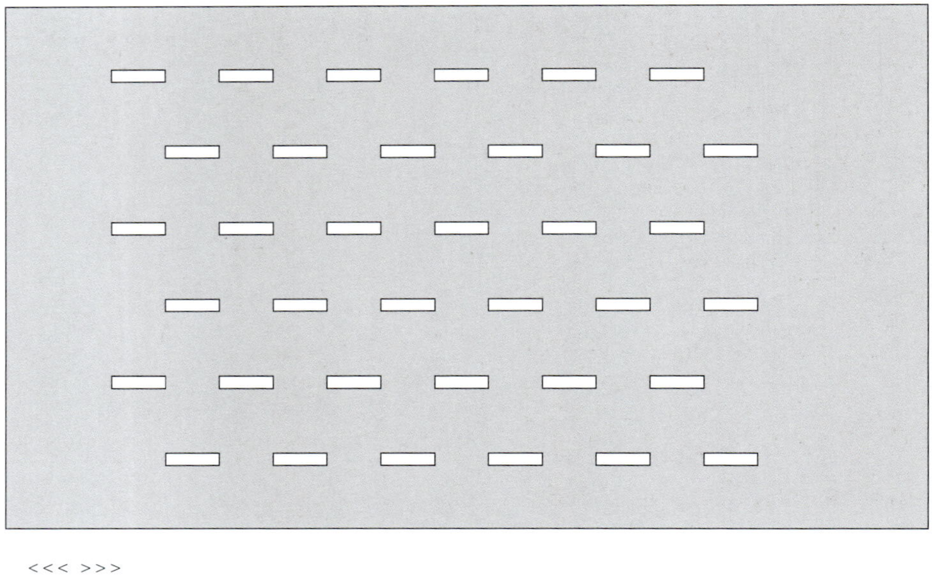

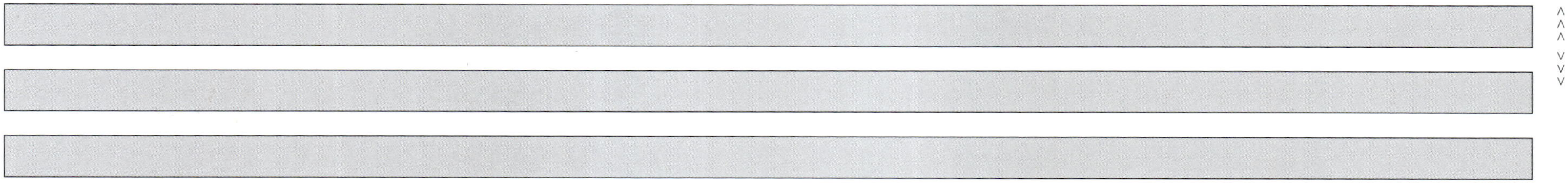

Lisa Gardiner, Surrey Institute of Art and Design

NOVELTY "SAFE" CARTON 1

Originally designed to represent a safe for the sale of padlocks or other security devices, this demonstrates how a two-piece carton can provide an attractive and interesting method of containment. Requiring no adhesive, each carton is constructed to fit one inside the other, creating a double-walled internal compartment with tuck-flaps contained by an outer cube with a framed entrance and double-flap lid. This innovative design demonstrates the potential complexity of integrated two-piece packaging; with interesting opportunities for use of different materials.

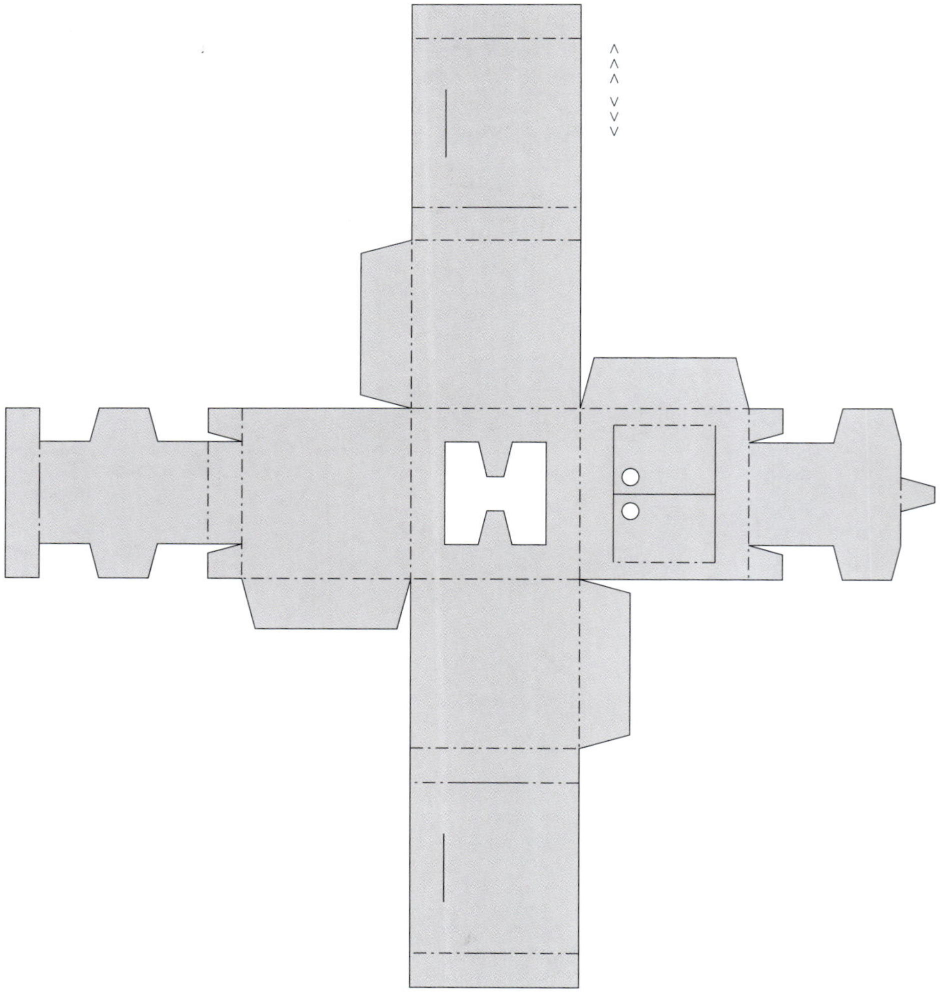

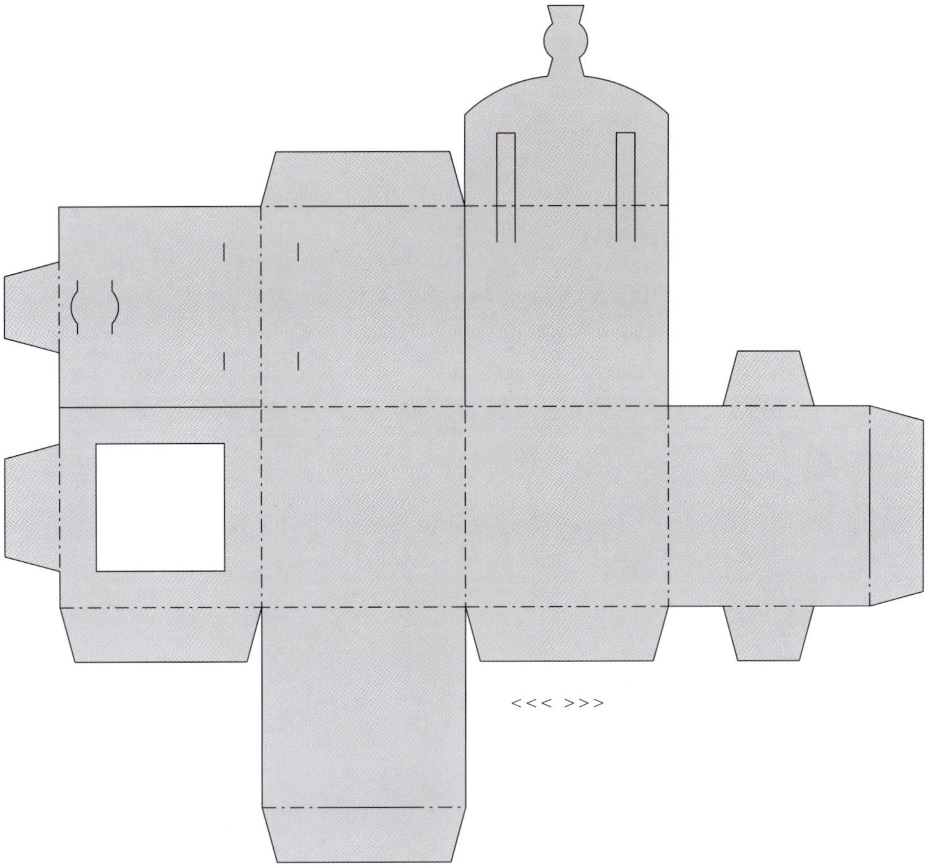

Neal Manning, Surrey Institute of Art and Design

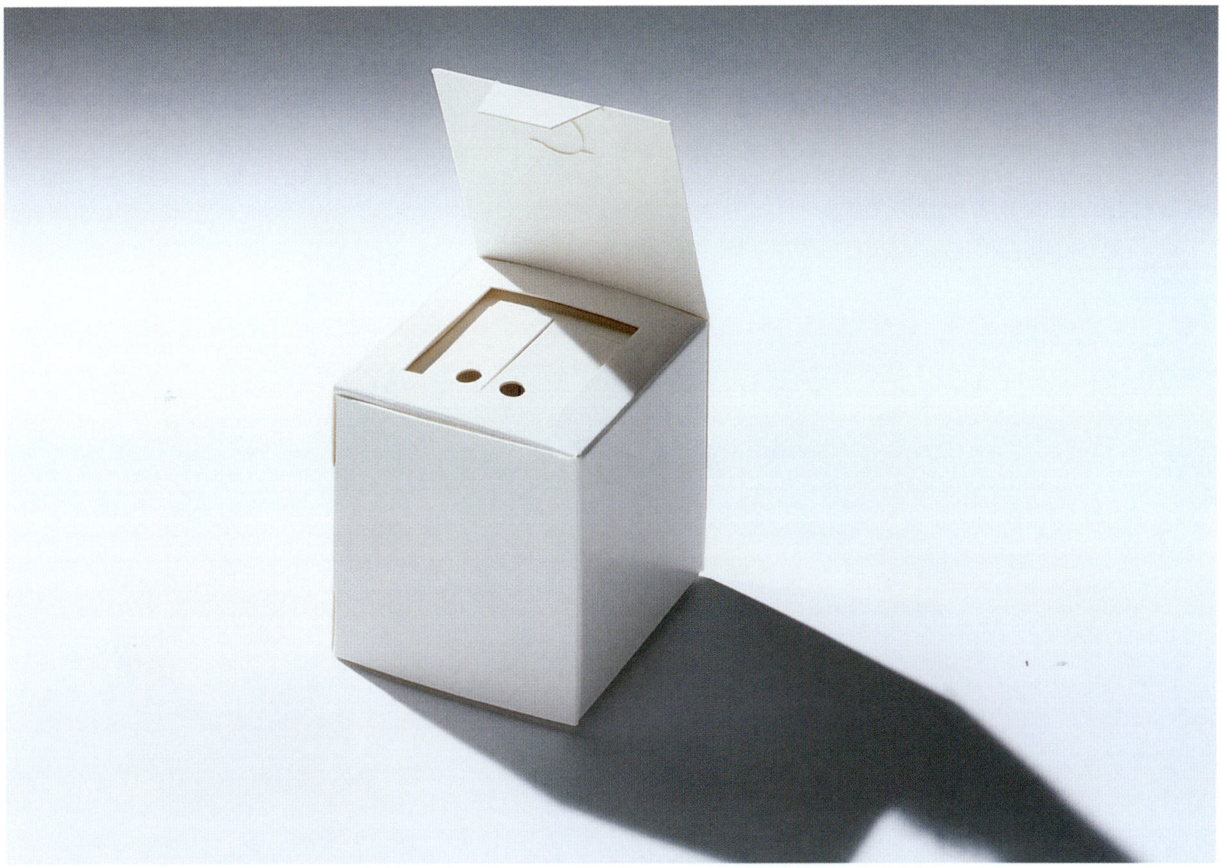

Patent Pending, Danapak

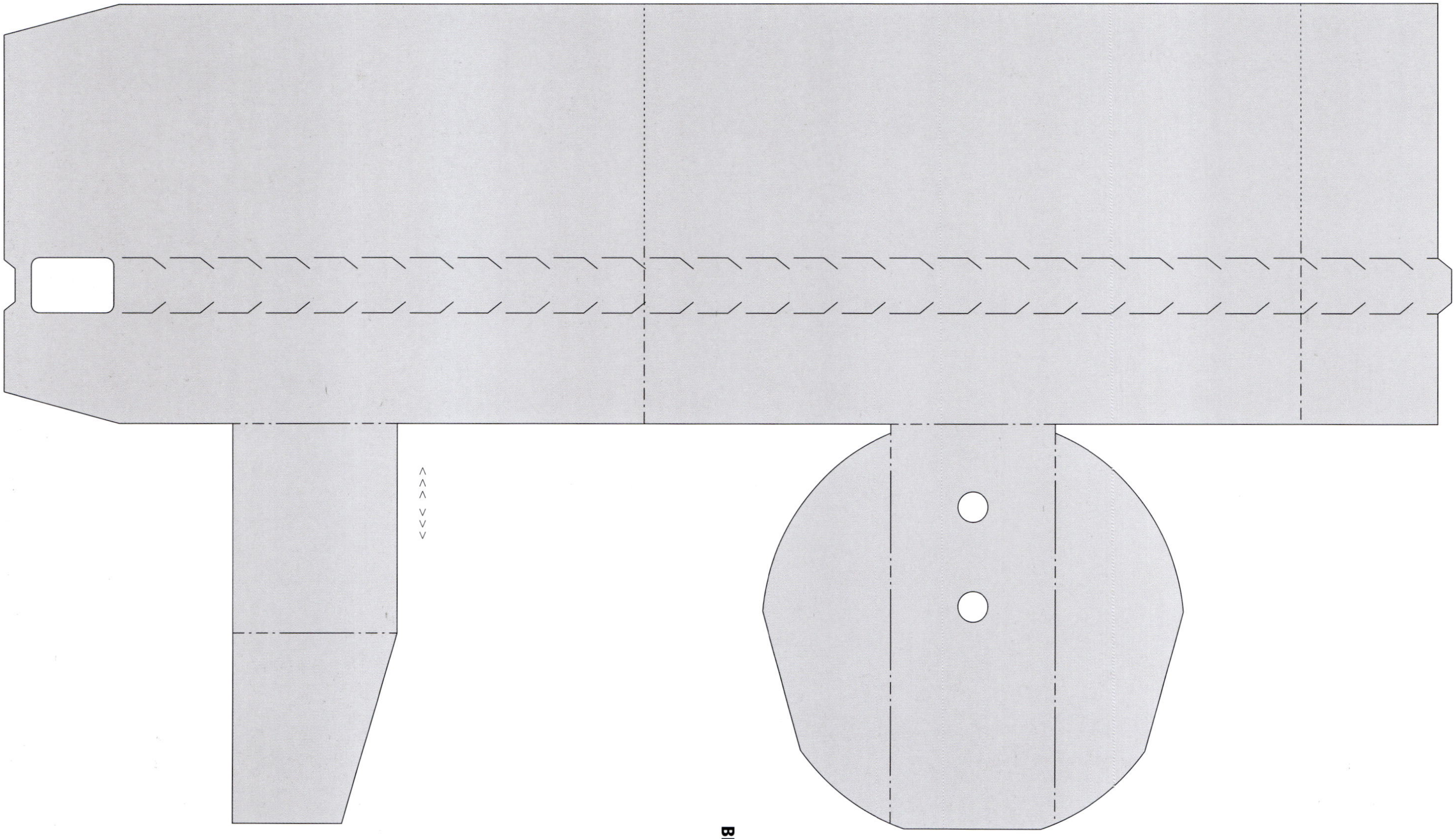

BIODEGRADABLE ROUND POP-UP TUB

This erectable tub can be used as a flower pot for seedlings, as an alternative to existing plastic pots. The flat-packed pot is easily erected and filled with soil. Once the seedling is ready to be planted, the base is removed by simply tearing off the pull-strip before placing the package in the ground, the sides support the seedling before biodegrading. This is a good example of responsible packaging design in tune with environmental demands. Alternative uses for the design could include easily-erectable holders for any cylindrical product or bulk-packed items, such as a bottle or pens.

HEAT-SEALED LIQUID CONTAINER

A classic liquid carton design made from a single sheet of laminated board. Through a series of simple folds this flat sheet constructs to form a fully-sealed rectangular block. This is an excellent shape for transportation purposes and shelf storage, as individual packs can be stored next to each other with no wasted space. To access the product the carton can be opened by tearing the seal on the top, thereby creating a reusable spout through which the contents can be poured. Adaptations to this design include a die-cut polythene-sealed hole in the top for a straw, or polythene-sealed perforations in the top for ease of pouring access.

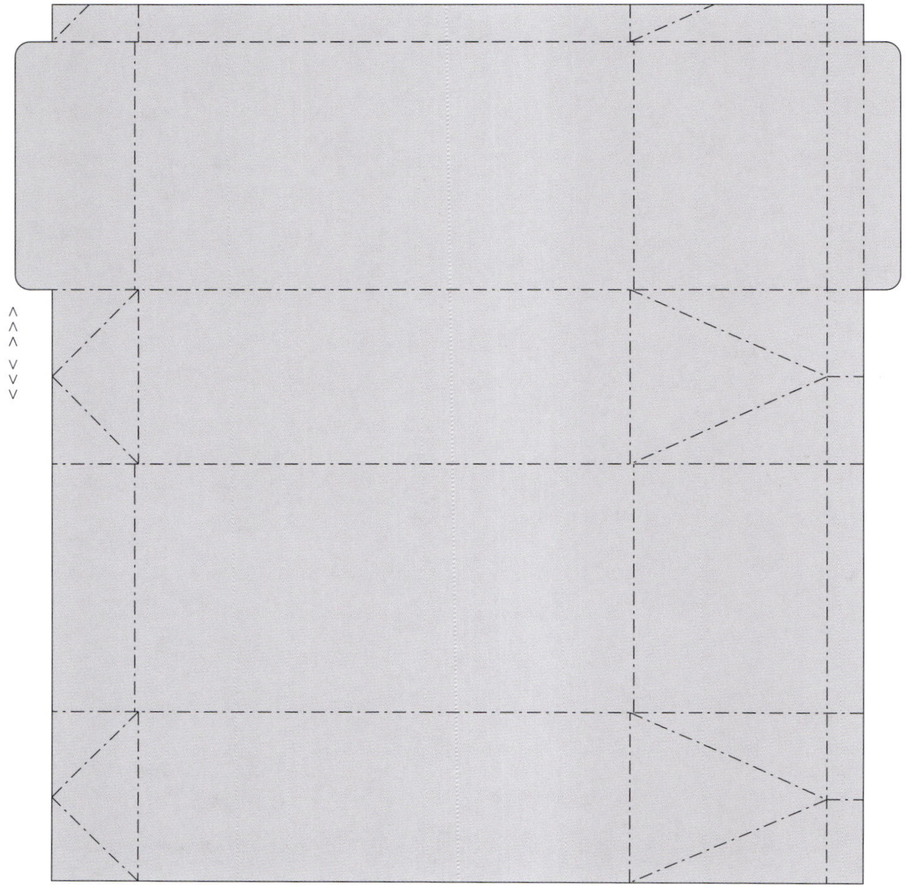

GABLE-TOP CARTON

This gable-top carton is a common design for confectionery or liquid products. It was made famous for containing liquids because of the easy-opening mechanism which gives access to the contents by pushing back the ridge on the top of the carton and pressing in the sides to break the adhesive seal. The internal tucks at the top of the carton also provide easy handling. The material used in this carton depends on the contents; liquid would require a multilayered, waterproof board.

Tetra Pak, Tetra Classic

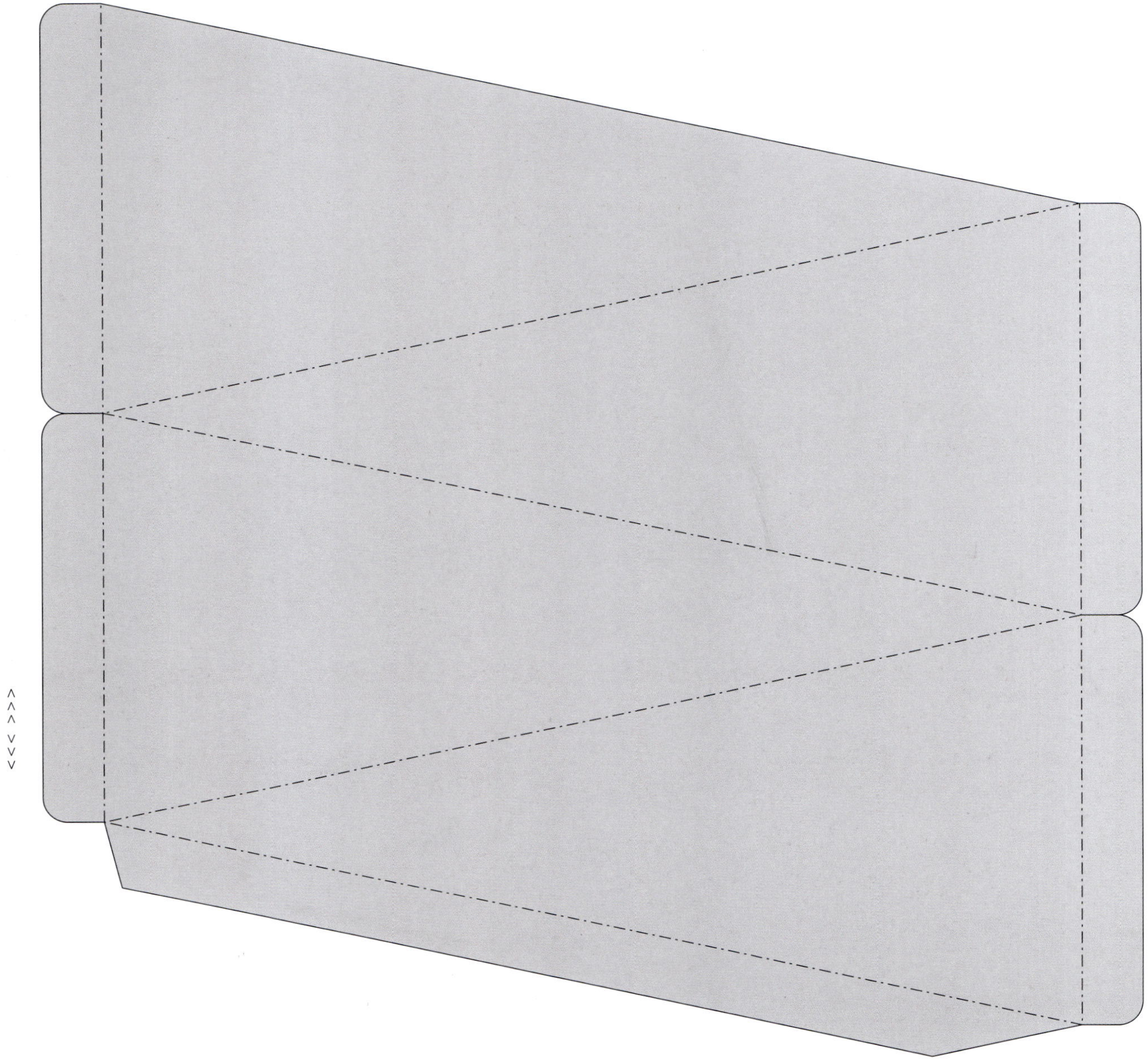

TE

…nd classic example of early card containers designed to hold liquids. …s still used around the world and it formed the basis for subsequent …ments. Laminated carton board is required for waterproof qualities …ems include the filling and dispensing of the liquid. This is a …yet complex design classic.

BOTTLE-NECK SUPPORT CARTON

Cartons can be used in conjunction with the visual appeal of the product on display. This carton is a good example of a method of grouping products without detracting from the product itself. It can be used to display families of products or special-edition product groups.

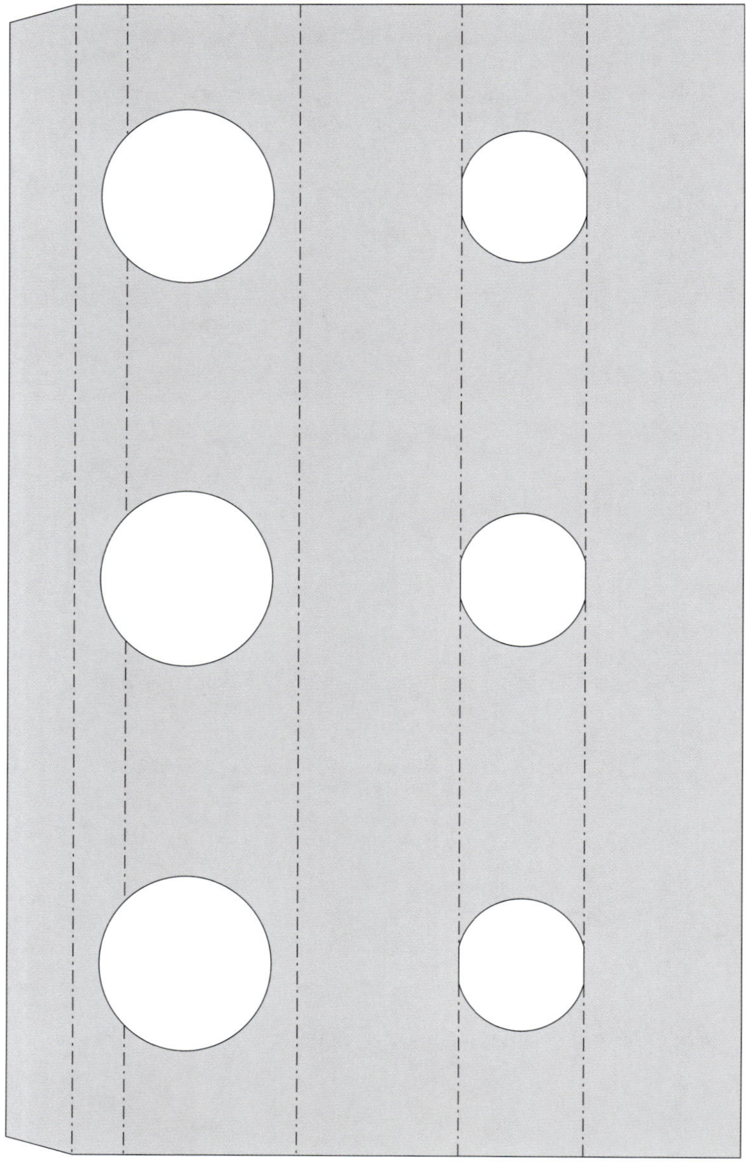

<<< >>>

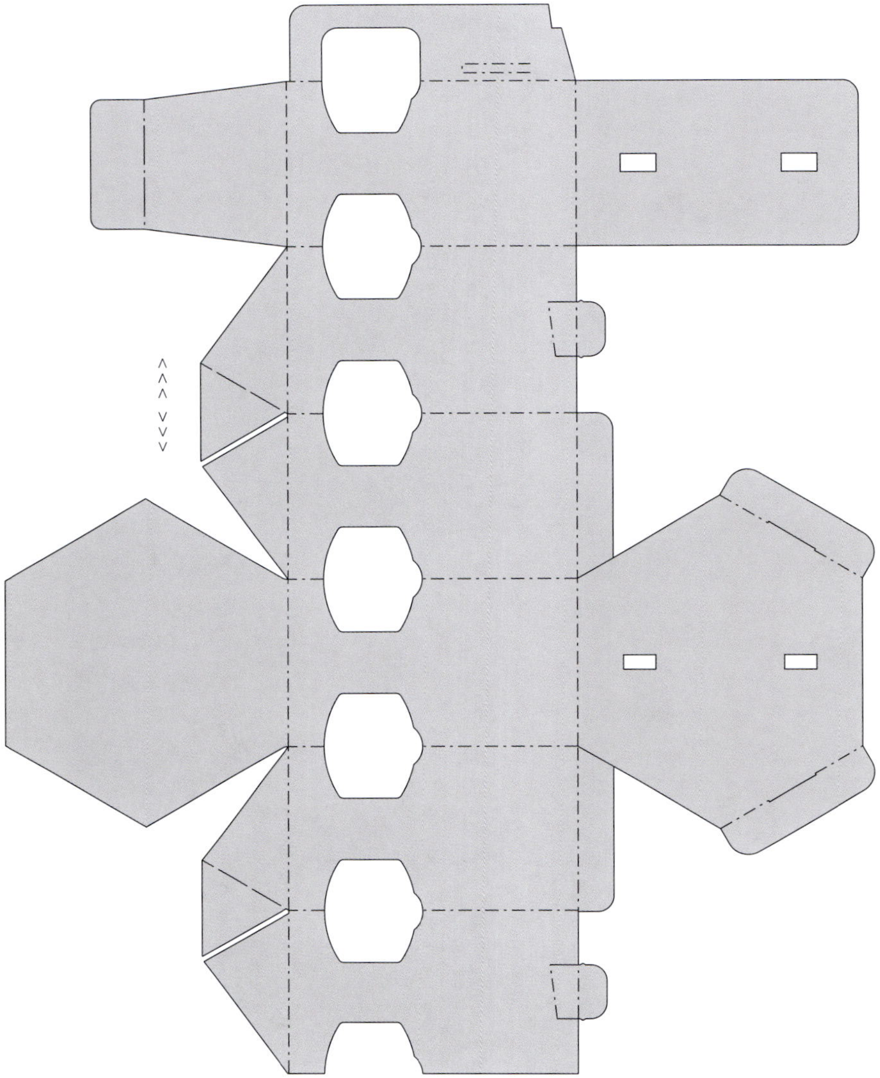

HEXAGONAL SIX-PACK CARRYING CASE

This shows how an unusual-shaped carton enhances the product it encases while providing stability and secure containment. The product labels are framed by the die-cut holes, this might allow a more inventive use of point-of-sale space on the side panels. By increasing the number of edges an octagonal version would be possible, depending on the strength of the material encasing the product. The case requires gluing along the base and the handle is optional.

BOTTLE CARRIER

An erectable carton for holding bottles, this features a crash-lock base and a one-piece design which is erected by sliding open the carton so that the diagonal cuts form the dividers for the individual bottles. The simple configuration allows any number of bottles to be stored in a single package by simply increasing or decreasing the number of diagonal cuts. This convenient carton features a handle in the central dividing section for easy carrying.

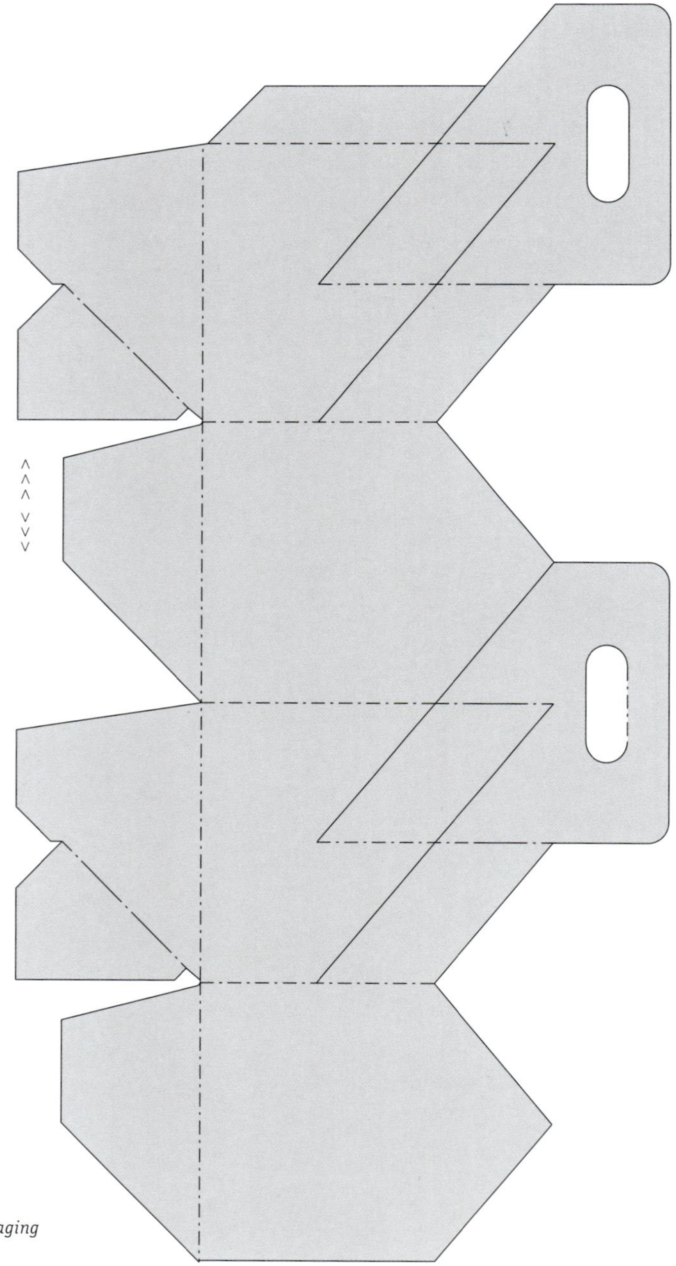

Cullen Packaging

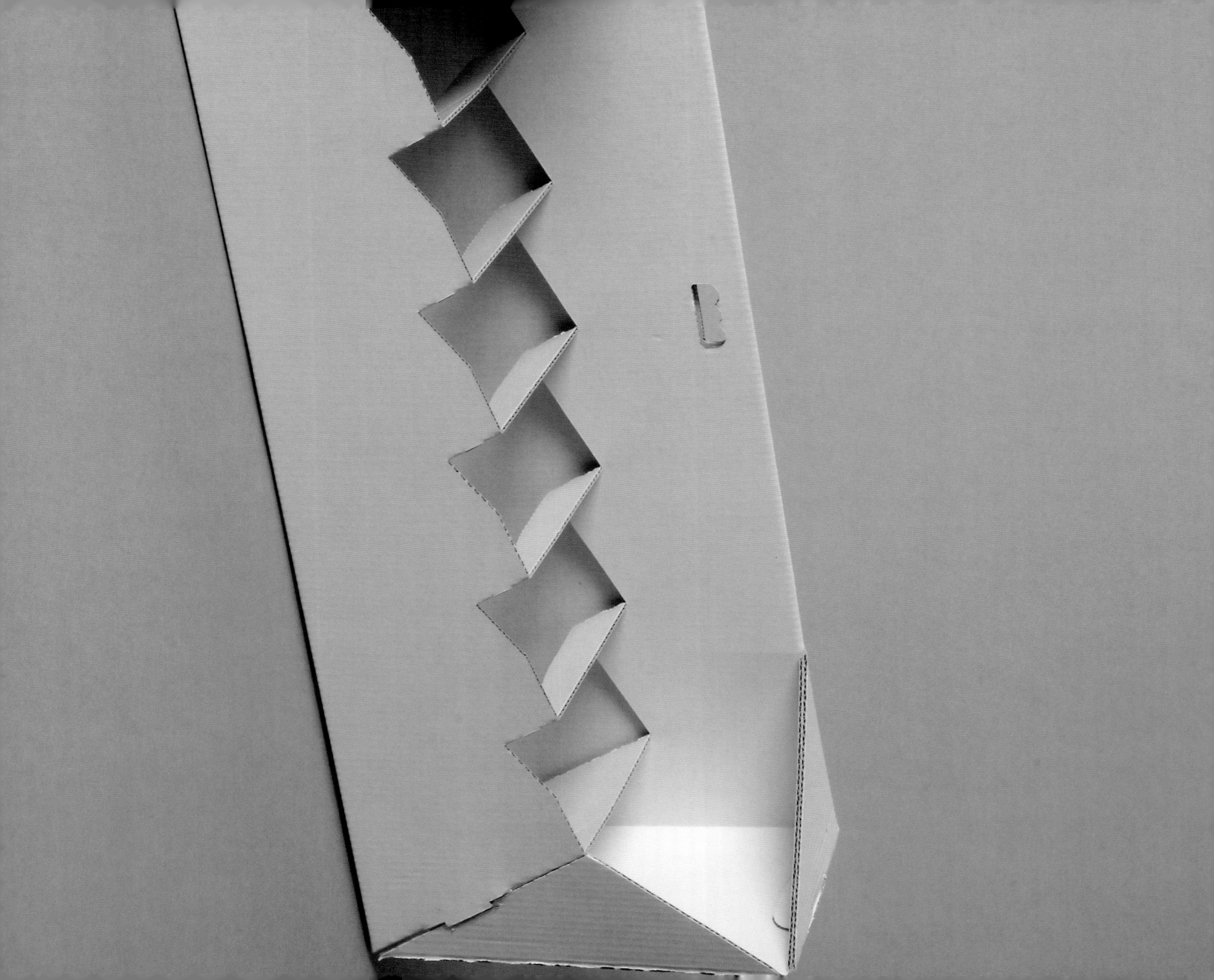

CPC Packaging

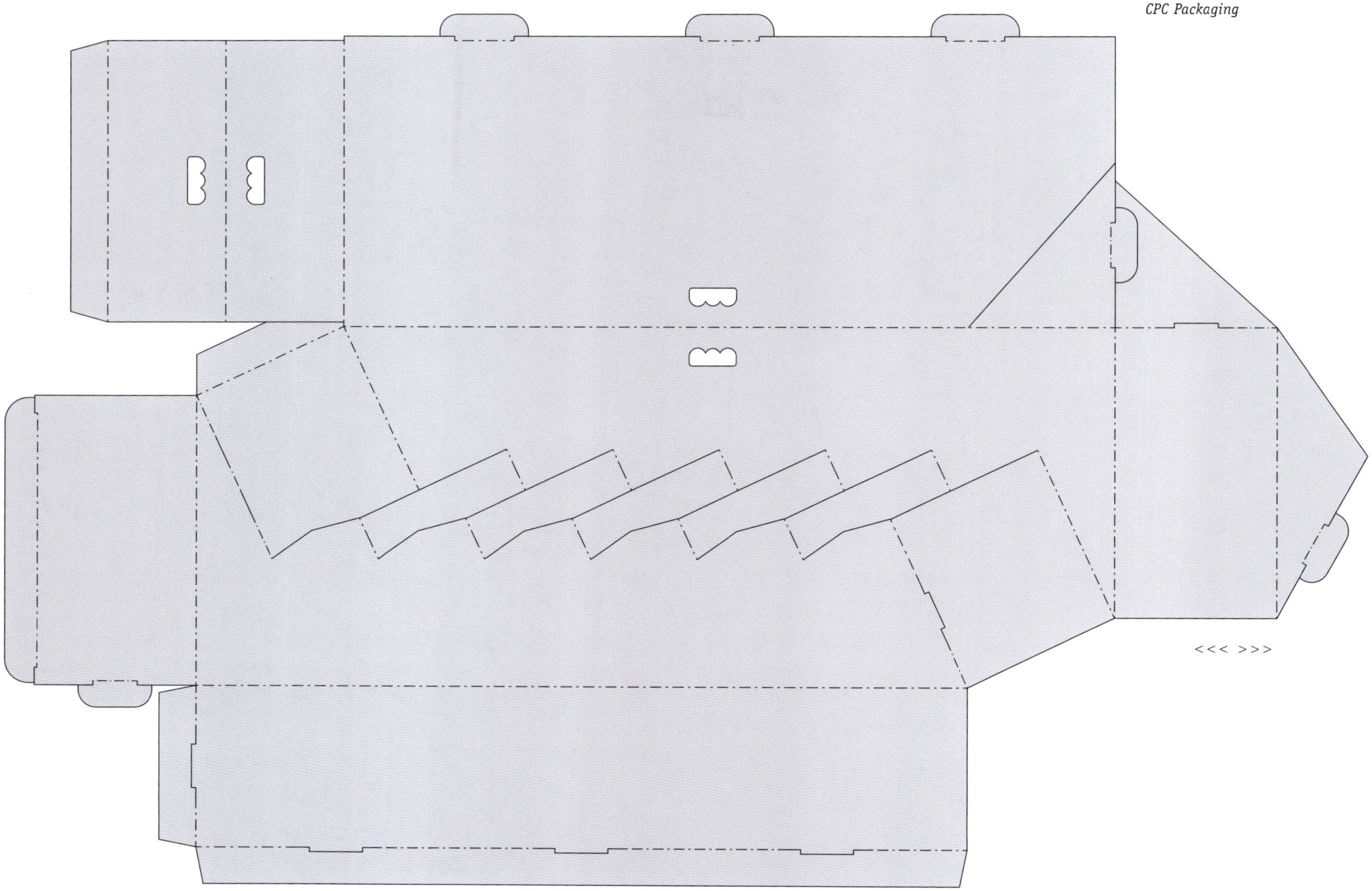

HANGING PARTITIONED POINT-OF-SALE

Euroslots on the length and width of this pack allow it to be hung in two different orientations. Diagonal slots and creases in the center of the net, which is cut from a single piece of corrugated board, create the partitions. Careful consideration to the design of each end provides structural rigidity to the overall pack.

FREESTANDING POINT-OF-SALE STACK

Fabricated from a single piece of corrugated board, the net of this point-of-sale device reveals an effective and efficient use of materials. The angled cuts provide 3D and rigid structural qualities, while the flaps at the rear of the pack provide a central internal separator and, together with the insertion of flaps and tabs at the top and base, complete the pack's construction.

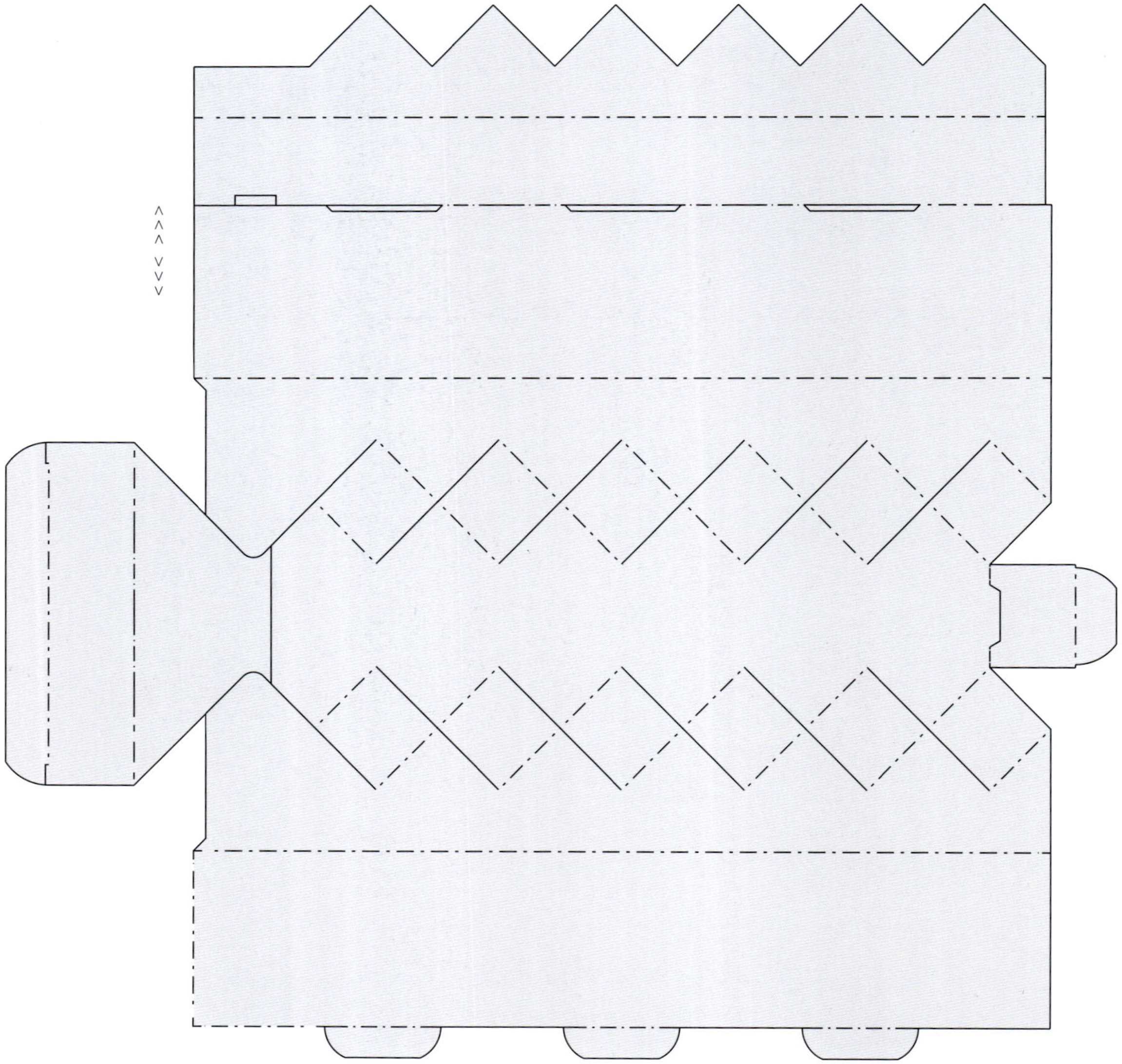

CPC Packaging

AssiDomän Frövi

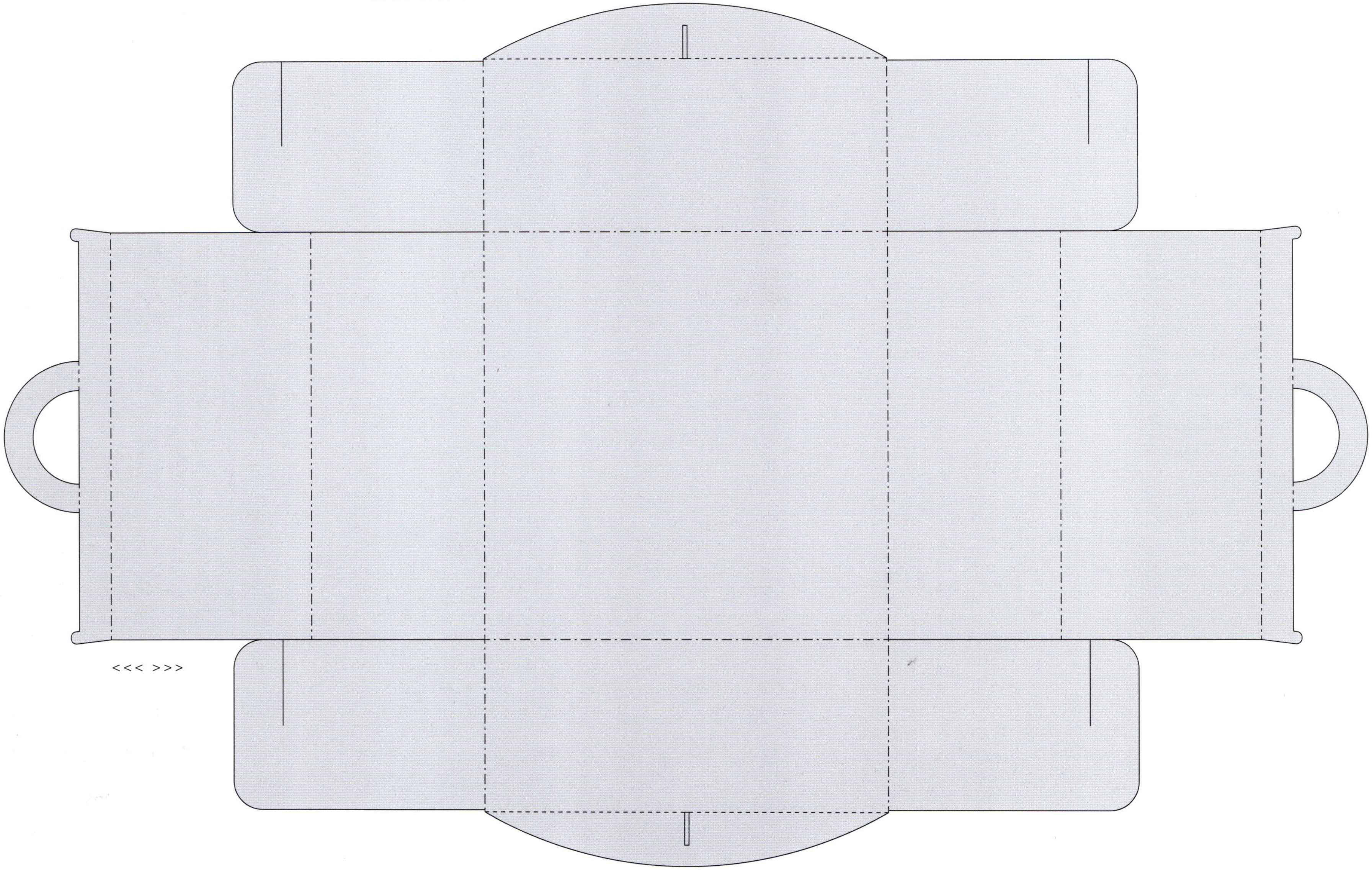

GIFT BOX WITH CARRY HANDLE

This large gift box is both easy to construct and offers excellent structural integrity. Each side is strengthened with interlocking internal flaps to give the carton a robust quadrangular form, while the lid's design, which uses two flaps and a handle device that meet in the middle of the carton and lock through slits at the side, provides a strong and effective method of carrying the constructed carton. These structural properties, together with the carton's flat-pack design, make it ideal for the packaging of fragile gifts, such as cakes.

TREASURE CHEST

A simple rectangular carton has been adapted slightly to incorporate a rounded top and curved sides, replicating the form of a treasure chest. The rounded lid, which is slightly wider than the carton's body, is formed by wrapping it over the curved sides and locking it to the front of the carton. The lock is formed using a tab on the lid that fits into a slot at the front of the carton.

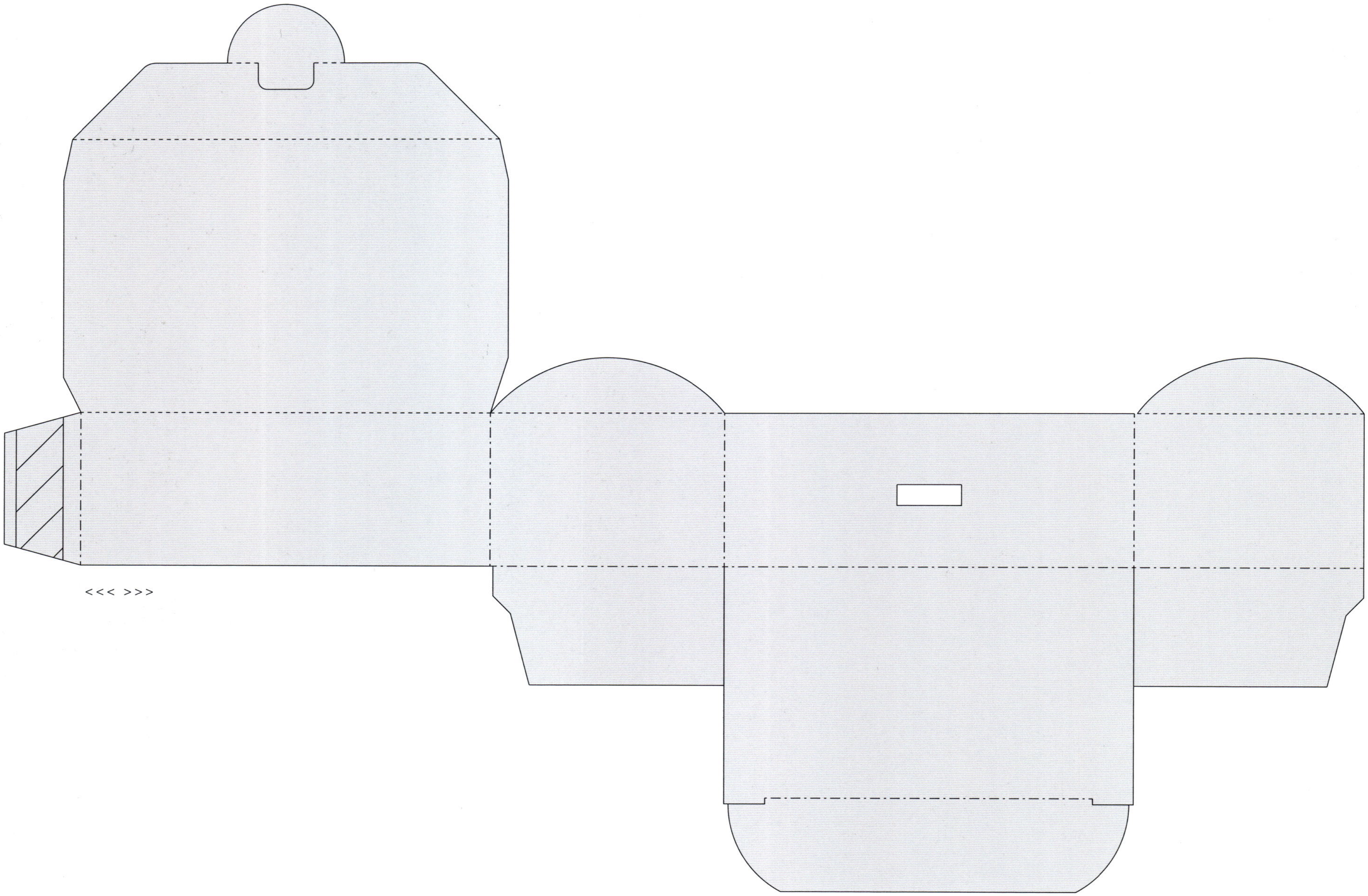

AssiDomän Frövi

AssiDomän Frövi

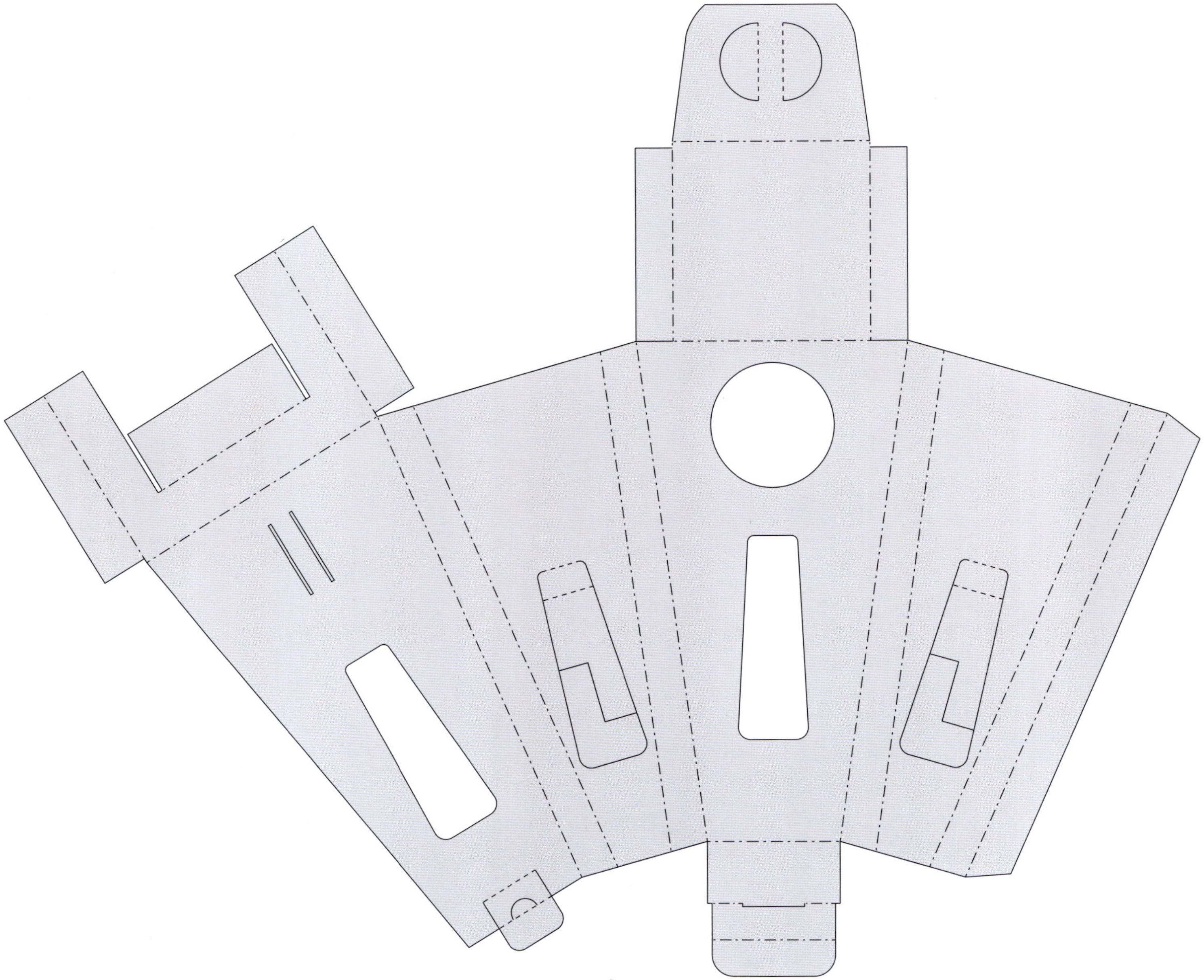

TWO-PART DISPLAY CARTON

Providing physical protection and visual display to two differently shaped products within a single one-piece carton poses many design problems. Here, the carton's tapered form reflects the chosen solution, which houses the wider, shorter product in the base and the taller, thinner product above. Tabs, taken from the die-cut windows at the sides of the carton and folded inward, separate the two products. The carton is sealed using a standard tab-lock at the top and a butterfly-lock at the rear of the base.

BOTTLE CASE WITH INTEGRATED NECK-LOCK

Many glass bottles are packaged in standard tall rectangular boxes with crash-lock bases and tab-lock tops. This follows a similar design, but with the addition of an adaptation to secure the neck of the bottle to prevent it from moving around in transit. In the past, this has often been achieved by adding a separate square insert with a round hole through which the neck passes, but this design incorporates the neck-lock into the tabs at the top of the carton, which fold internally, allowing the whole carton to be produced from a single sheet of board.

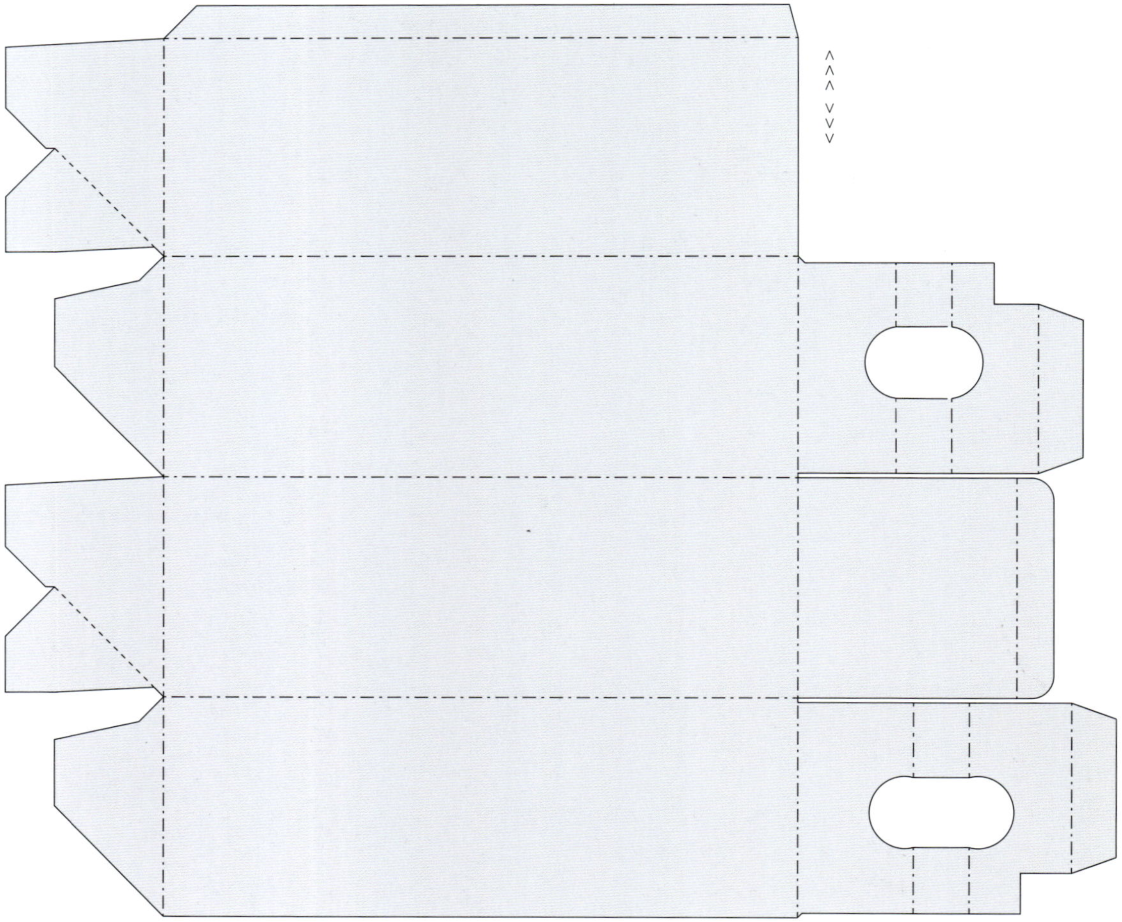

AssiDomän Frövi

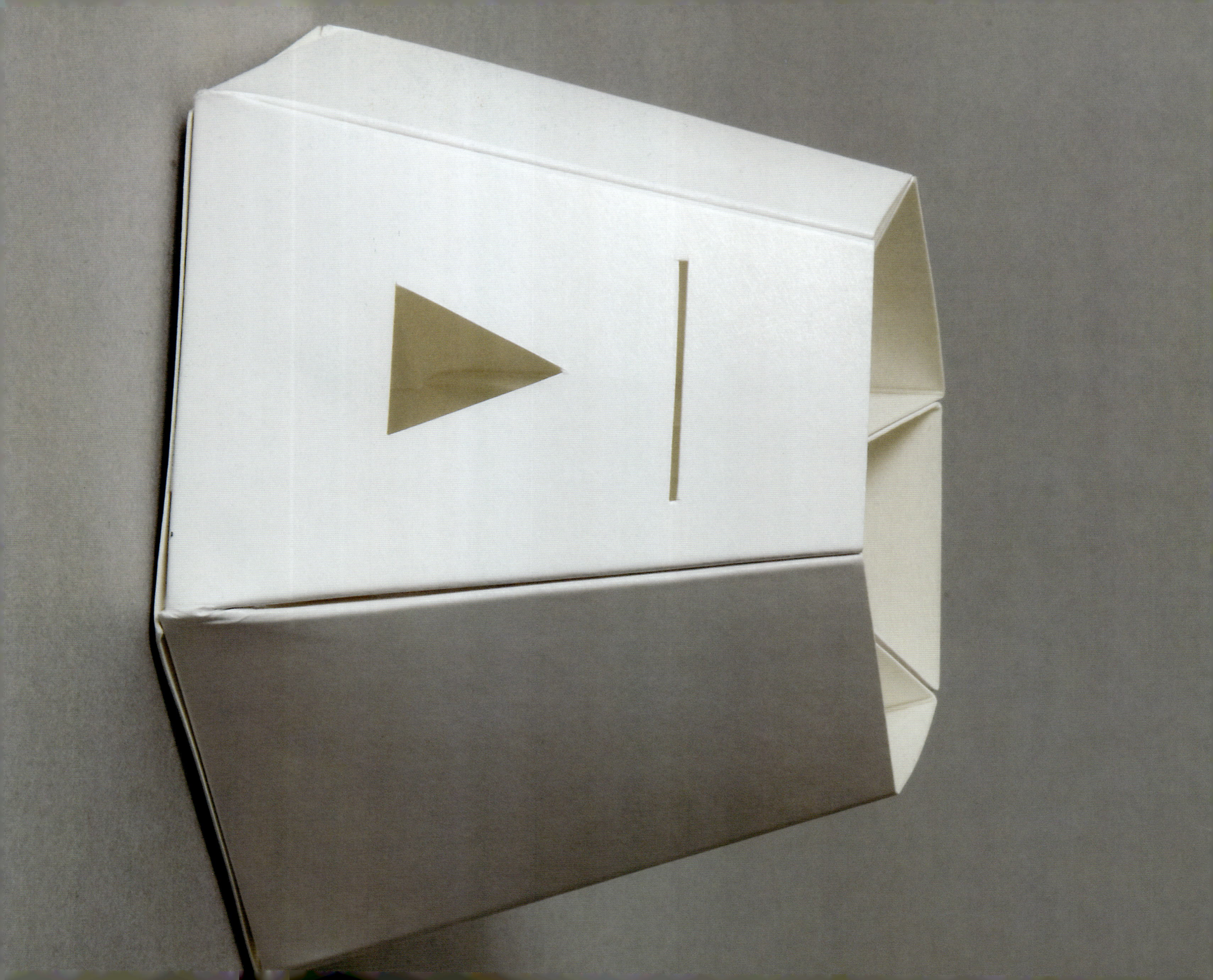

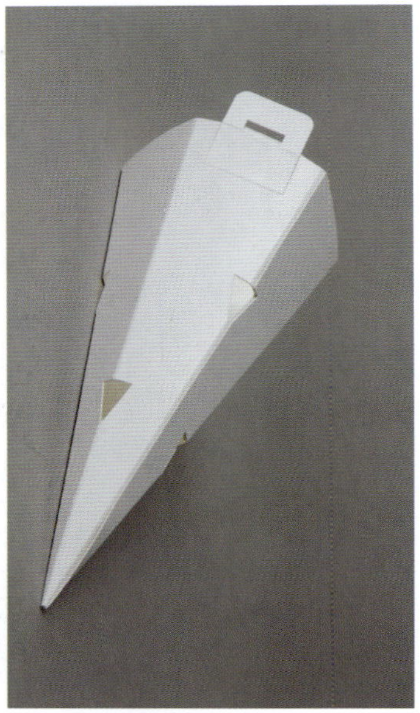

AssiDomän Frövi

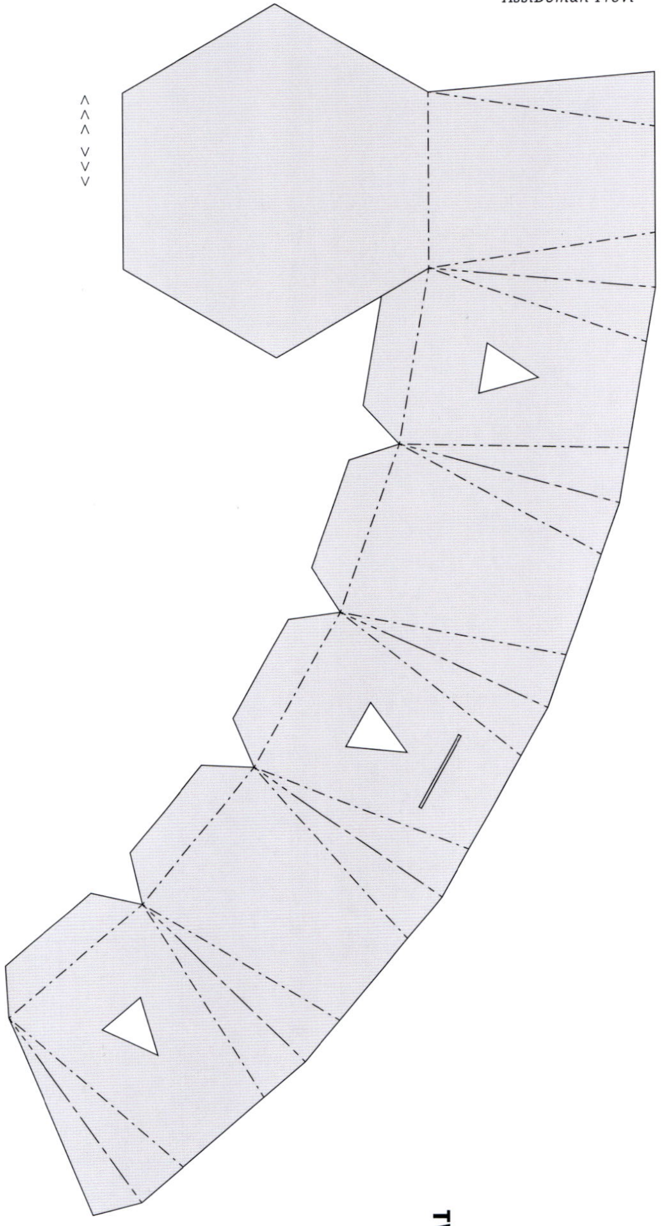

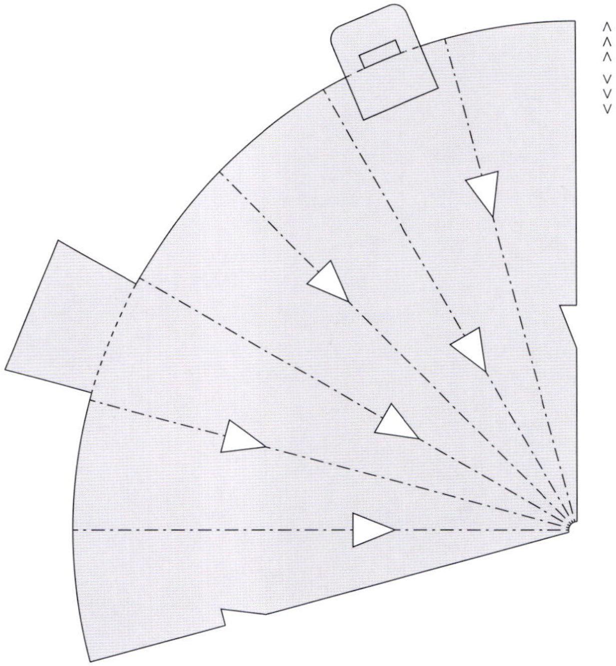

TWO-PART HEXAGONAL CONE

Two components are used to create this hexagonal cone. The top is a regular six-sided conical form, glued down one edge and designed to sit on top of the base, secured by a tab-lock. The base is constructed using a hexagonal piece of board that is glued to tabs at the bottom of each side. Webbed corners allow the base to expand when the top is removed.

TABBED-HANDLE CARRY CASE

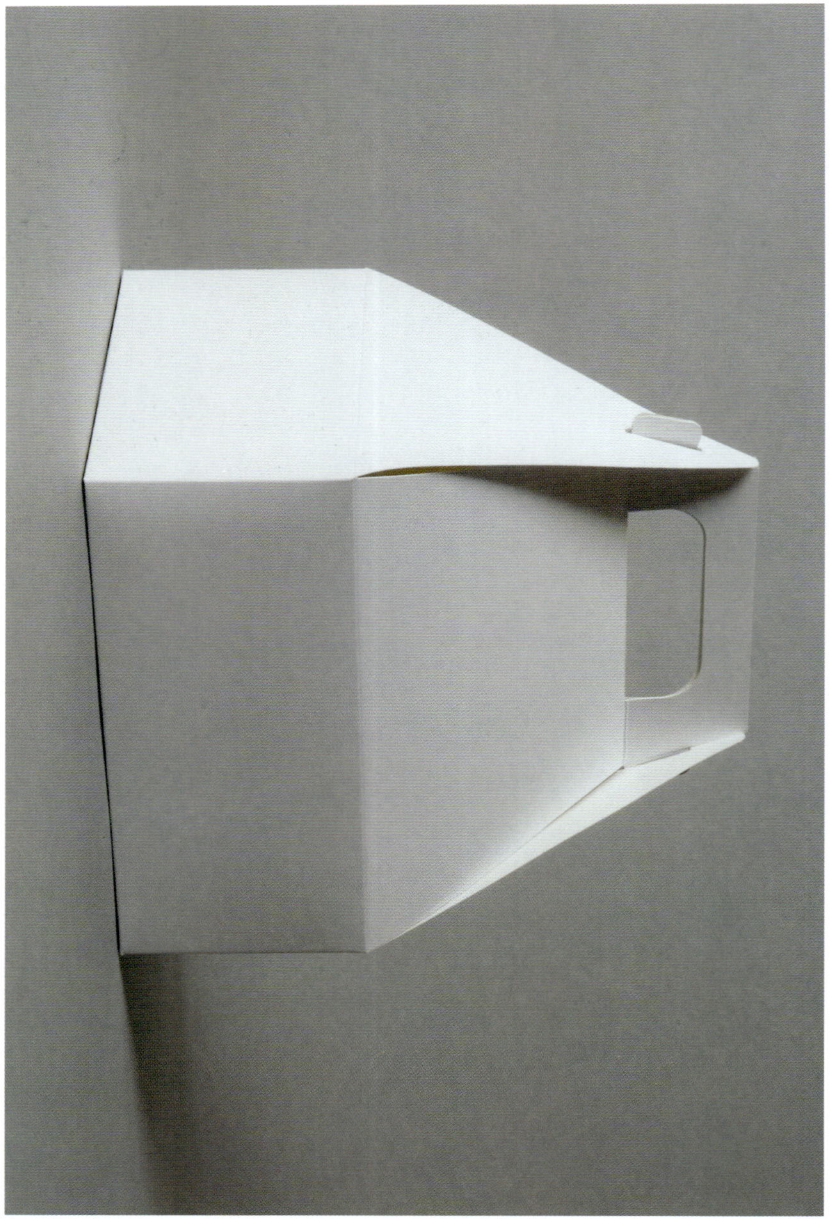

Unlike the design of other carry cases, where the handles on opposite sides of the case interlock with slits on adjacent sides of the case to provide an effective means of closing the package, this design has used protruding tabs that interlock with slits on the opposite sides to provide an efficient but less-fastened means of closure. The advantages of this design are that it provides a greater internal volume for the amount of material used and it is easier to erect.

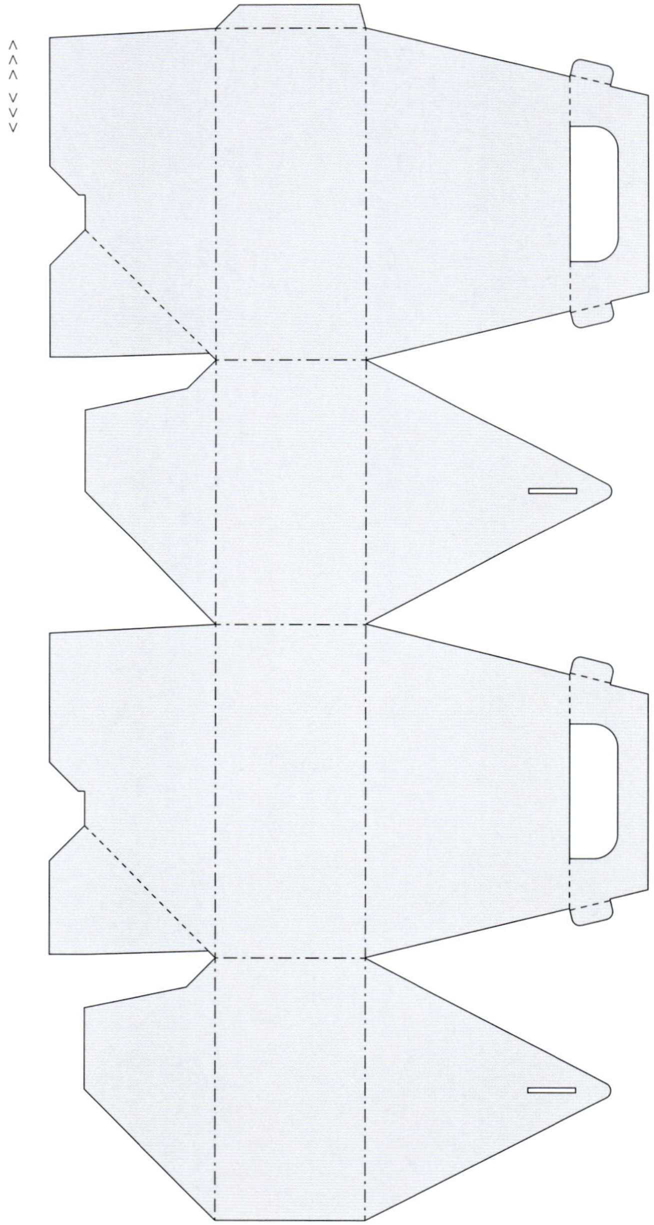

AssiDomän Frövi

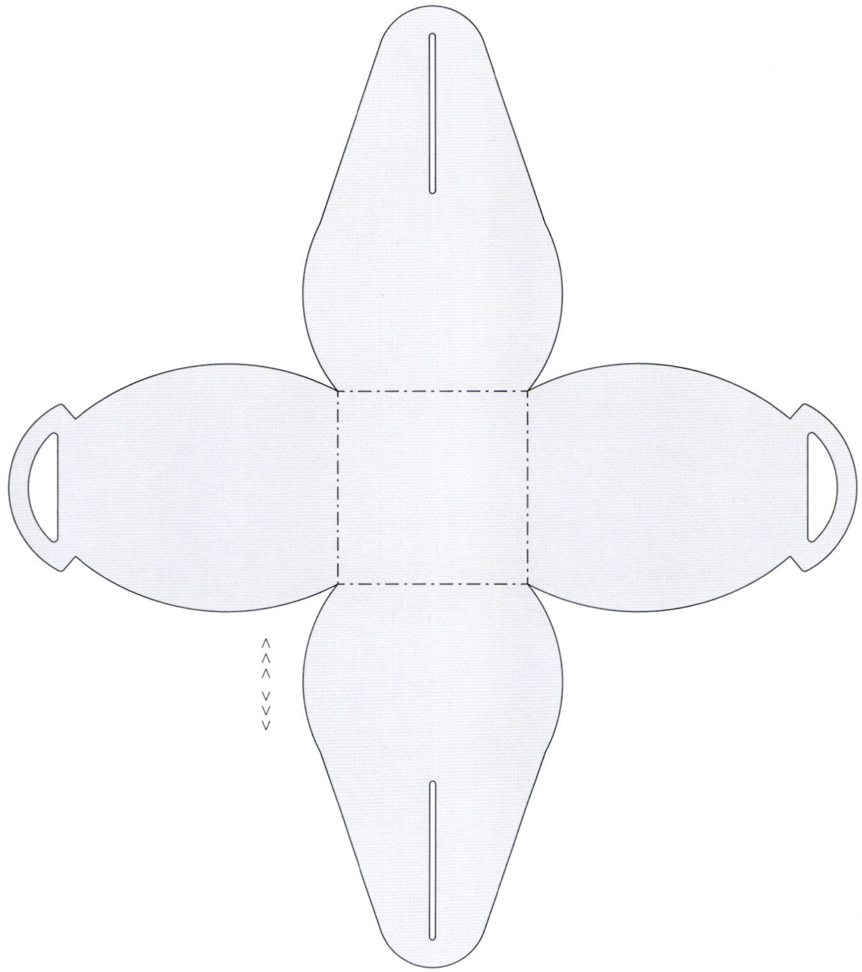

FOUR-SIDED BULBOUS CARRY CASE

The curved sides of this carton serve as an attractive and protective casing that is held together by handles on opposite sides of this four-sided pack. These handles pass through slits in the top of the adjacent sides, making this carton very easy to erect from a flat piece of board and providing structural integrity.

RECTANGULAR CARRY CASE

Rectangular cartons with angled corners require different design solutions from cartons with rounded sides and few angled corners. This carry case requires a crash-lock base to give it a rectangular form, but its method of closure at the top is similar to that of rounded carton-board carry cases, where the handles on opposite sides interlock with slits on adjacent sides to form a secure closure.

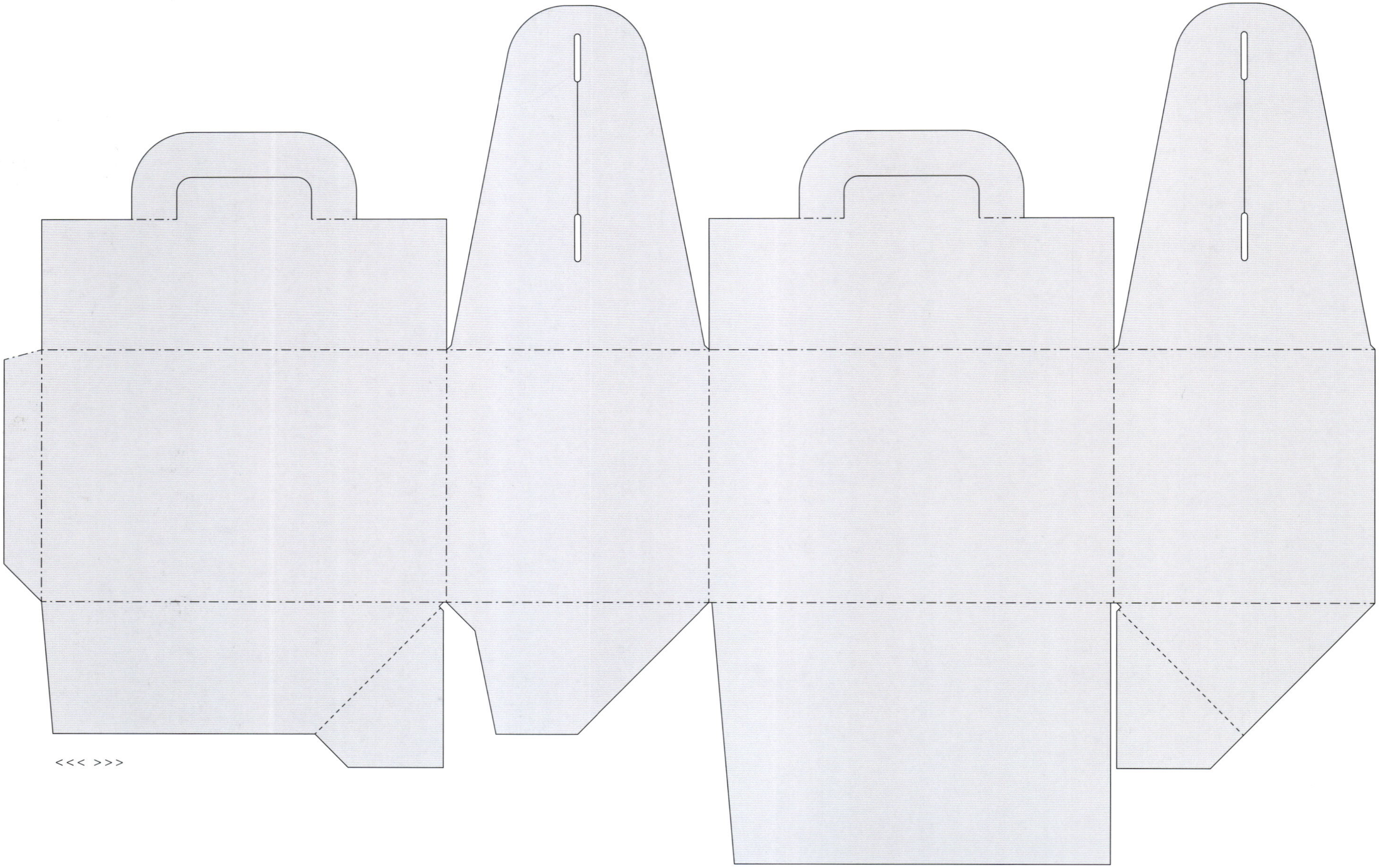

Benson Box

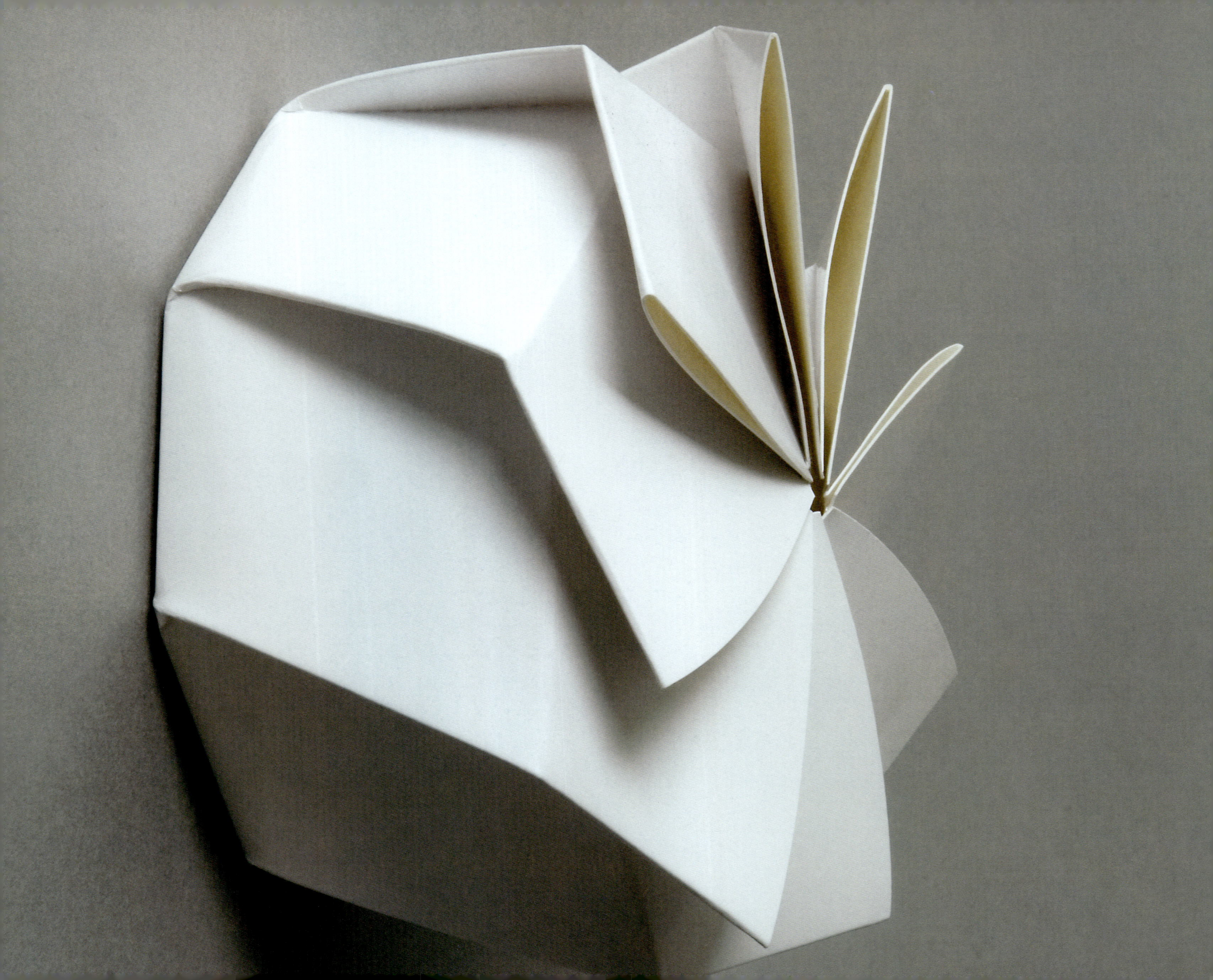

Mayr-Melnhof Packaging/Lindt & Sprüngli

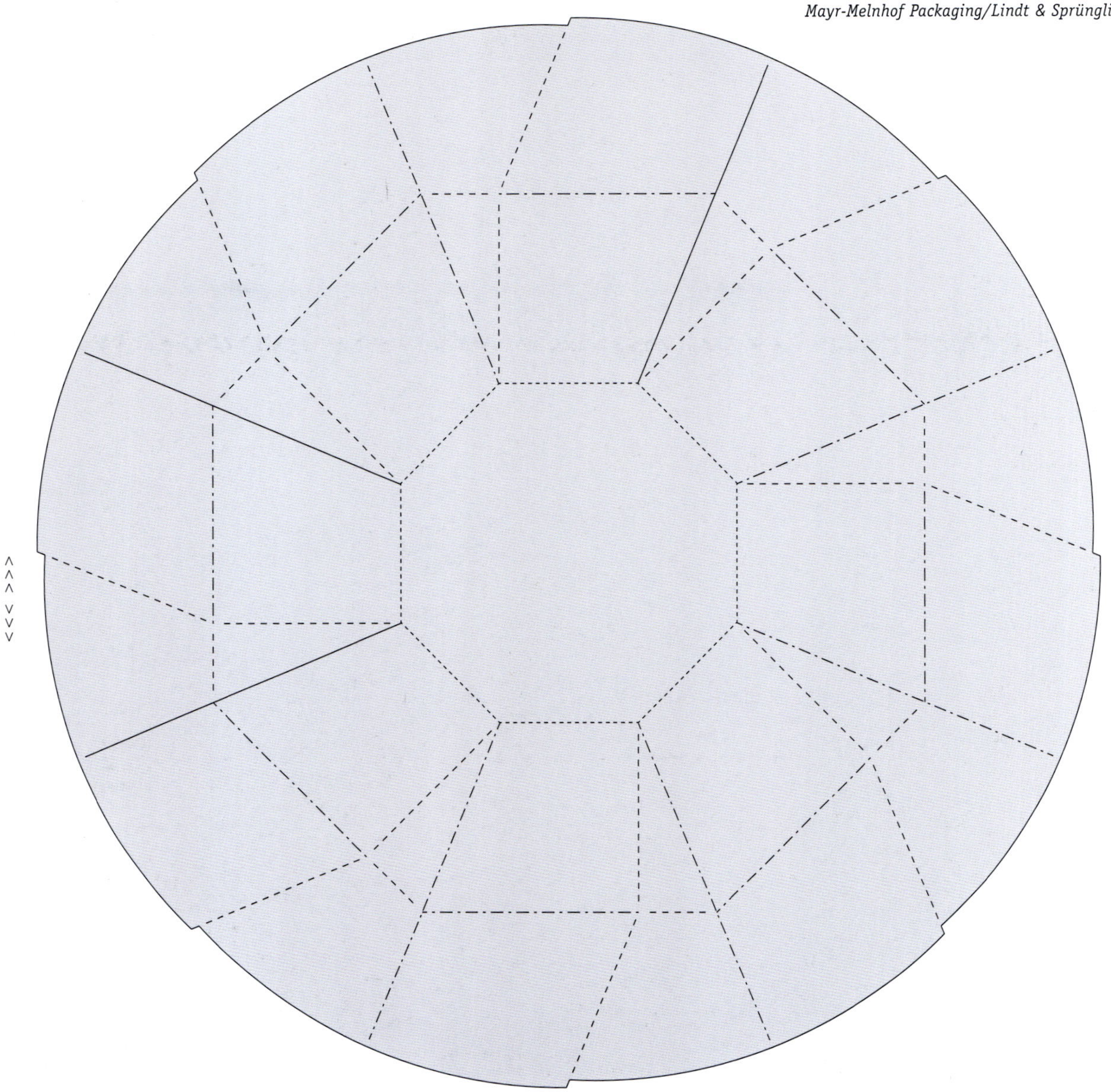

FRILLED OCTAGONAL CARTON

Sophisticated folds transform a circular piece of board into a highly attractive octagonal carton without the use of glue or any other fixative. The carton erects easily by gathering the sides together along the folds and creases, which naturally form the carton's structure as it contracts. The frilled exterior serves as an appealing and eye-catching feature, with a variety of shapes and surfaces on which to display product information.

TEN-SIDED CARTON WITH EXPANDING BASE

Unglued webbed corners have been used in the base of this carton to create an expanding display tray when the lid is removed. Subtle design considerations, including rounded corners on each side of the base, and the method of folding the sides of the lid to give the internal sides the same surface finish as the exterior, further improve the carton's appearance and performance.

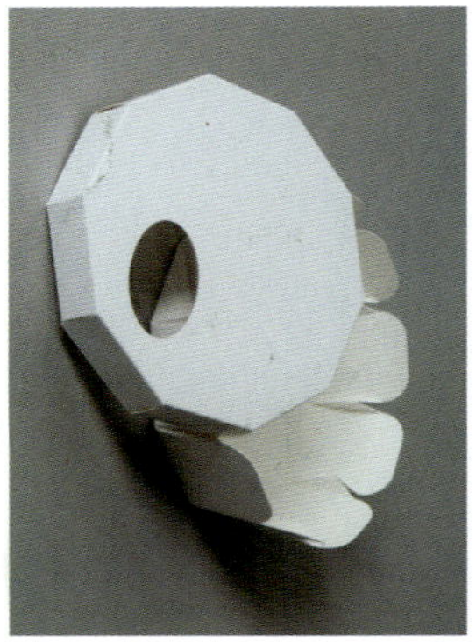

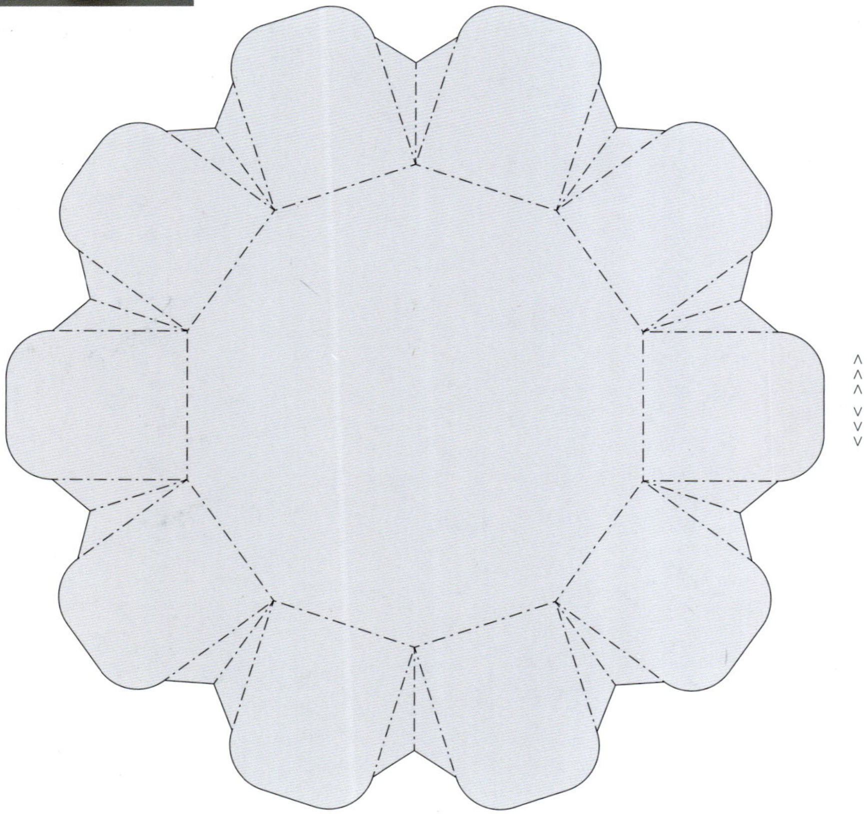

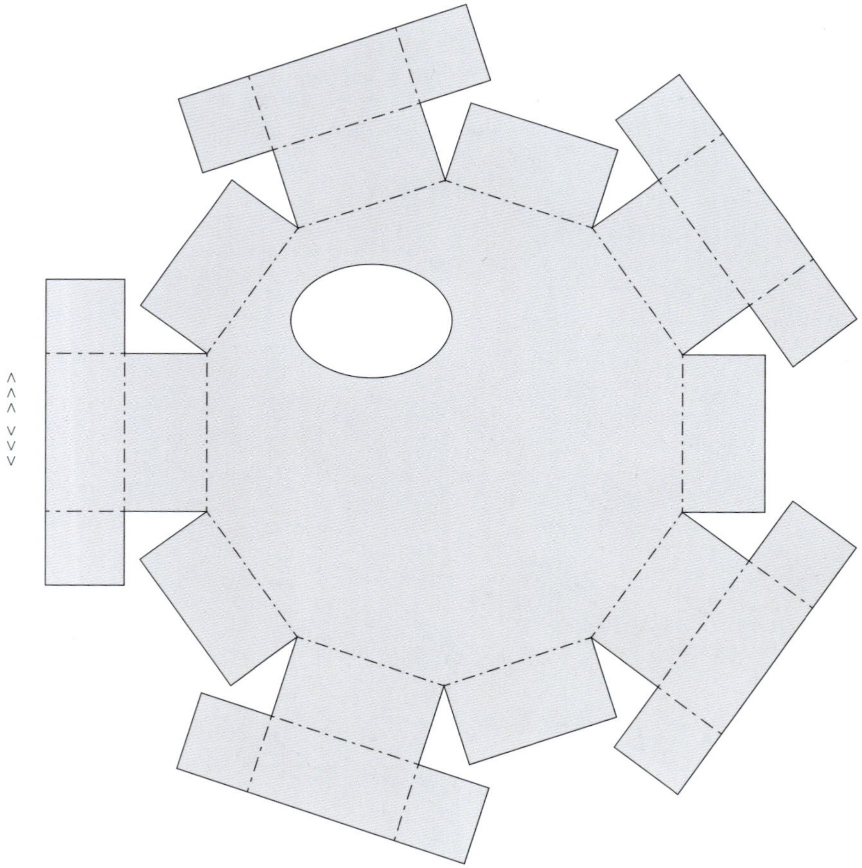

Cadbury/E. Wedel

Mayr-Melnhof Packaging/Alfred Ritter

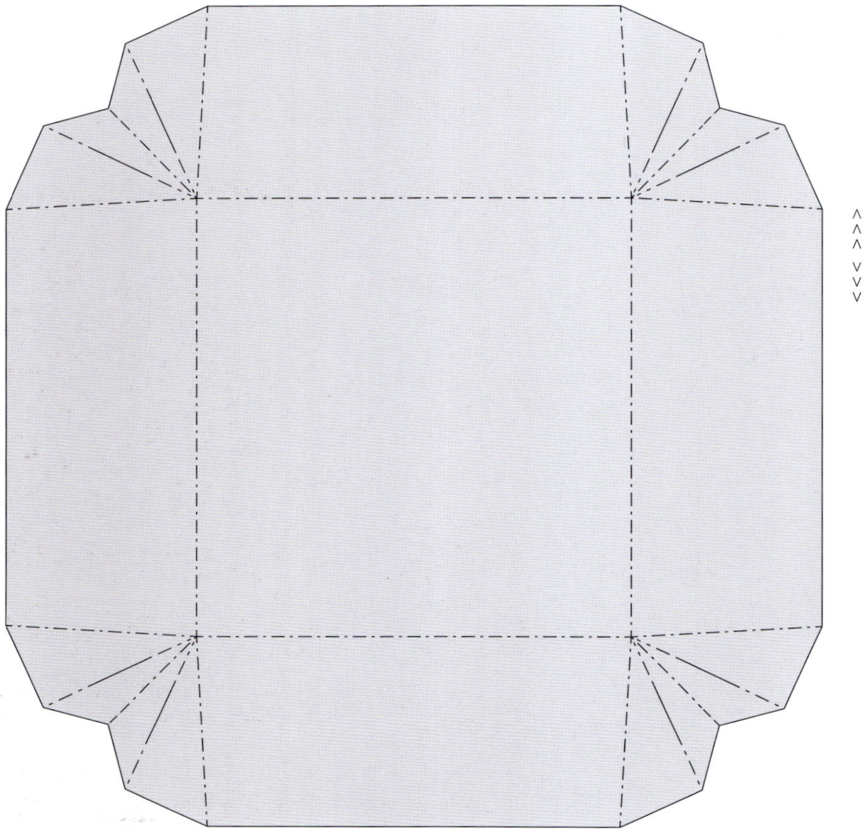

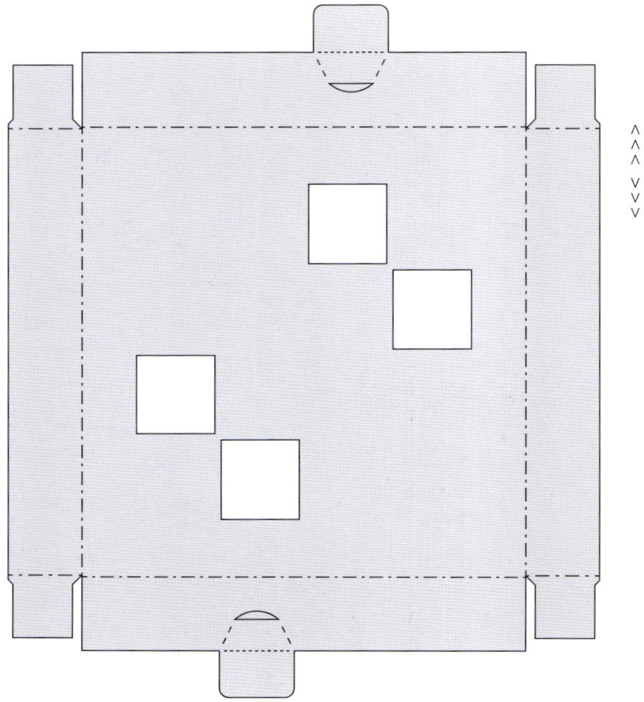

WEBBED-CORNER CARTON

Minor alterations to the common webbed-corner tray have given this box added value and improved functionality. By gluing only a segment of each web in the corners of this carton's base, the sides fall outward when the lid is removed to present the product within. This allows the base to become a standalone presentation tray that can be reclosed easily by replacing the lid.

BARREL-VAULTED DISPLAY CASE

Designers can intrigue the consumer by exploiting the interdependent mechanical properties of two-part cartons. Here the two components comprise an inner tray that contains the product and an outer sheath that conceals it. The product is revealed by sliding the tray from the sheath. Careful attention to detail has given the package a feeling of quality that is expressed most obviously in the rounded surface of the sheath, which has been created by a series of creases along its length.

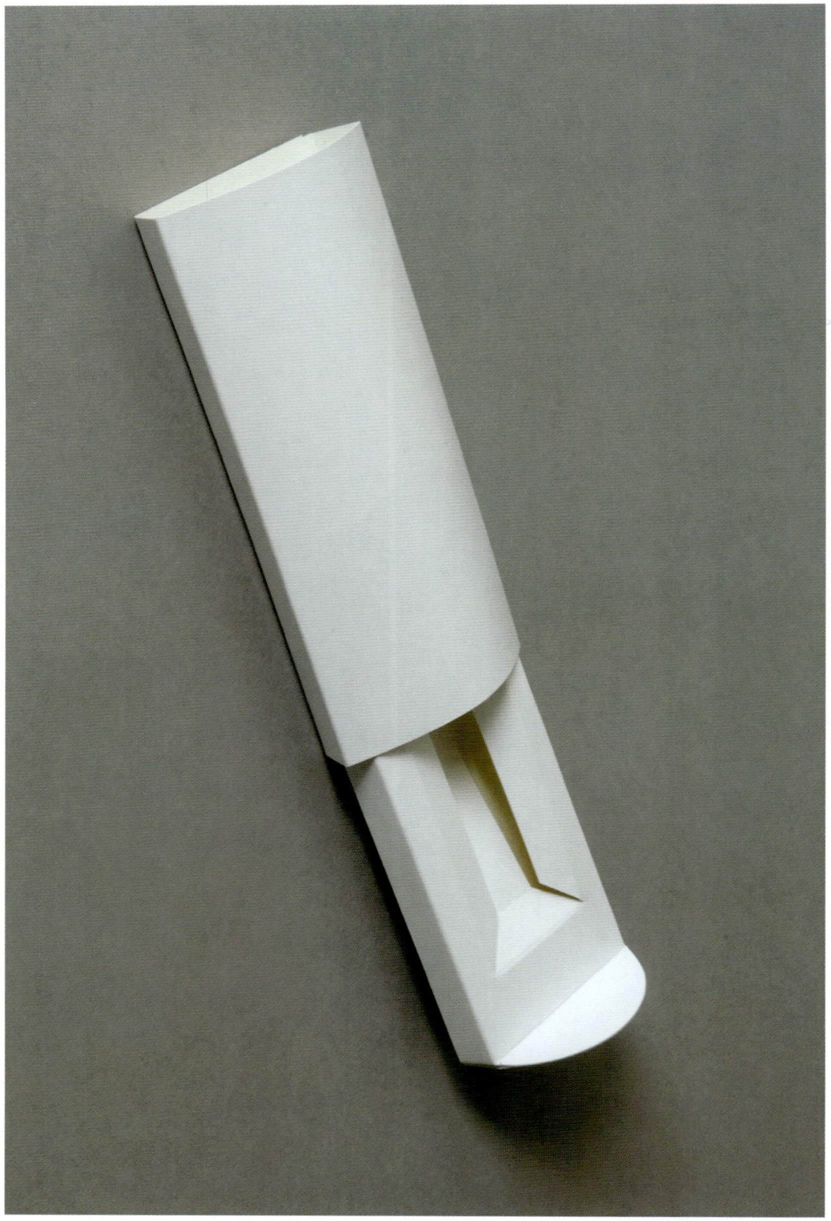

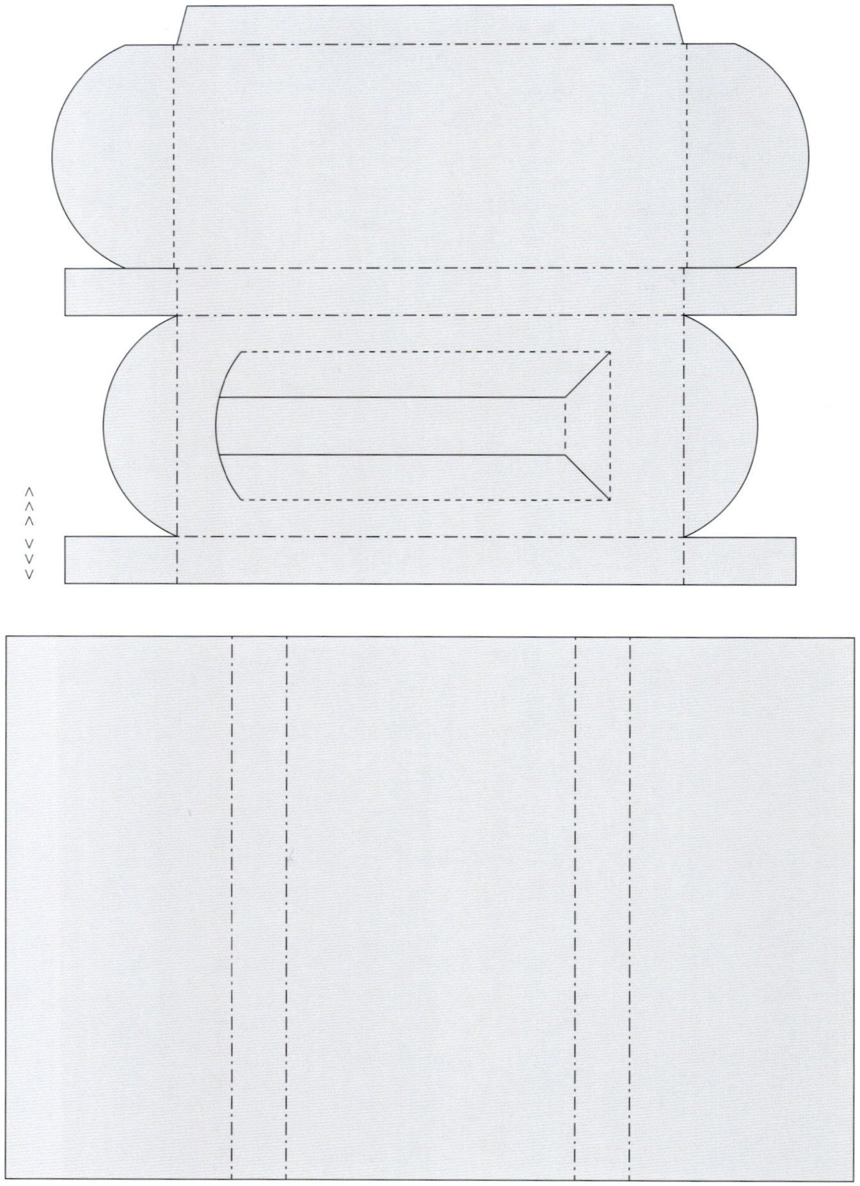

Mayr-Melnhof Packaging/Palmers

<<< >>>

Rob. Leunis & Chapman

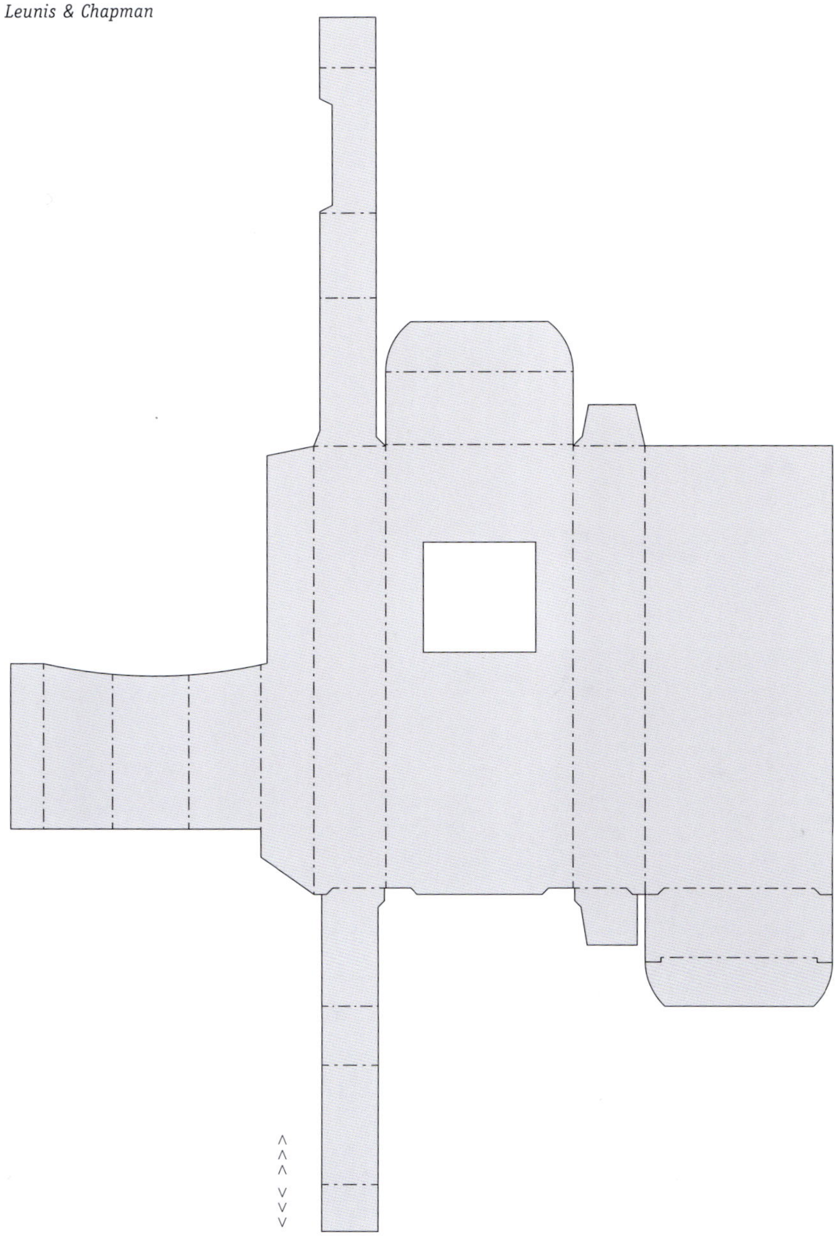

Simple rectangular cartons often belie more sophisticated design characteristics concealed inside the package. This flat-pack design contains an insert which, when the carton is erected, forms a series of partitions in which different products can be contained separately. A design adaptation to the folded flaps at the top and base of the carton also provides further division of internal space and helps to house the different-sized products securely.

EXPANDING PRESENTATION BOX

By incorporating folded corners into a hexagonal container and gluing the sides to the base, this elaborate carton expands to reveal the product when the lid is removed. This simple adaptation of a standard hexagonal carton creates a mechanical feature that offers a surprise to the user. The intelligent choice of design for the elaborate lid echoes the carton's overall idiosyncratic character. The lid is shown here partially open so that its construction can be seen clearly.

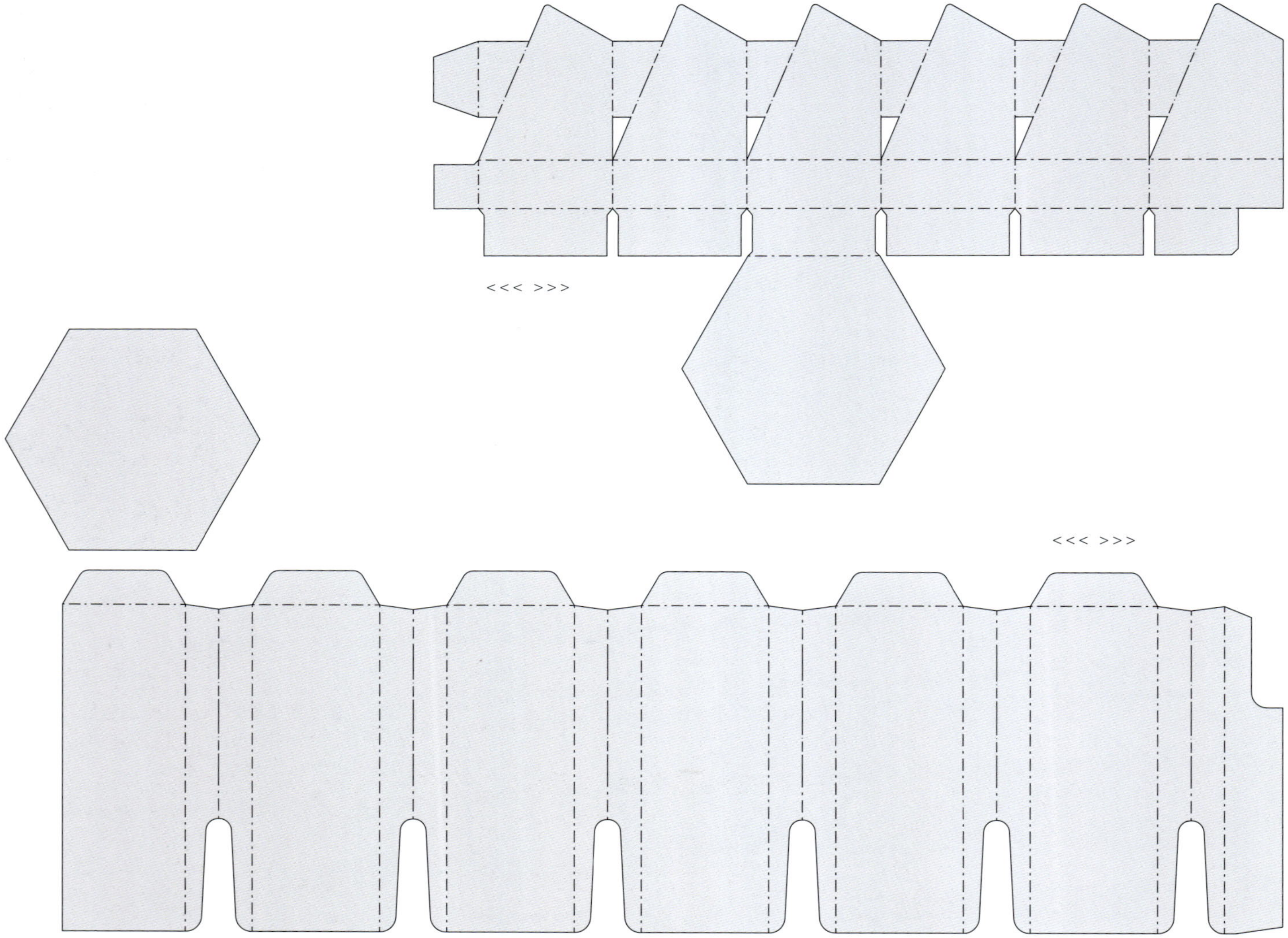

Mayr-Melnhof Packaging/LU

Rob. Leunis & Chapman

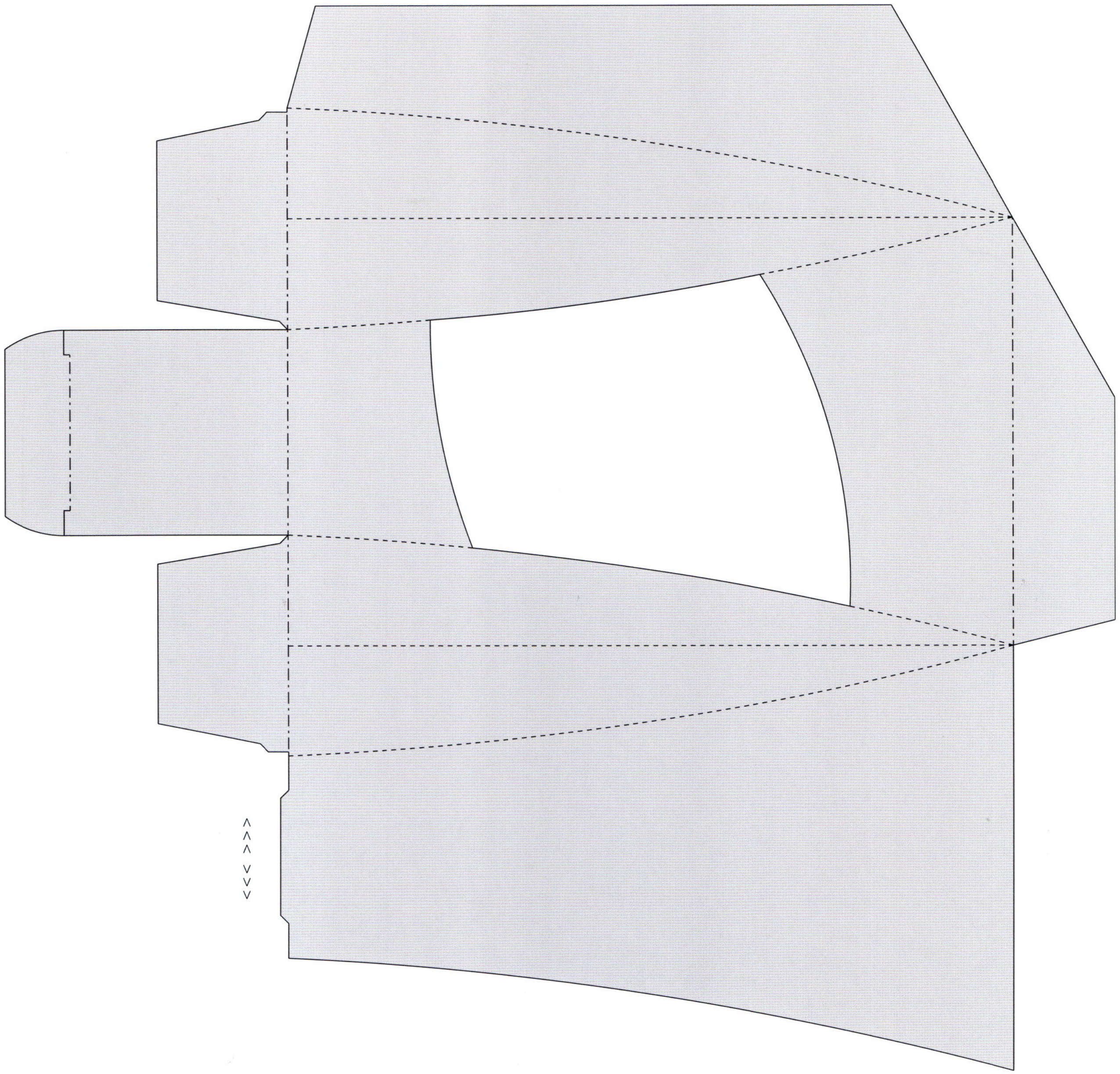

PINCHED-SQUARE CARTON

The square footprint of this carton has been pinched toward its top to provide a large display front and slender side profile. The carton's shape reflects the form of the product inside; this would be suitable for end-filled plastic bottles. The crease lines down each side of the carton allow it to be flat-packed prior to filling.

HIDDEN-LOCK DISPLAY CASE

The creative art of carton design is at its best when the structure arouses the curiosity of the user while providing exceptional functionality. The design of the base of this simple display case incorporates an internal-locking mechanism that prevents it from being separated from the main body of the carton. As the base is withdrawn, the cover flap concealing the front of the carton is released to reveal the contents inside. The superior design and careful choice of materials make this carton appropriate for luxury items.

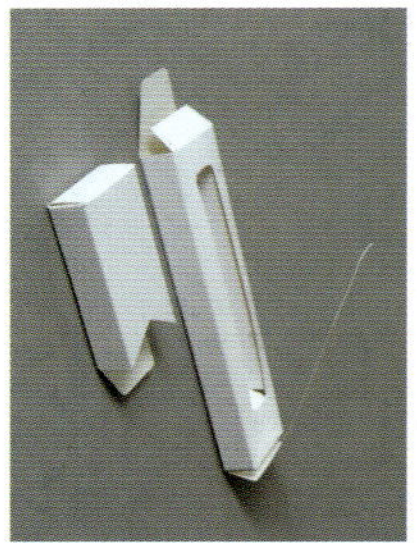

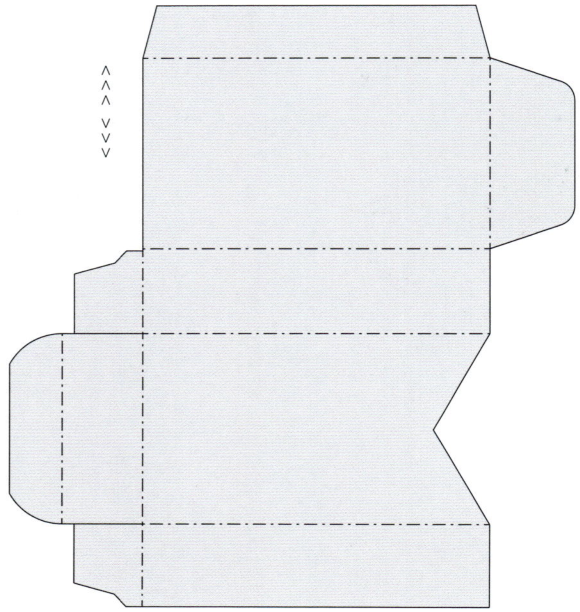

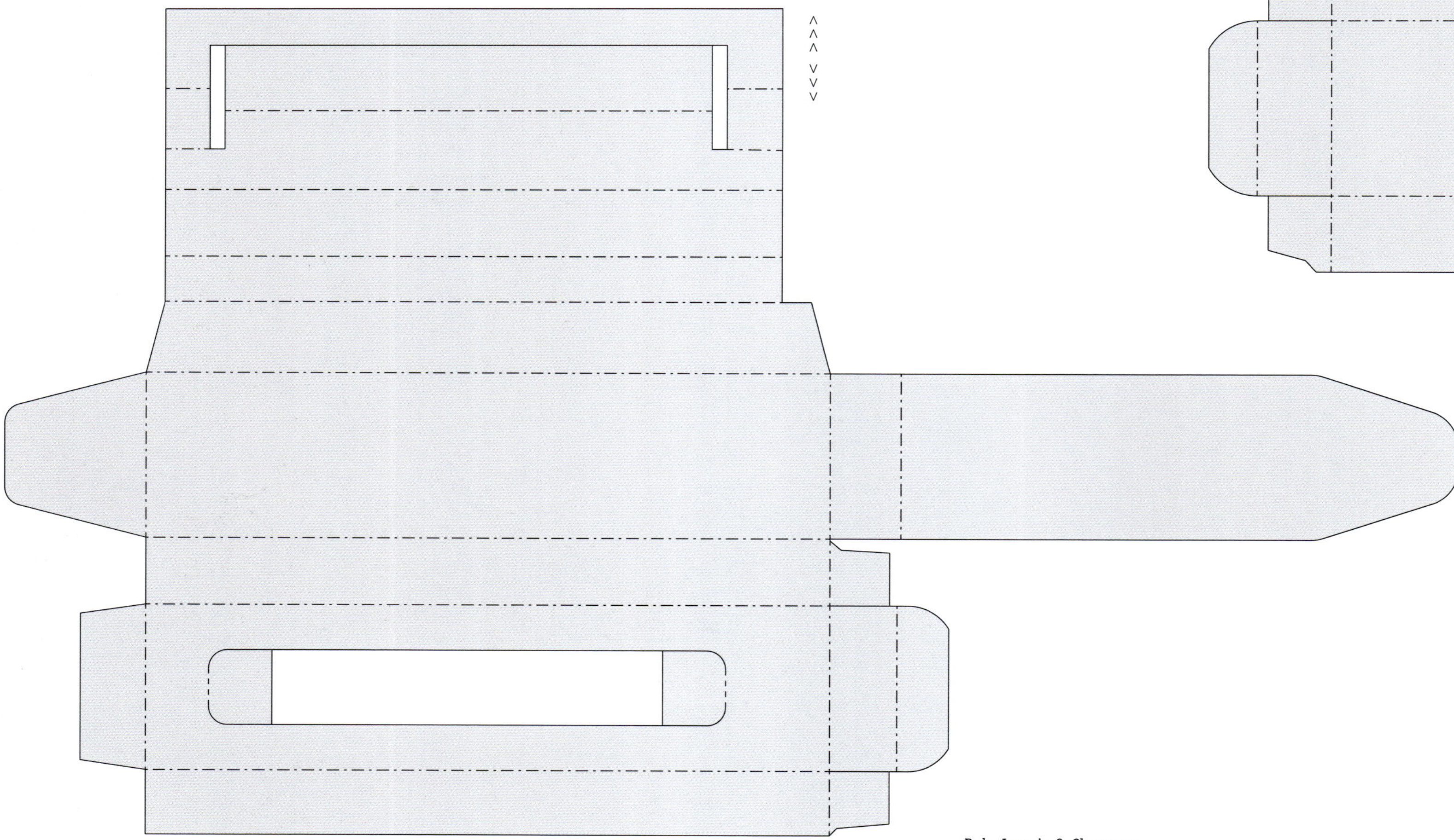

Rob. Leunis & Chapman

Rob. Leunis & Chapman

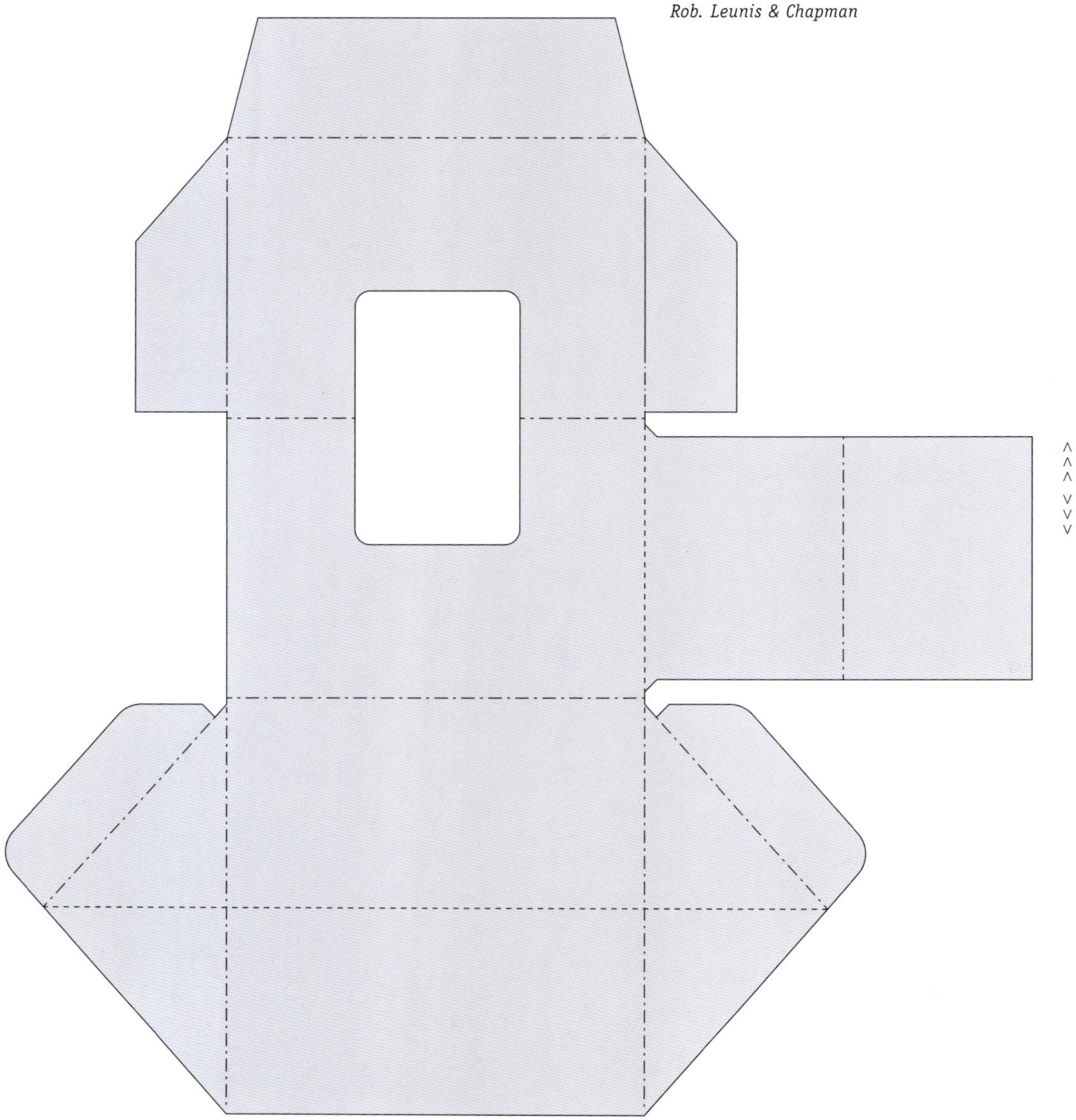

TRIANGULAR CARTON WITH WINDOW

This triangular design uses the two shortest sides of the container as the primary display surfaces. The front corner, which contains a display window, doubles as a fold-line that allows the carton to be flat-packed for transportation. Tuck-flaps at the top and bottom provide structural integrity to the package when it is filled.

CHILD-RESISTANT CARTON

Child resistance was once considered to require more sophisticated packaging materials than carton board. However, improvements in design and manufacturing have ensured that these highly desirable and, in some cases, essential characteristics can be obtained with carton board. This design incorporates an internal locking device that prevents the inner tray from being detached from the outer sleeve and also, most importantly, boasts a highly innovative mechanism for ensuring that young children cannot access the product, making it ideal for medicines. This mechanism relies on two holes on either side of the container being depressed at the same time as the inner tray is withdrawn using the tab.

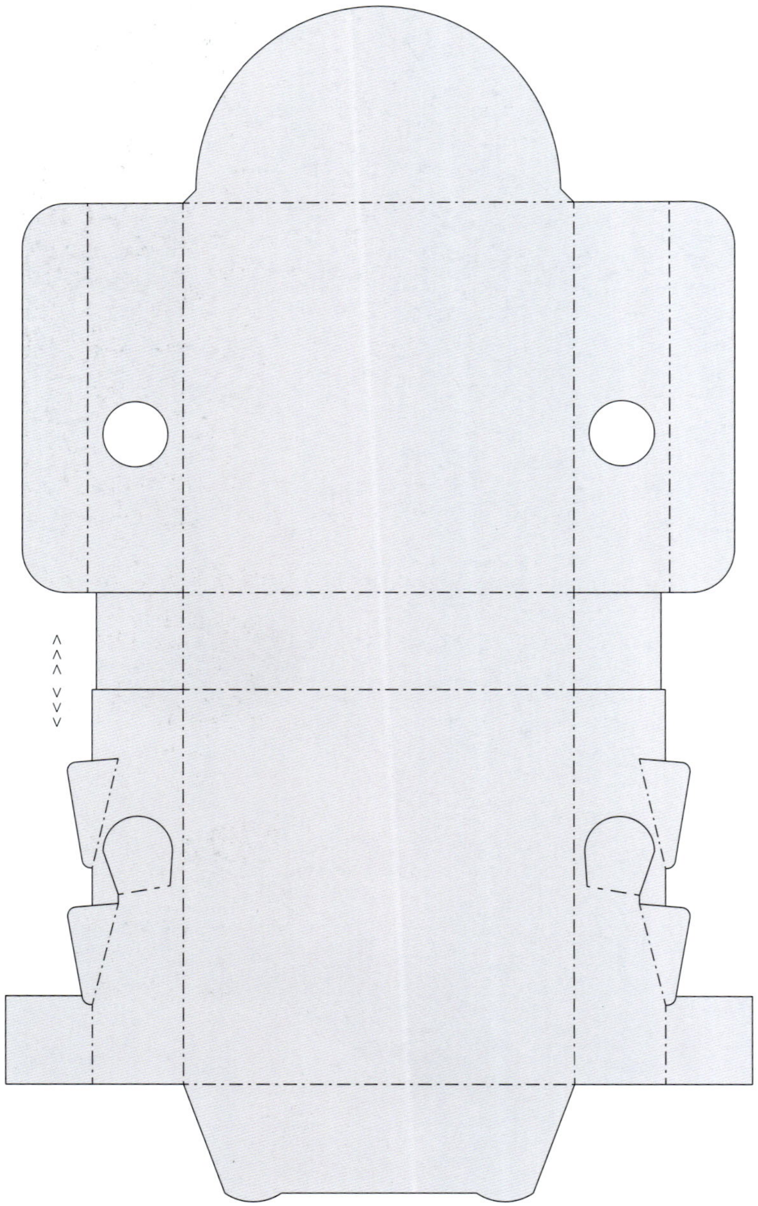

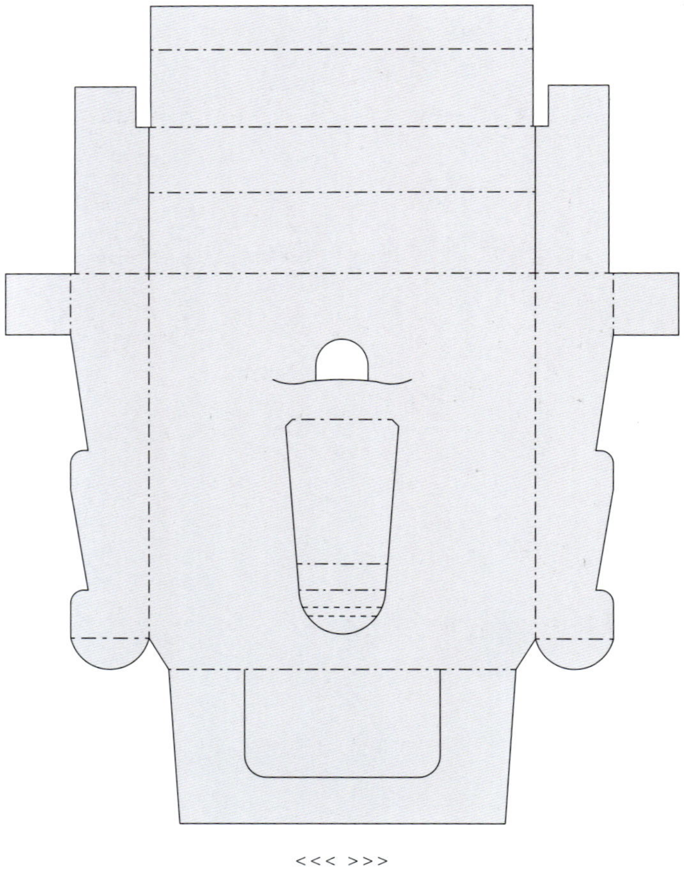

Stora Enso

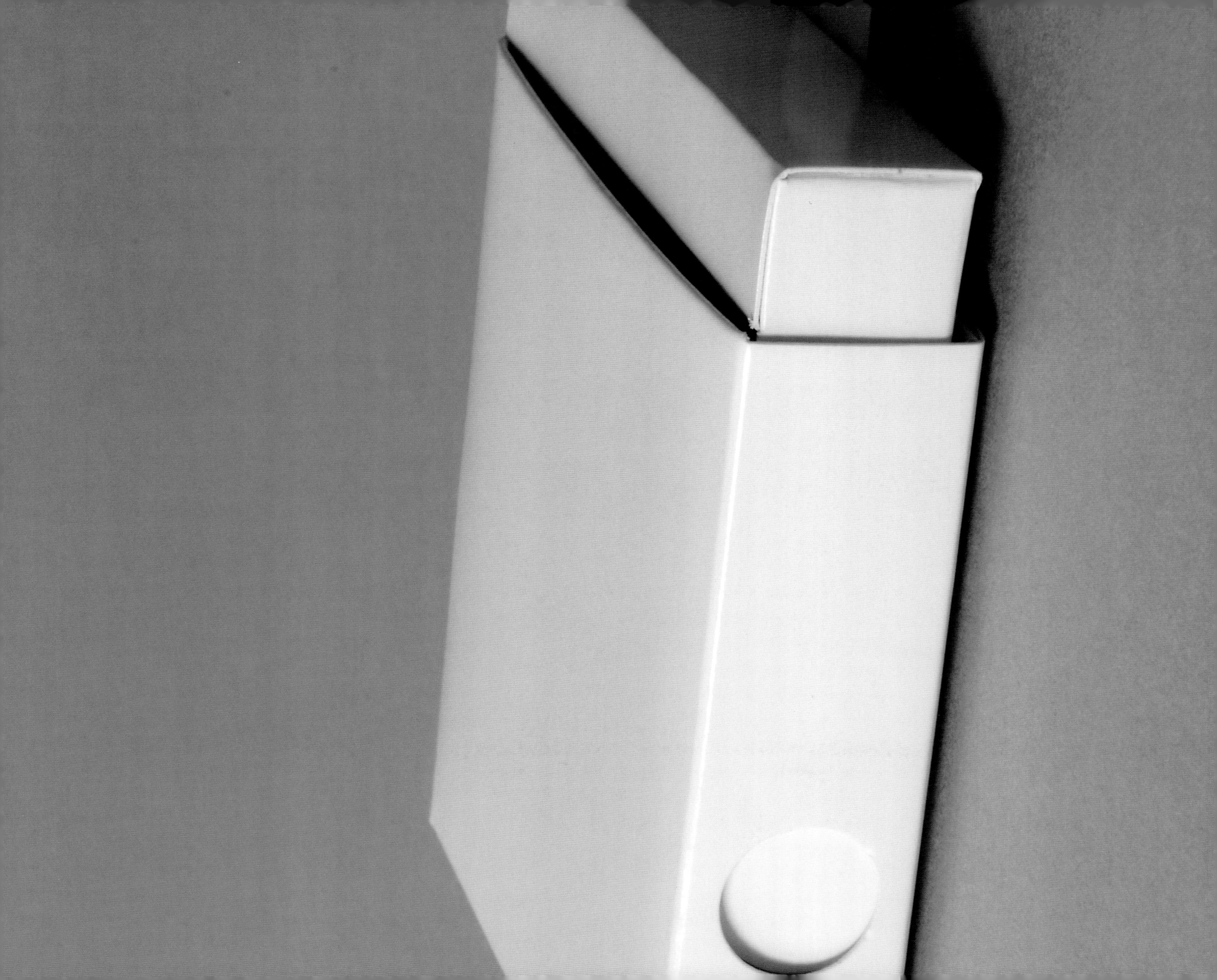

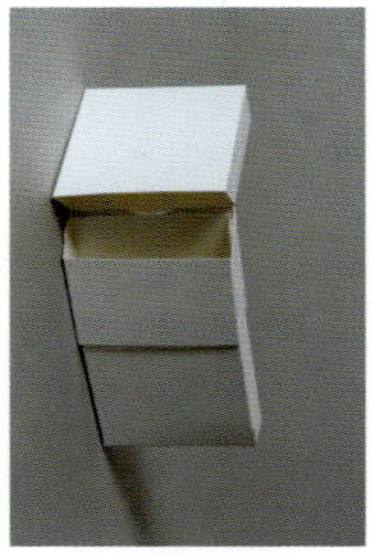

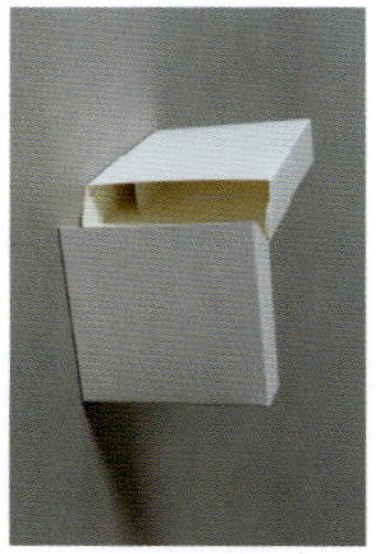

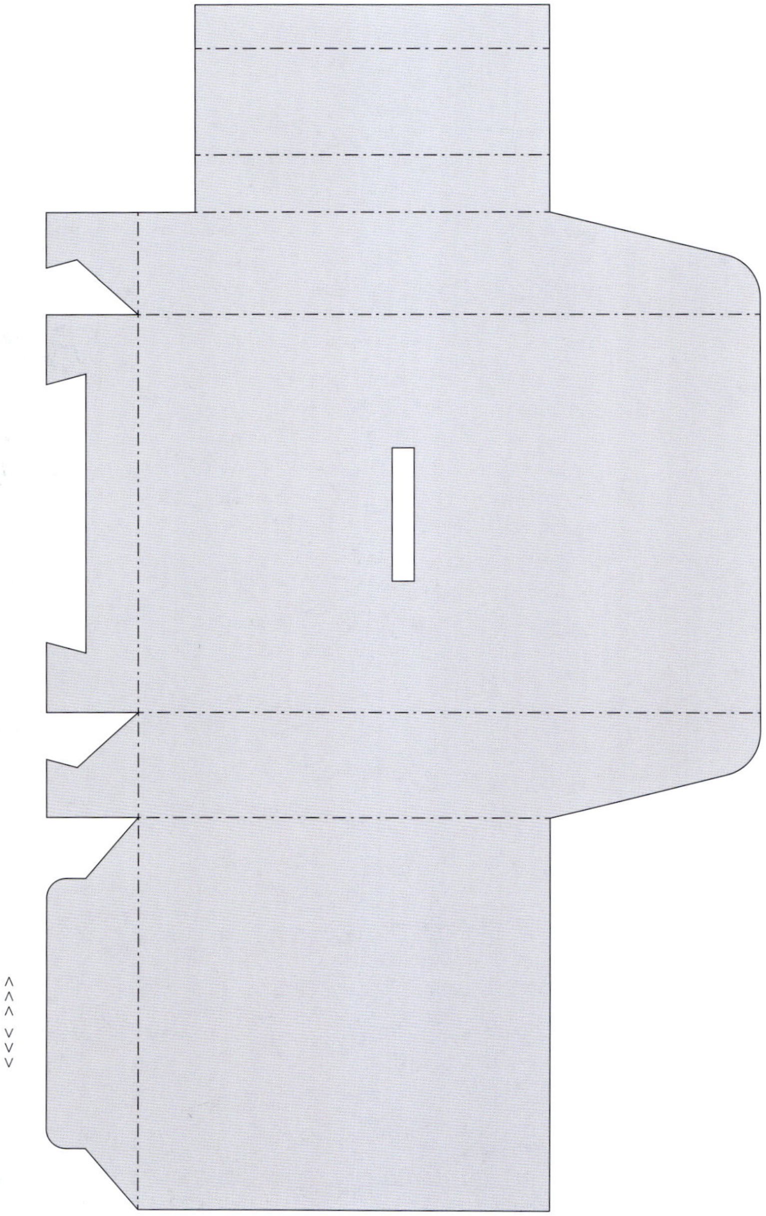

Rob. Leunis & Chapman

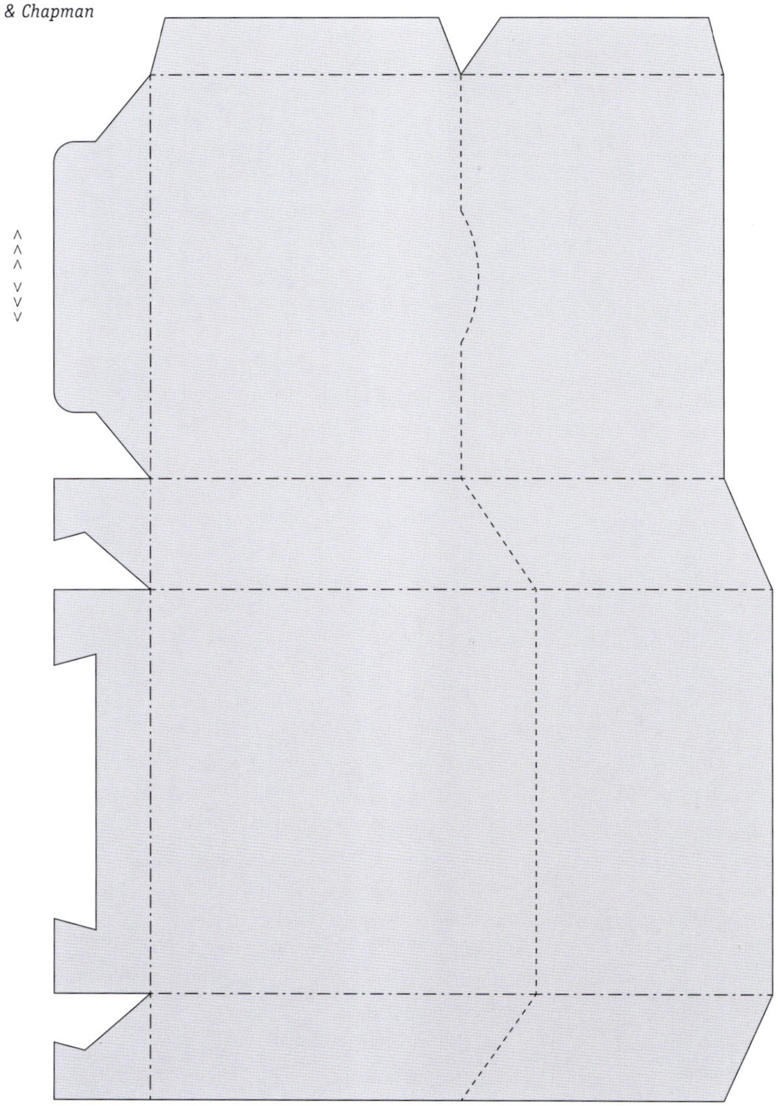

DESKTOP-DISPLAY CASE

The likelihood that packaging will be thrown away as soon as it has been opened is common to many products. This short life span is often reflected in the utilitarian nature of many packaging designs, which seldom warrant highly specialized treatment for such a limited shelf-life. This pack, however, is intended to last as long as the product it contains. Designed in two parts, this desktop-display case comprises a main body to house the product, and an outer sleeve. Subtle features of the design of the outer sleeve, especially the tear angles and the tap and slot, allow it to function both as an efficient component in transit and as a display stand throughout the life of the product.

COMPACT-DISC SLEEVE

The proliferation of digital media over the last decade has caused a minor revolution in the types of containers for digital storage devices such as compact discs. This design allows a compact disc to be safely secured to an internal card that slides from the outer sleeve to reveal the product. A concealed lock on the inside of the sleeve prevents the insert card from being completely removed. The innovative design is further enhanced by delicate detailing that provides tactile interface with the user to suggest functionality, such as the die-cut finger slot and depressible tab at the rear.

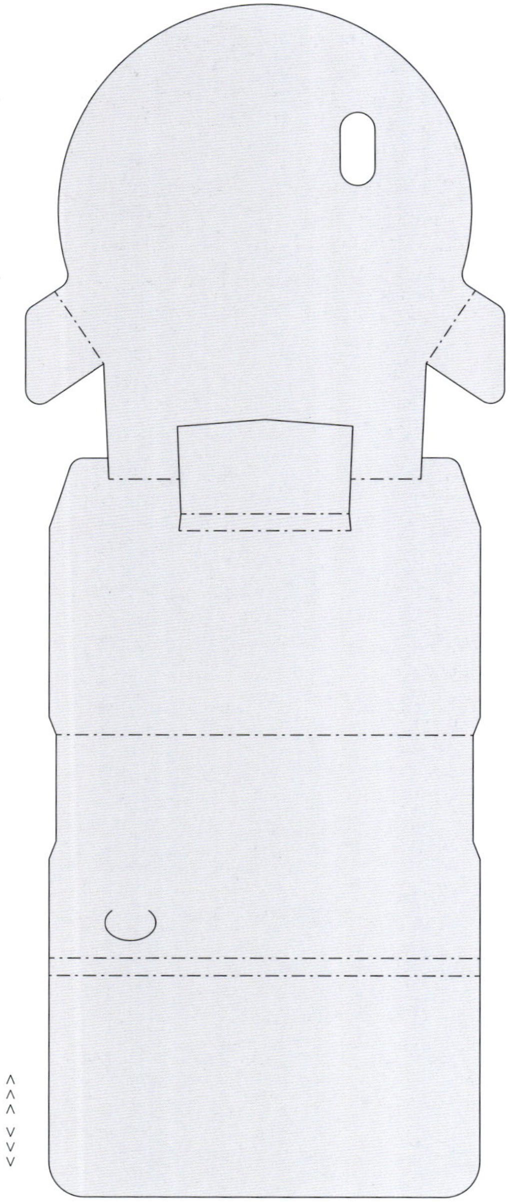

Stora Enso

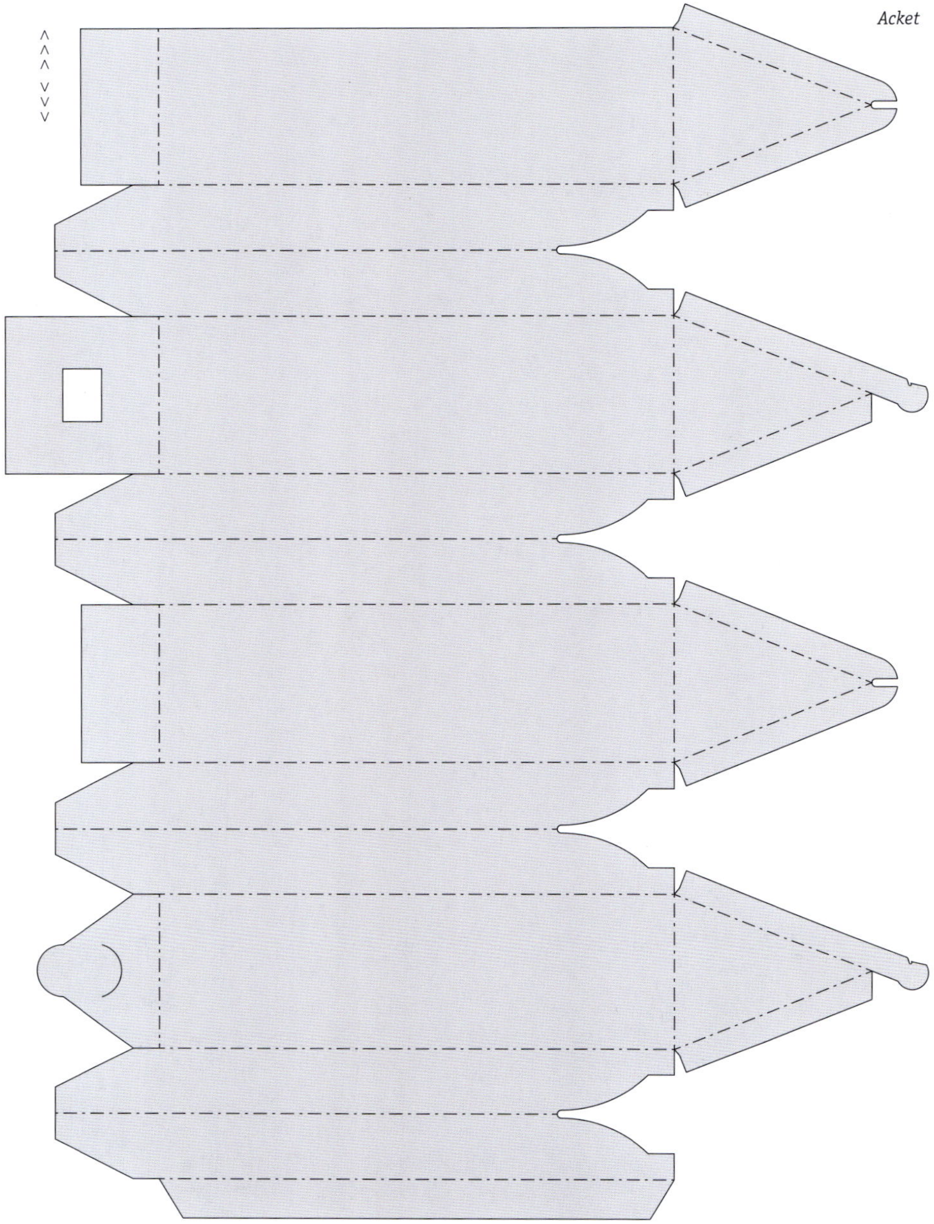

ROCKET CARTON

Replicating familiar shapes and forms in packaging design is a method employed, often humorously, by designers to make associations with the type of product contained within. The rocket-shaped form of this carton presents an opportunity to package products that have an association with outer space. The folded corners create legs that elevate the carton as though "ready for lift-off," while the rocket-like upward thrust is further accentuated by the carton's pointed top. The concept behind this particular package is notable for actually traveling on a manned space mission as part of an experiment to compare the growth rate of seeds in space and on earth.

WEDGE CARTON

There are many ways to package a triangular or wedge-shaped product. This design solution allows the carton to be supplied to the producer flat-packed and is exceptionally easy to erect and fill. The folded sides ensure that the surface finish on the outside of the carton matches that on the inside. The product itself helps to provide structural integrity to the package, and is secured using an adhesive strip over the top of the pack.

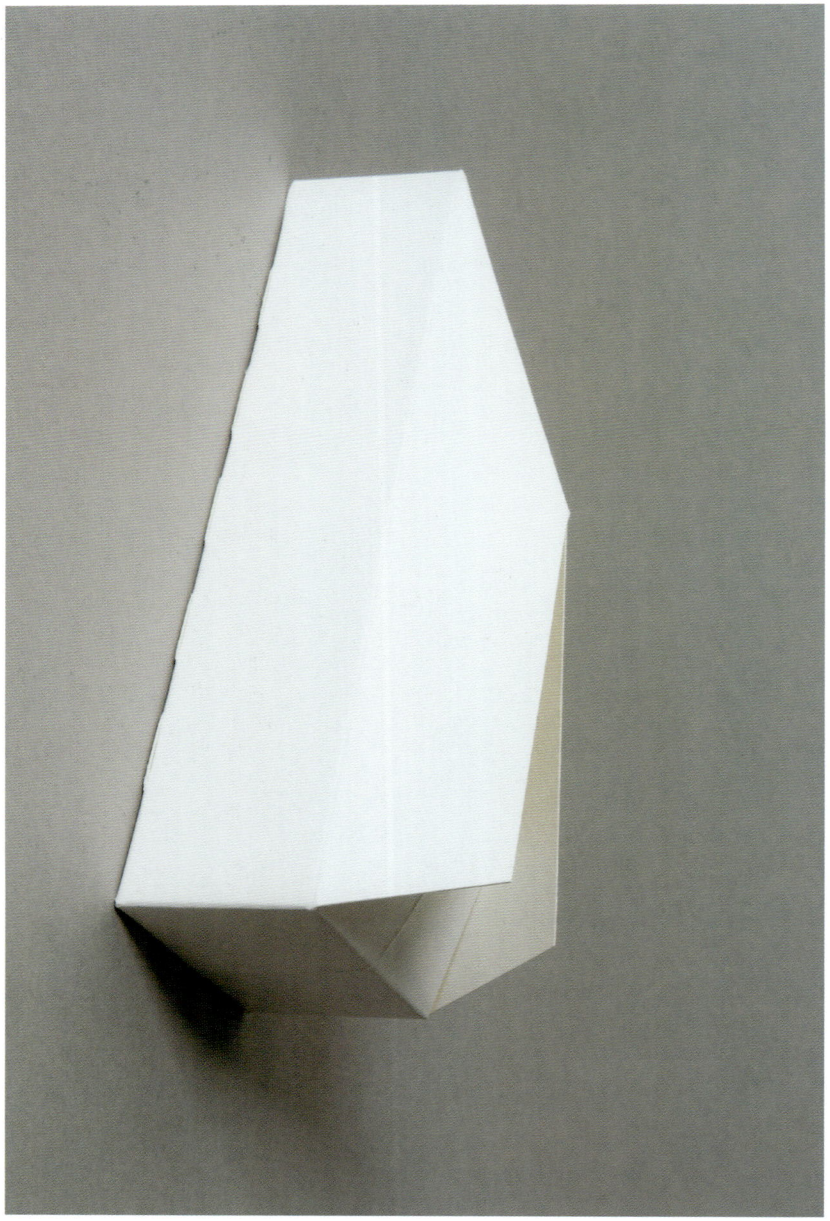

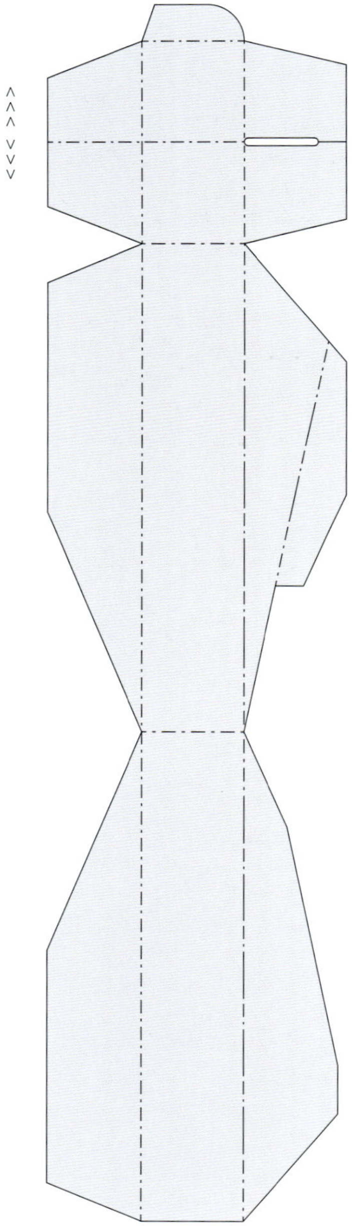

Acket

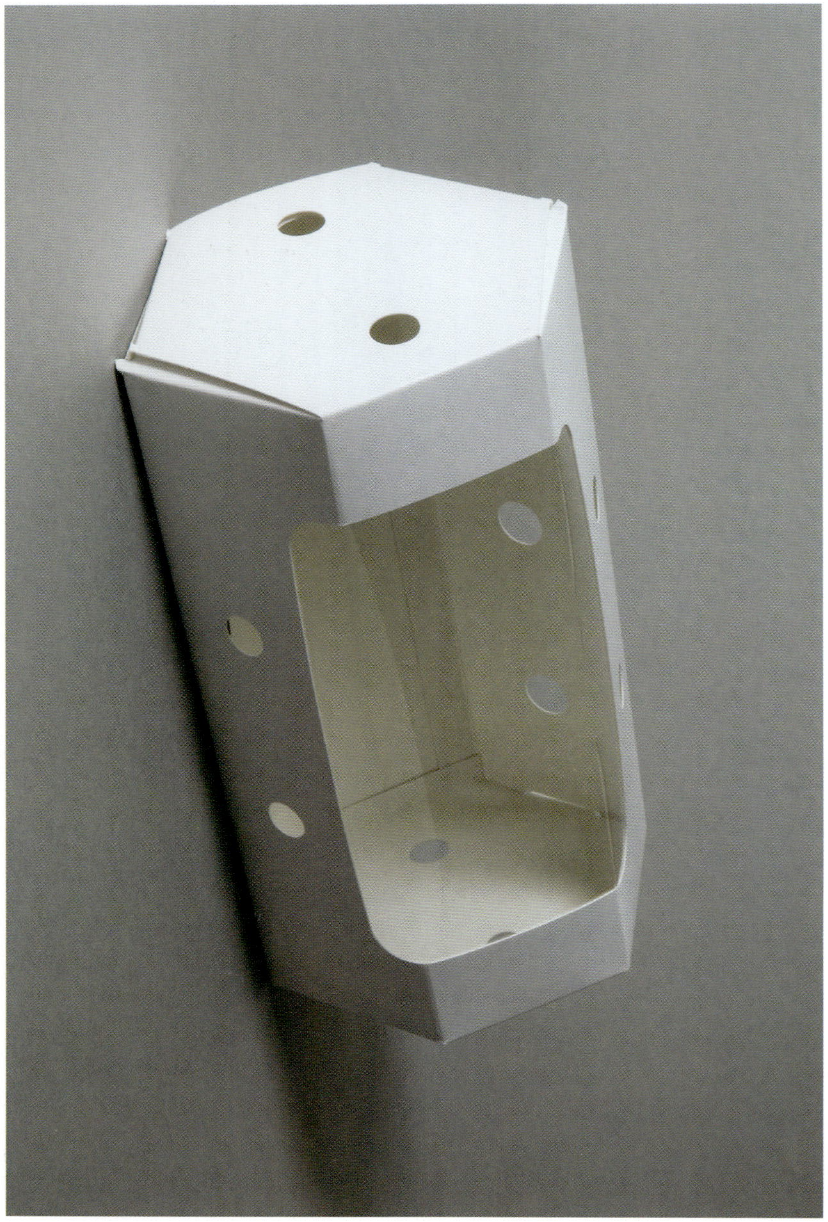

SCENT-ENCOURAGING HEXAGONAL CARTON

Acket

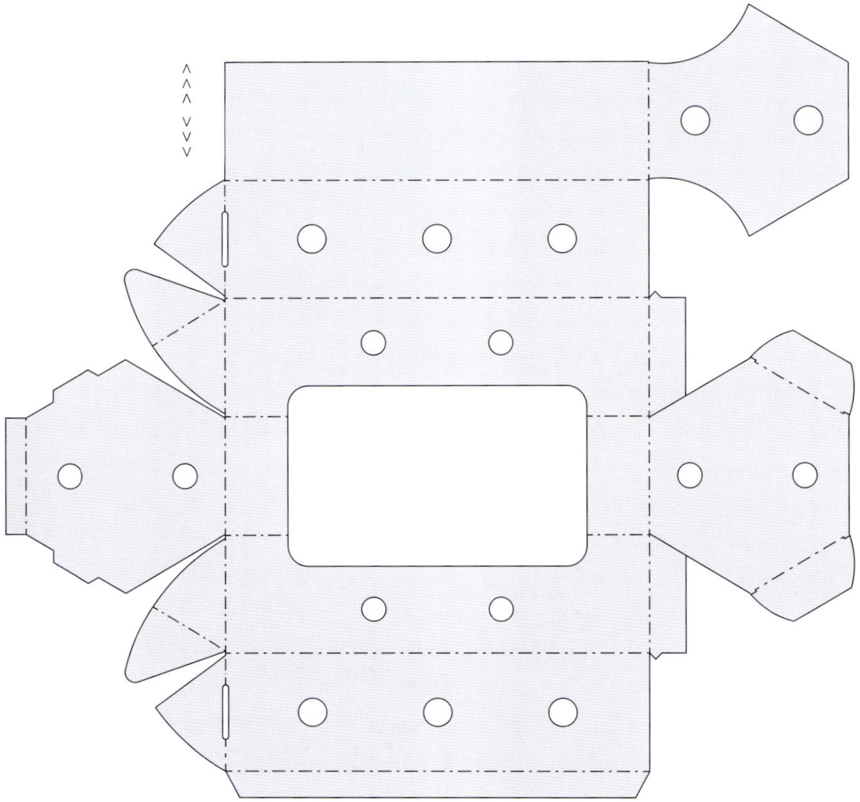

Packaging is often used to encourage sensory interaction with the product, either visually (with a transparent or die-cut window, or through graphic representation), or tangibly where the product is not fully concealed by the packaging. It is relatively unusual for cartons to be designed in this way, to encourage the product's aromas to reach the consumer. The holes in this package allow the consumer to appreciate the smell of the product, while the transparent window allows the product to be seen—a more common mode of interface.

HEAT-SEALED SANDWICH PACK

This variation of the triangular sandwich pack allows for easy manufacturing and fast filling times. The carton is supplied in a flat-pack form that is easily erected and filled before the lid, containing a transparent window, is folded and sealed to the flanges on the sides of the carton's base to provide a hygiene barrier.

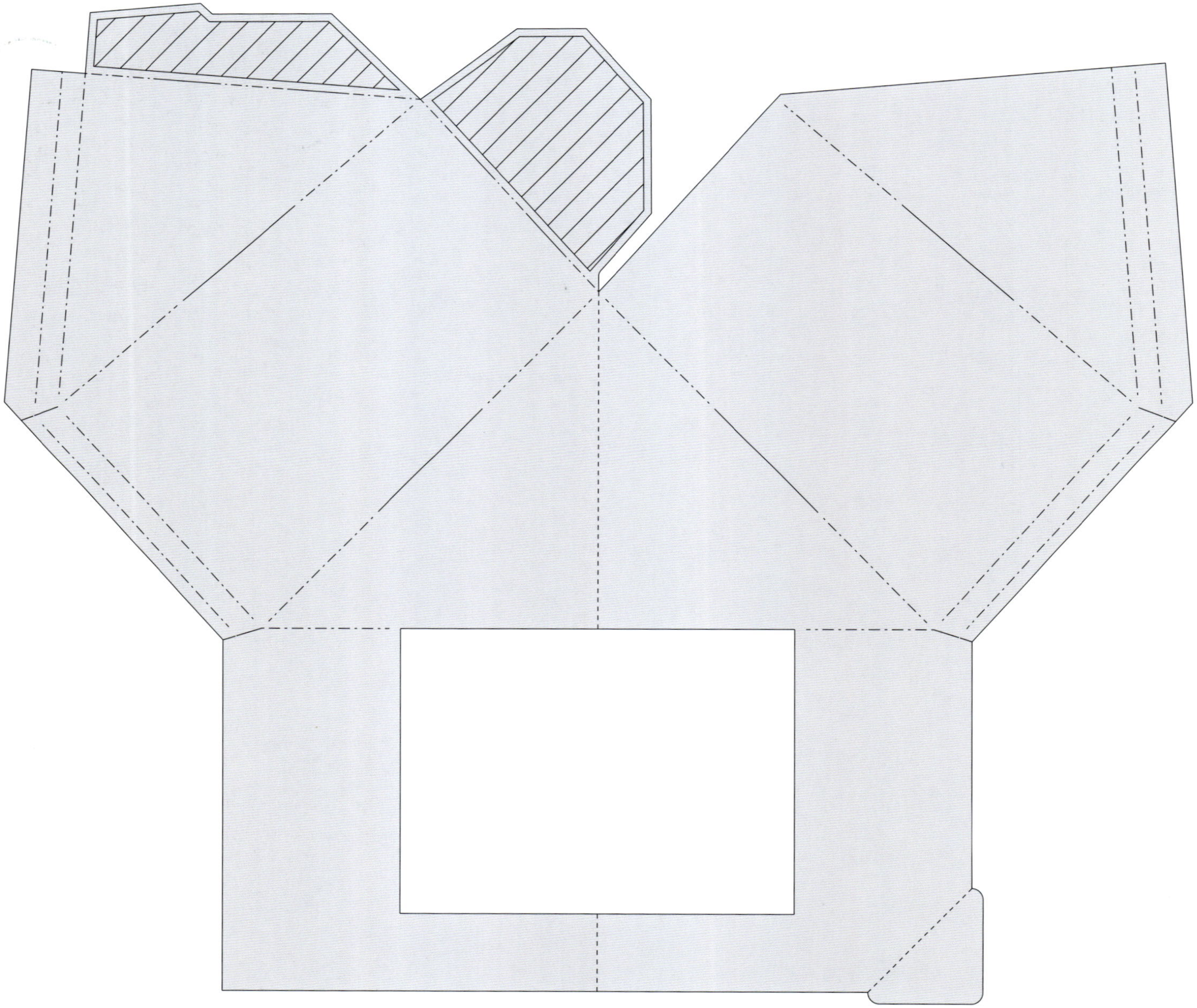

<<< >>>

RAP

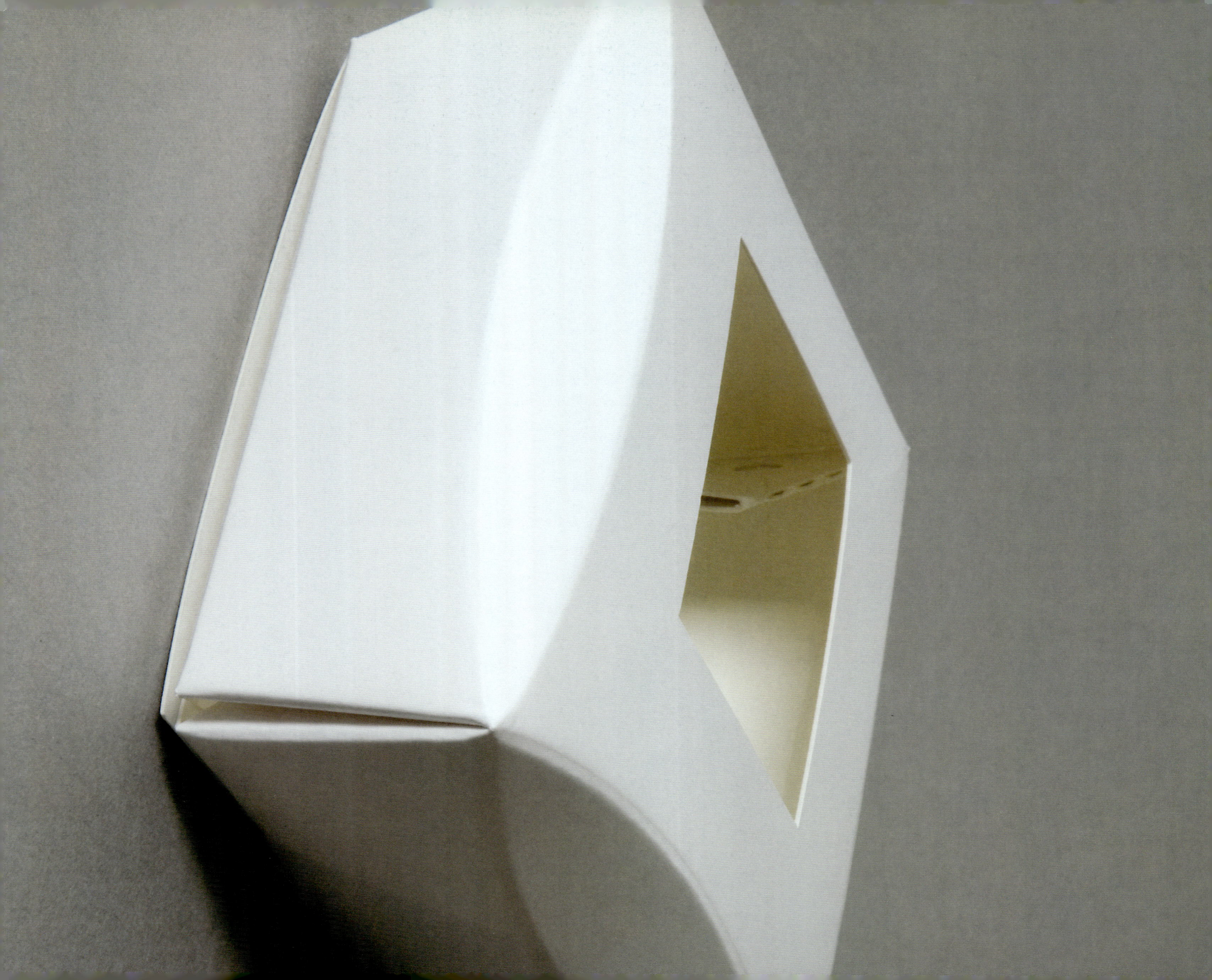

Model PrimePac

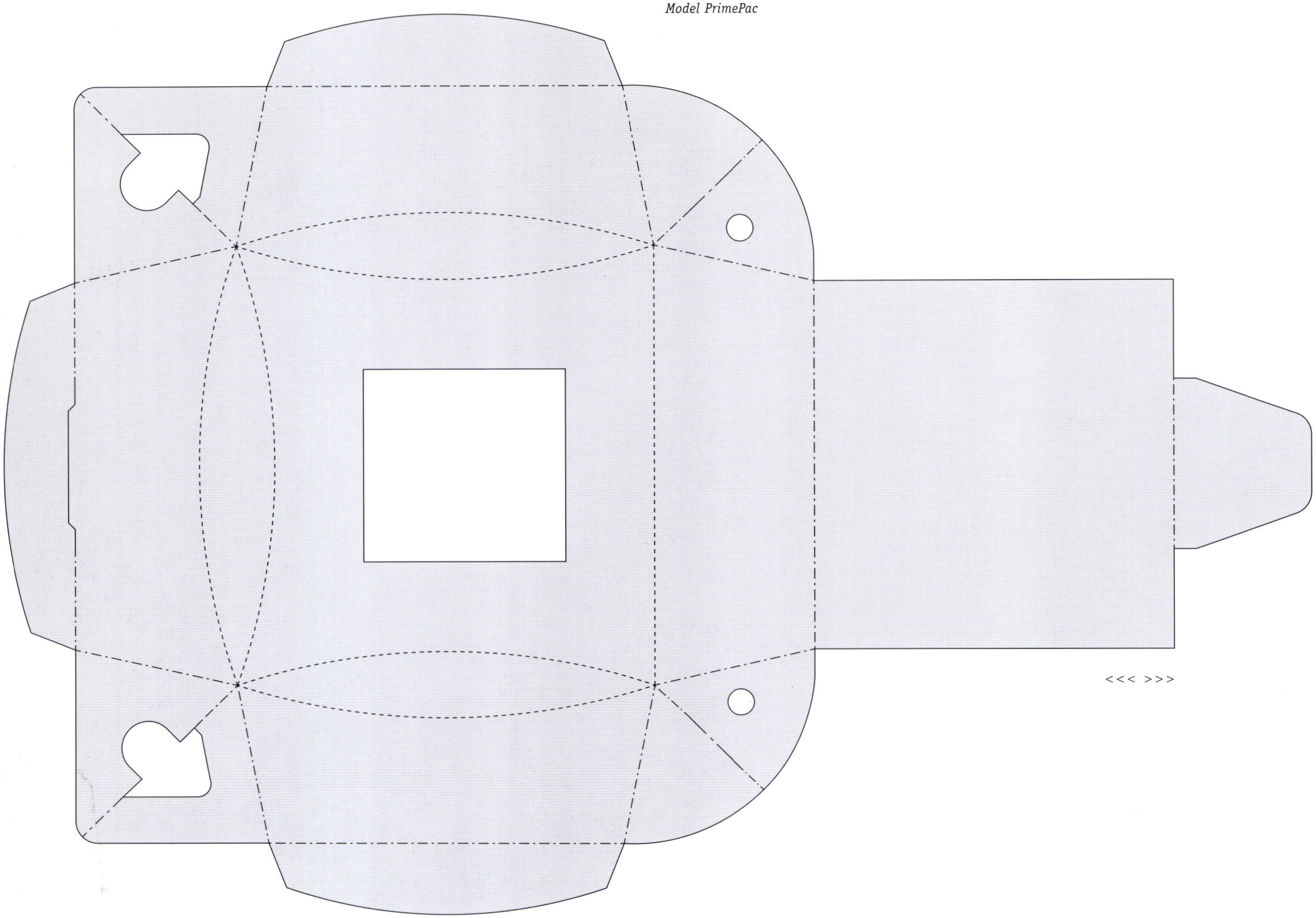

CHAMFERED-SIDED CARTON

A web-cornered carton has been given an interesting twist by bevelling the upper edges and tapering the sides. These two design changes give emphasis to the top of the carton, which is larger than the base and so provides a prominent platform for point-of-sale information. The carton is sealed with an adhesive tab on the base that is folded and fixed to the side. The pack is resealable using this tab and a lock-slot on the bottom edge of the carton.

CRACKER CARTON WITH INTERNAL-PRESENTATION DEVICE

A two-piece design is the secret to the success of this carton. The familiar and popular "cracker" design has been used for the exterior packaging, while a simple insert provides an innovative device for presenting the contents inside. The contents are accessed by removing the tear-strip on the exterior packaging, which cuts the pack into two portions. Once opened, the packaging transforms from a container to a freestanding presentation device. One end of the cracker forms the base of the display and the insert unfolds to reveal the contents.

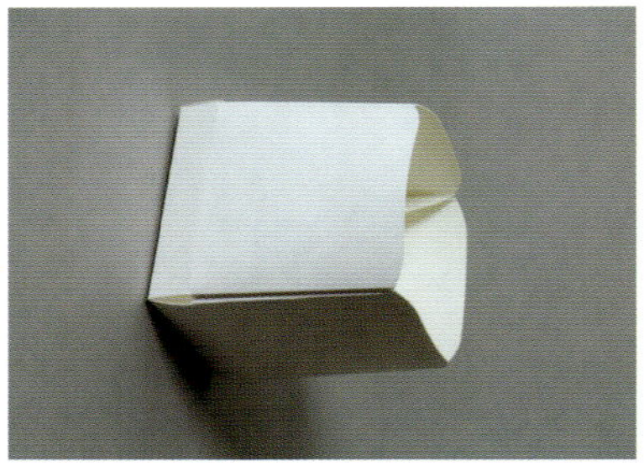

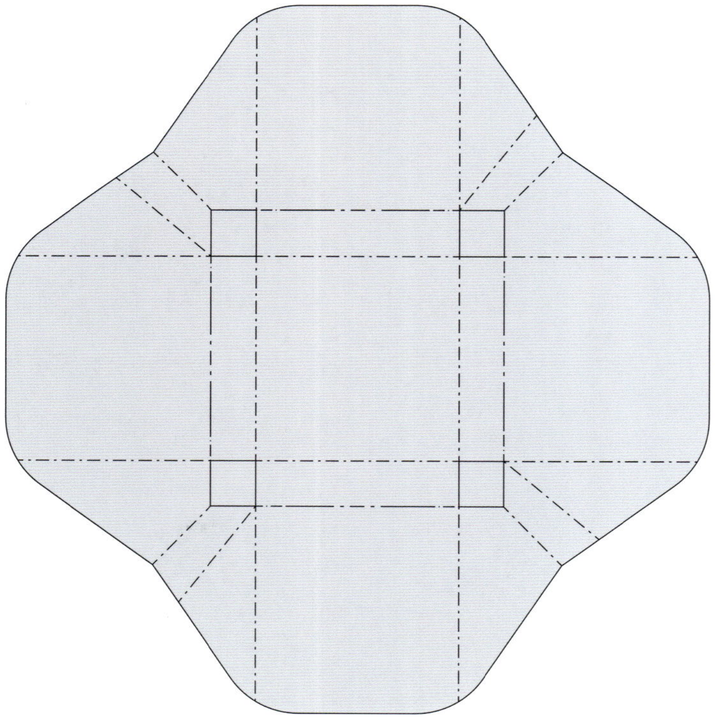

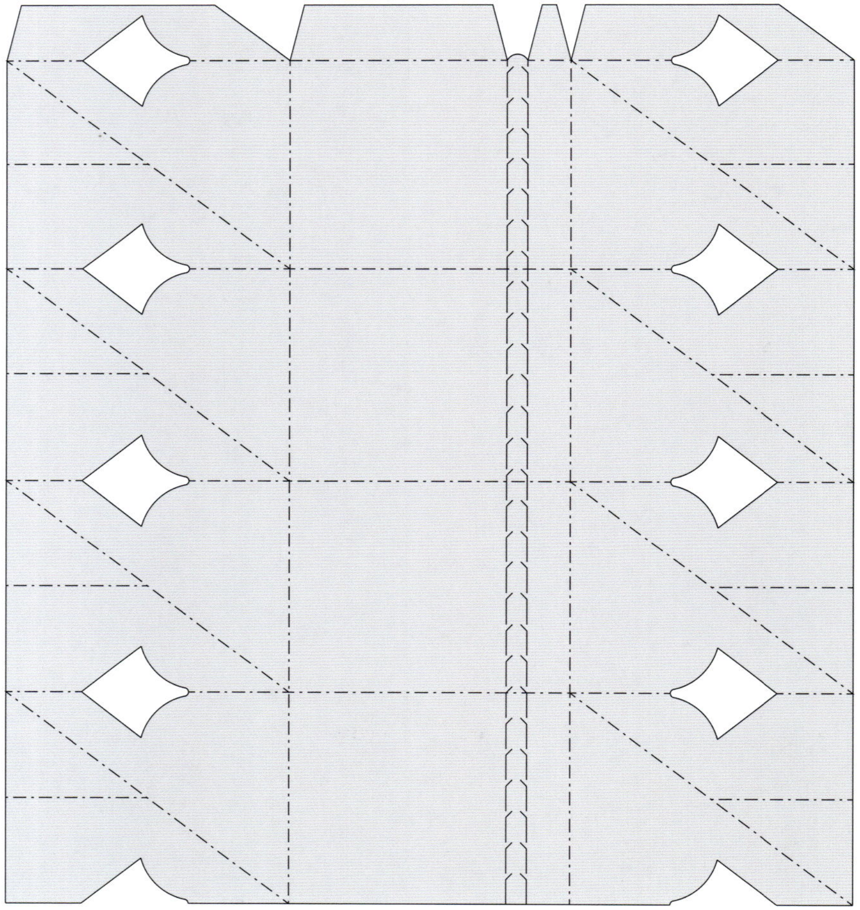

Model PrimePac

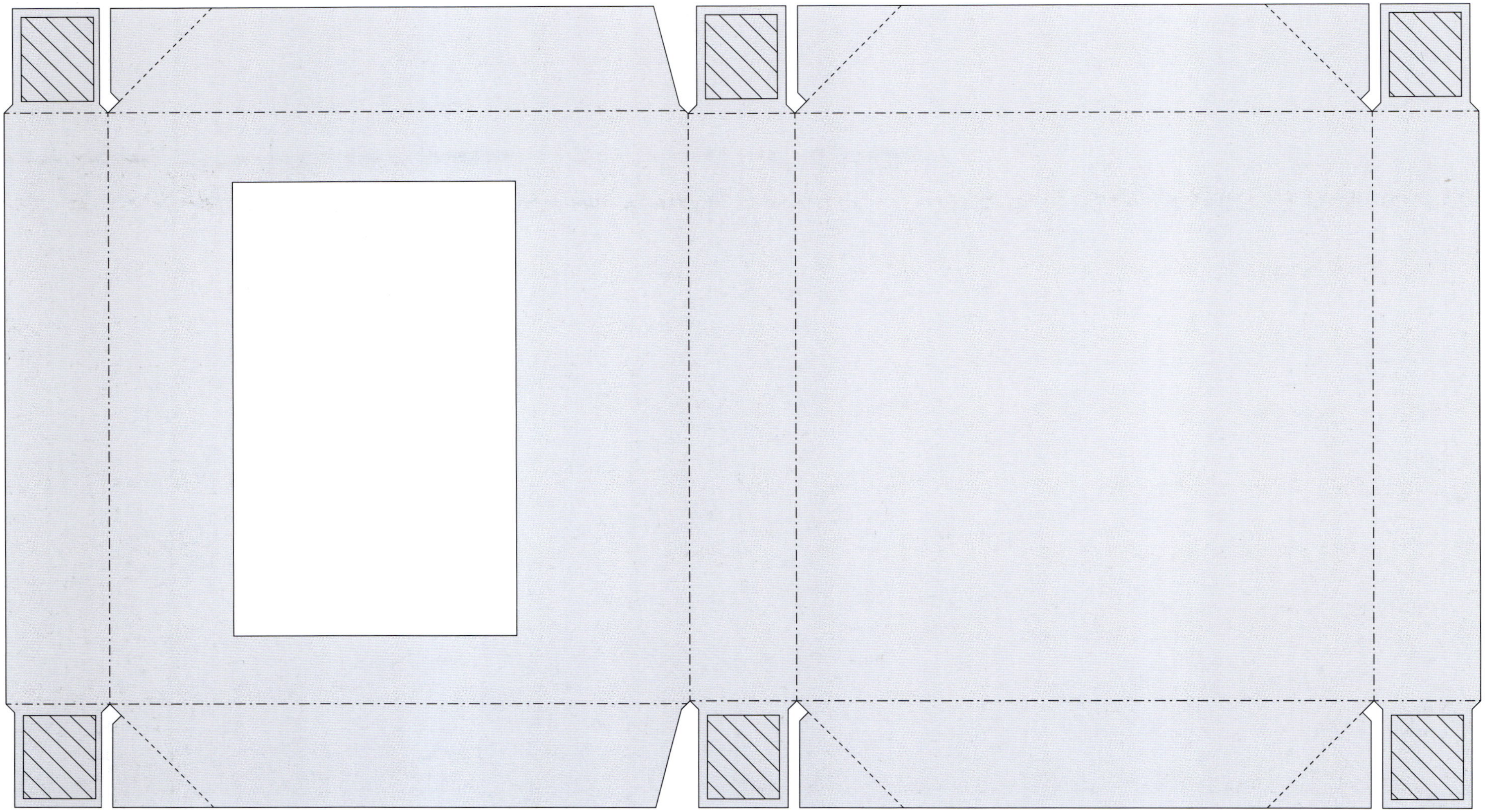

<<< >>>

SIX-POINT GLUED TRAY WITH INTEGRAL LID

Transporting the package to the product often poses more design problems than presenting the product to the end-user. This package represents a common solution for many standard packaging scenarios and can be adapted to meet specific requirements, such as the size and volume of the product. This carton is transported to the product in a flat-pack form allowing for high volumes and reduced cost of shipping, and is easily erected and sealed. The carton is glued at each corner on its base and at the front two corners of its lid. The lid's large surface area provides a good platform for presenting product information or providing an attractive display.

FAST-FOOD WALLET

The proliferation of packaging waste from fast-food outlets led to a major rethink of packaging design in this sector in the 1990s. This pack is one of a range of packaging solutions that have helped to improve standards of design, delivery, and customer experience, while also substantially reducing packaging waste. This simple wallet design is easily erected at the point of sale and uses minimal materials for the transportation and delivery of fast food.

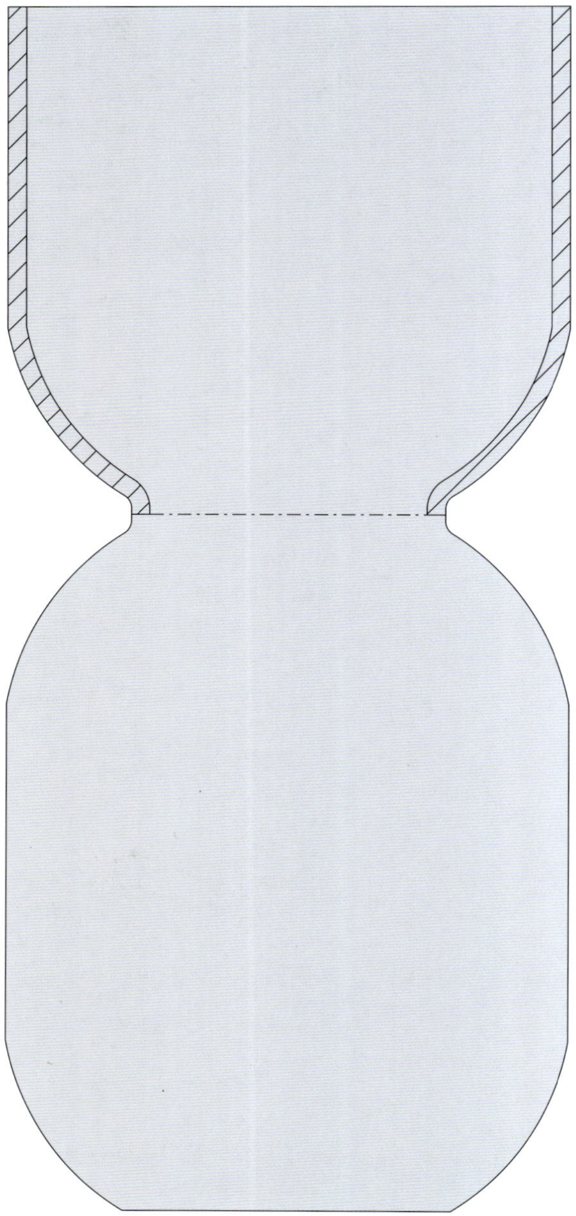

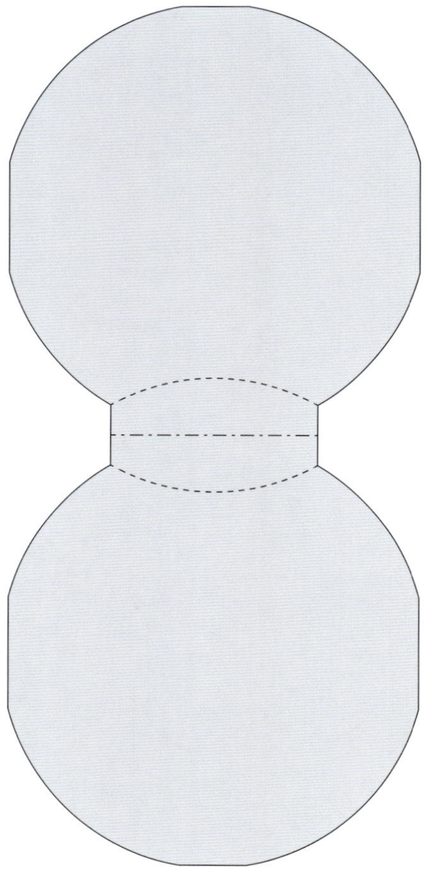

<<< >>>

RAP

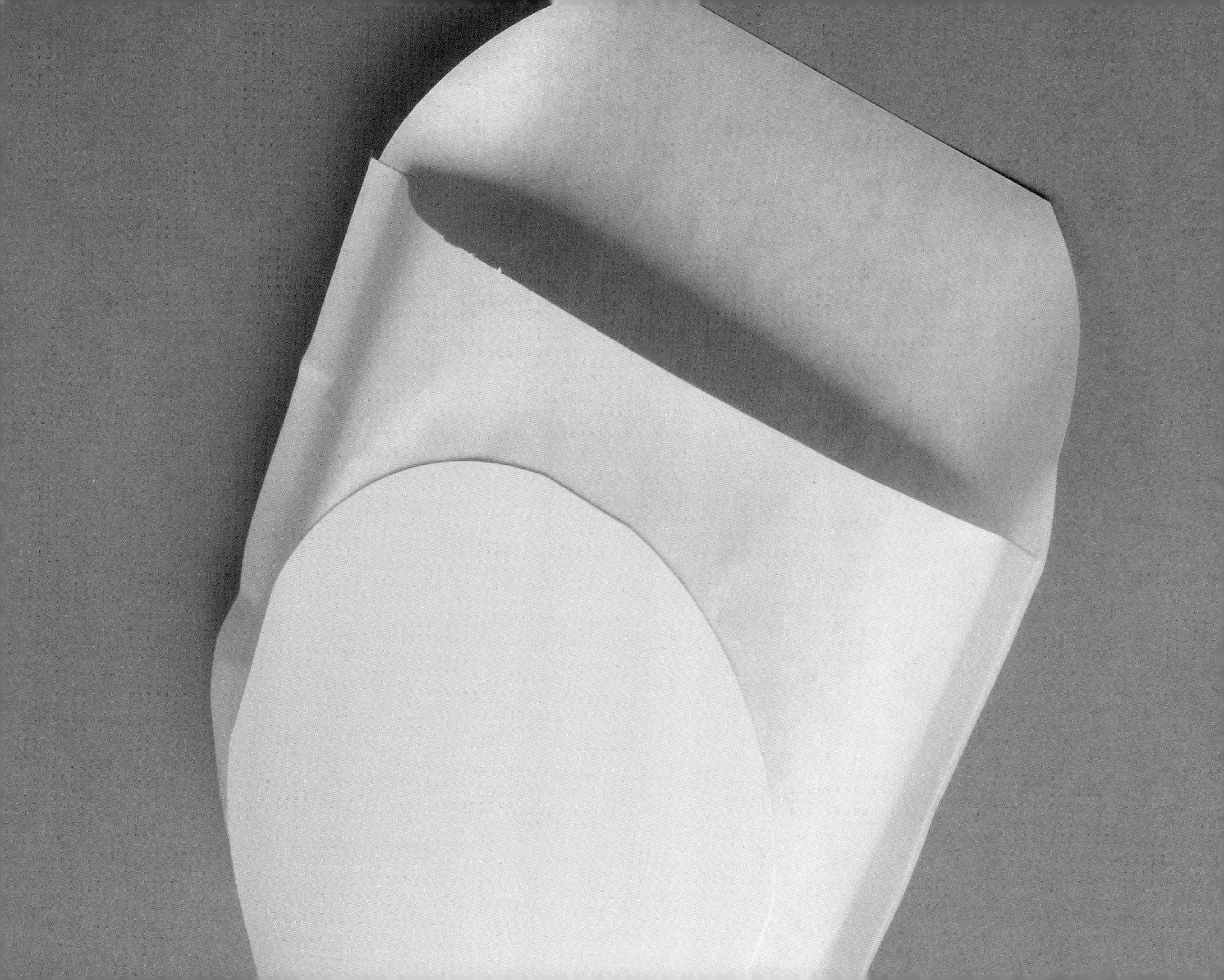

FAST-FOOD SLEEVE

This pack relies on the use of two different materials to provide a lightweight, easily erectable fast-food package. Carton board is used to provide structural integrity and strength, while a lighter composite paper is used to wrap the product. The simplicity of the design allows it to be flat-packed, filled quickly, and resealed using a tab-lock.

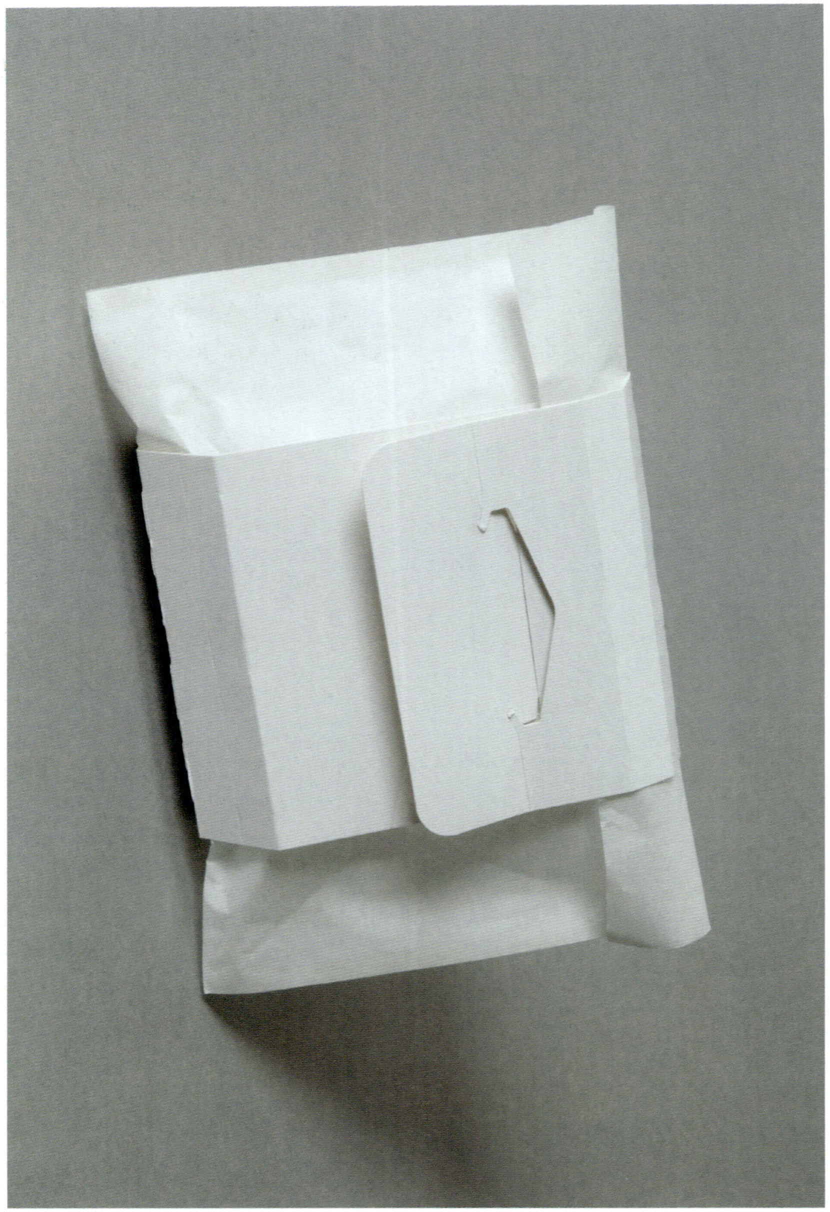

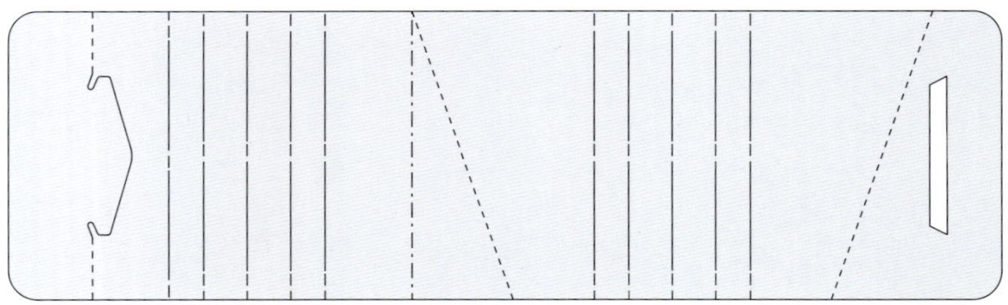

RAP

<<< >>>

Rokov Design

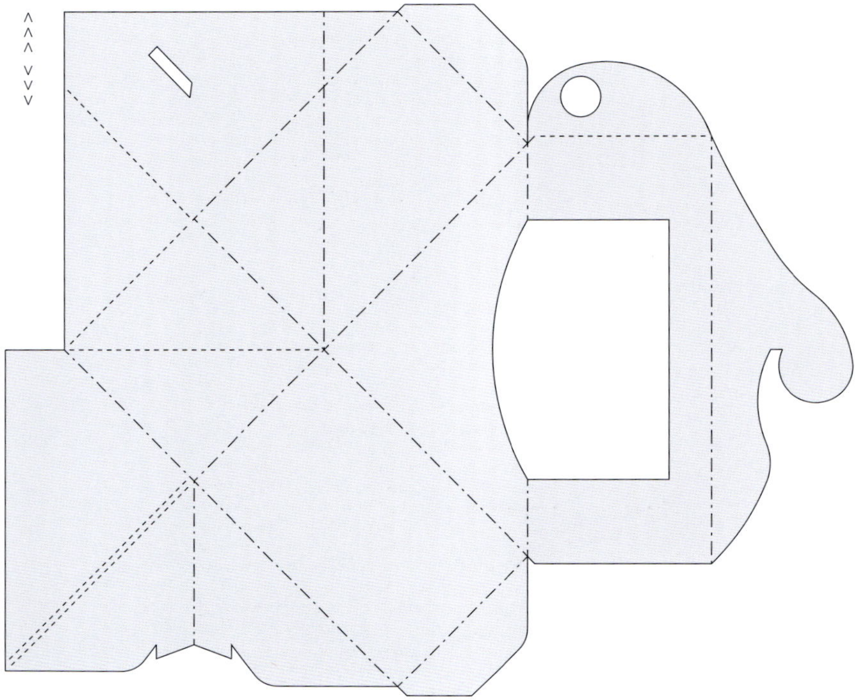

GLUELESS SANDWICH PACK

Fast food has helped to stimulate design innovation in what has become the most competitive packaging sector. With billions of packs sold each year worldwide, the lucrative rewards for succeeding in this market have driven innovation to remarkable extremes. This highly inventive design uses a series of complex folds to create a triangular sandwich pack that is easy to erect and keeps the product fresh, but also has several added characteristics. These include an internal pouch for sauces or a napkin, an unfolded plastic window that offers a flawless view of the product, a finger hook for easy carrying, and a design that, when fully opened, serves as a plate.

Rokov Design

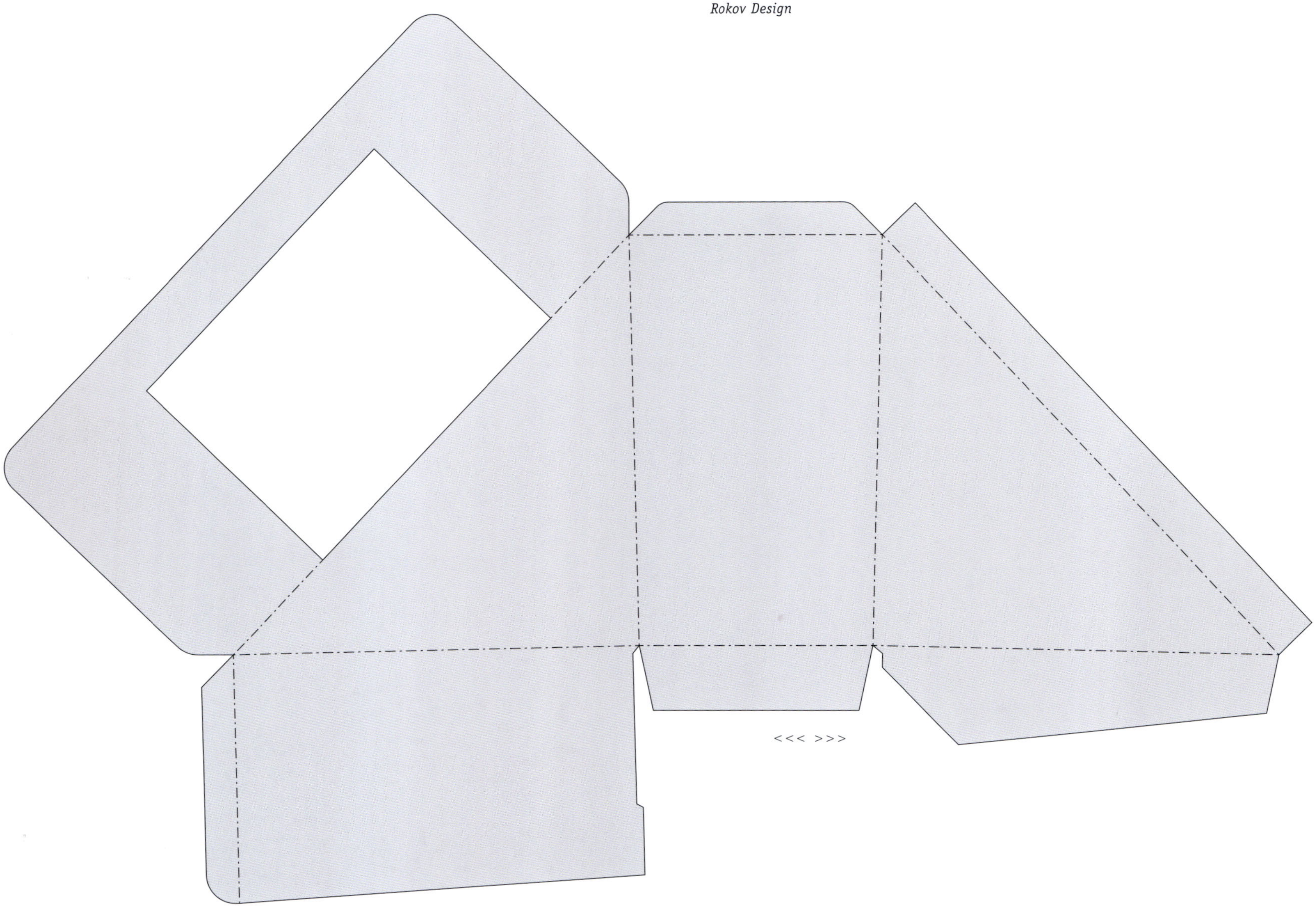

HEAT-SEALABLE SANDWICH PACK

Packaging of fresh produce for periods of longer than one day poses a wide range of problems that have to be overcome through design, manufacturing processes, and use of materials. This pack keeps produce fresh for longer periods than manually sealed packs and is constructed using preglued tabs along two sides while leaving the lid open. After filling, the lid is folded and heat-sealed to the flanges along the length and sides of the pack. The window, which is glued to the underside of the lid, provides a view of the contents without compromising the pack's functionality.

WRAP PACK

Increasingly diverse fast-food types present increasingly wide-ranging problems for packaging designers. One food type that has yet to receive the same level of attention as traditional bestsellers such as sandwiches, pizzas, burgers, and fries, is the wrap. Certain characteristics of the wrap, such as needing to reveal its contents and its shape, make it a particularly difficult product to package effectively and attractively. This pack employs a similar technique to the triangular sandwich pack to contain two halves of a wrap, end to end, so that the contents are revealed through the front windows. The package is easily erected and heat-sealed; it also stacks well in transportation and at the point-of-sale.

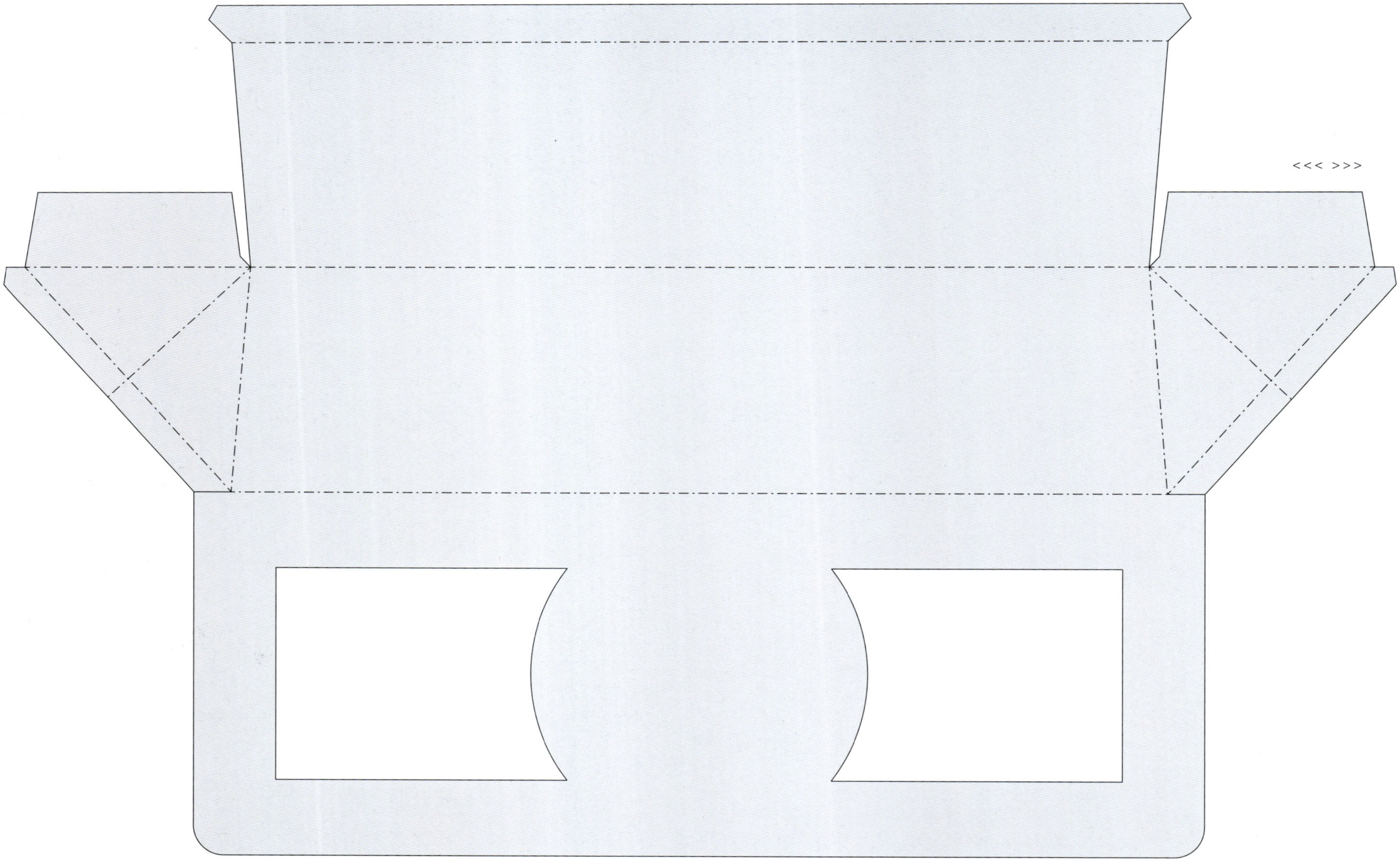

Rokov Design

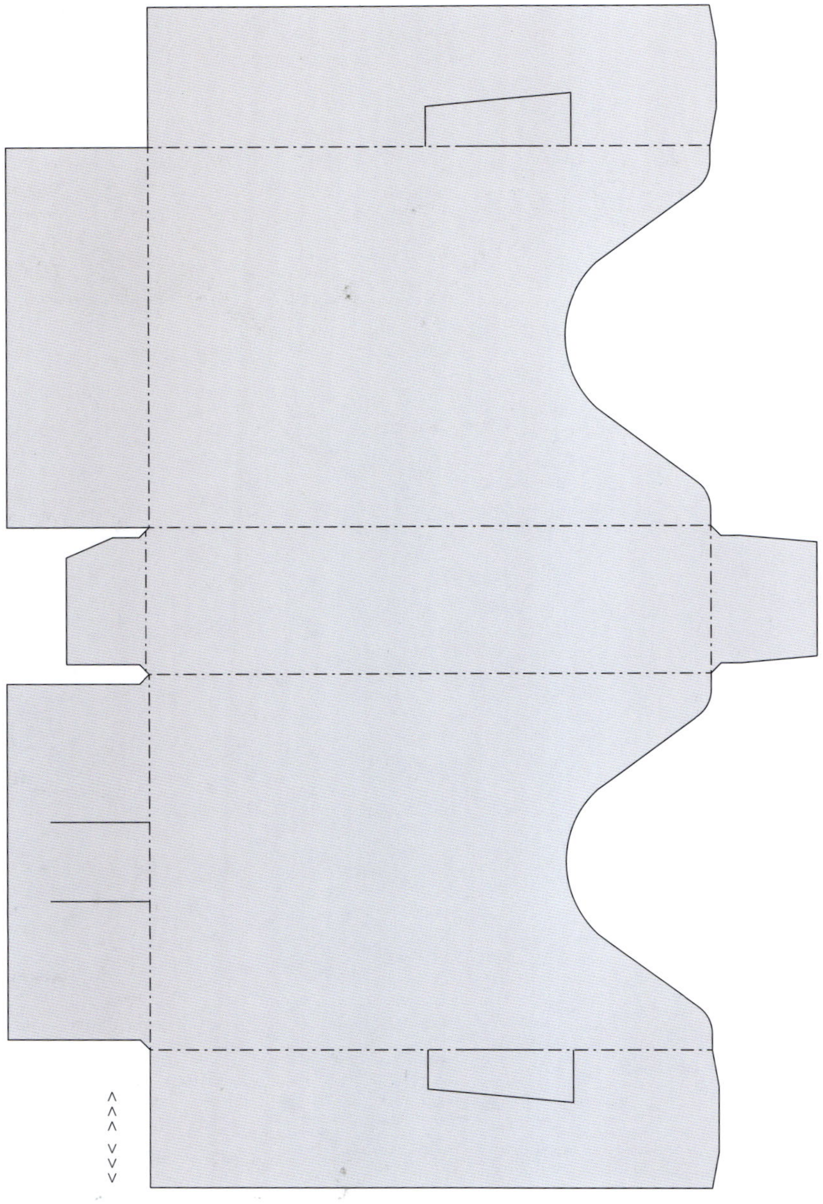

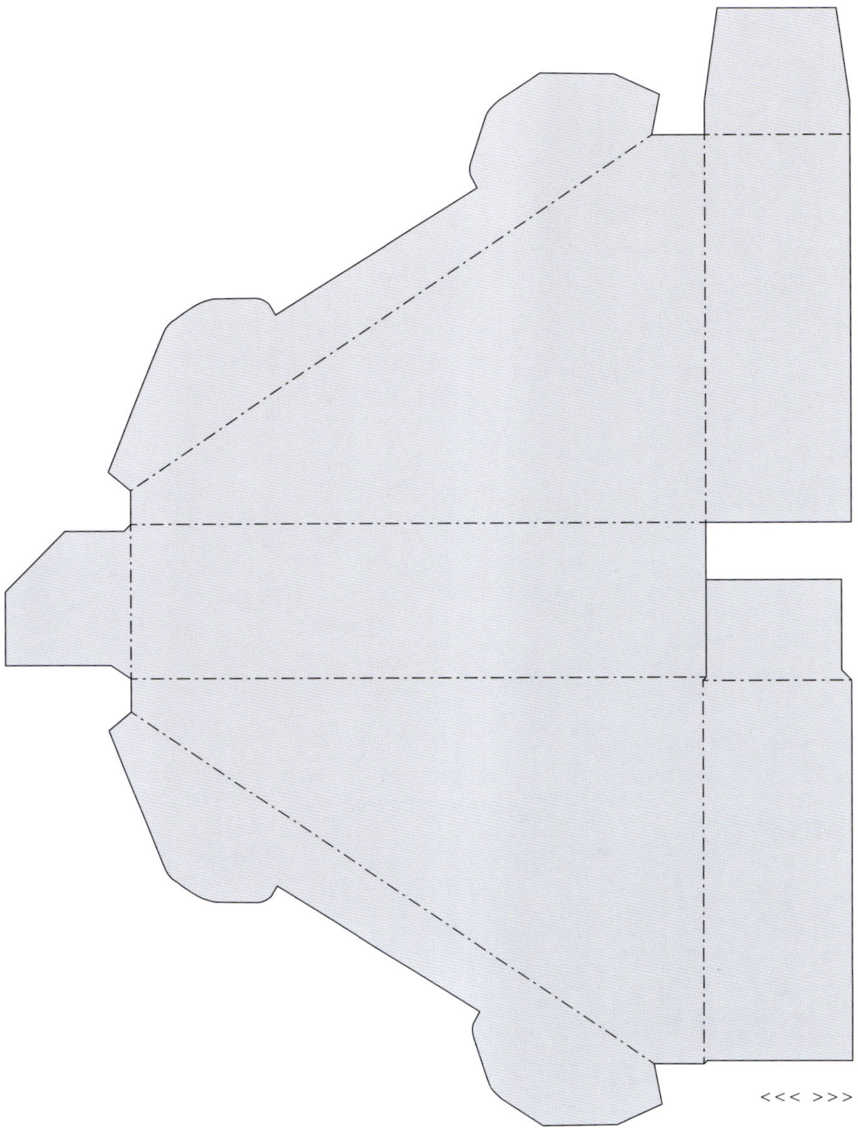

HINGED CARTON WITH INTERNAL LOCK

The user experience is often an essential part of a successful carton design, adding value by enhancing brand identity and product satisfaction. This two-part carton provides a memorable experience by opening and closing using a hinge on the outer casing, which pivots at the bottom corner of the carton where it is glued to the inner case. An internal locking device prevents the two components becoming separated.

DUAL-ACTION WALLET (BURGOPAK™)

By using one or more materials, the design and functionality of a carton can be significantly improved. However, designers should always consider carefully the impact of using multiple or composite materials: they are harder to recycle and may rely on nonrenewable materials. In this instance, the use of a plastic strip inside the carton produces an effect almost like a magic trick. When the internal sleeve is withdrawn from the outer wallet, another sleeve appears from the other end of the wallet. Neither sleeve can be completely removed from the wallet. When the sleeve is pushed back, the other sleeve also returns.

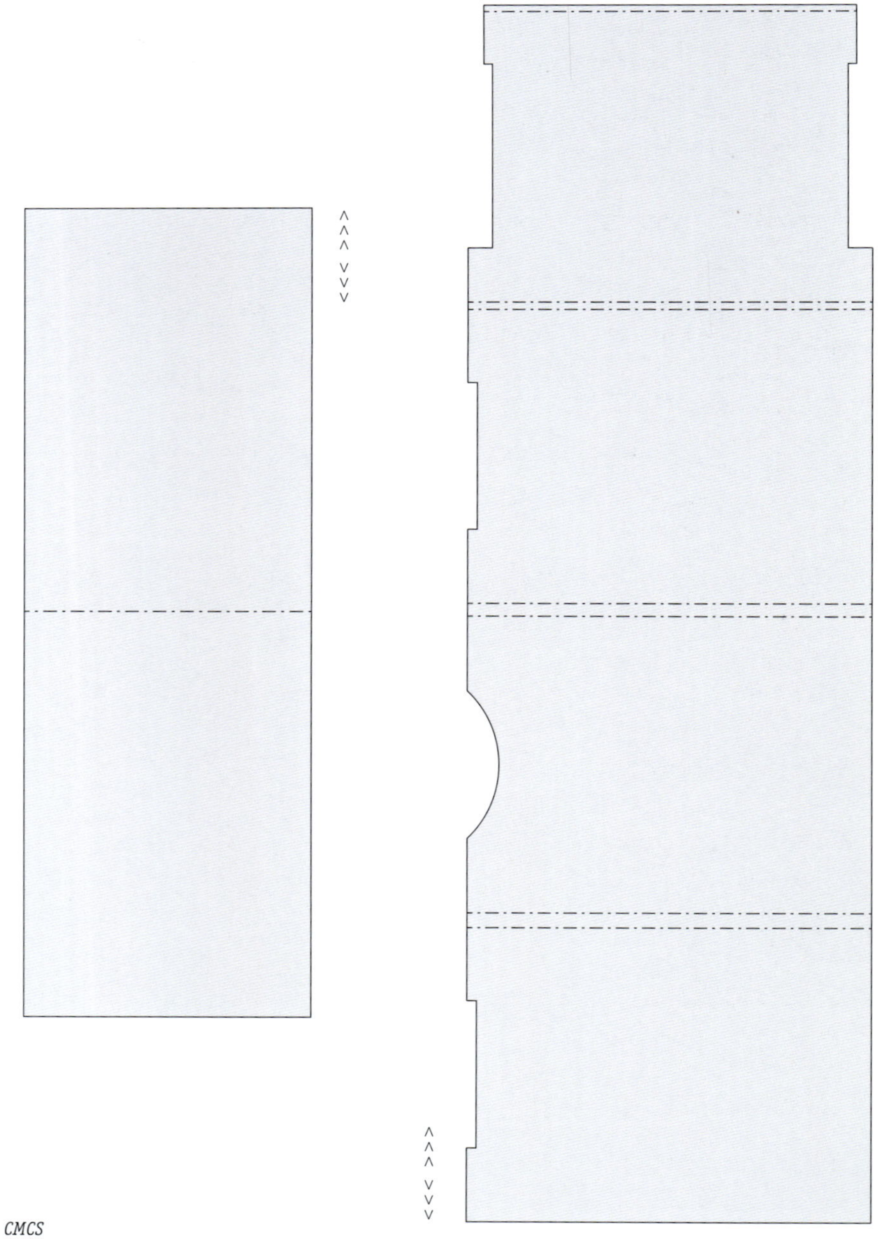

CMCS

Arden Packaging

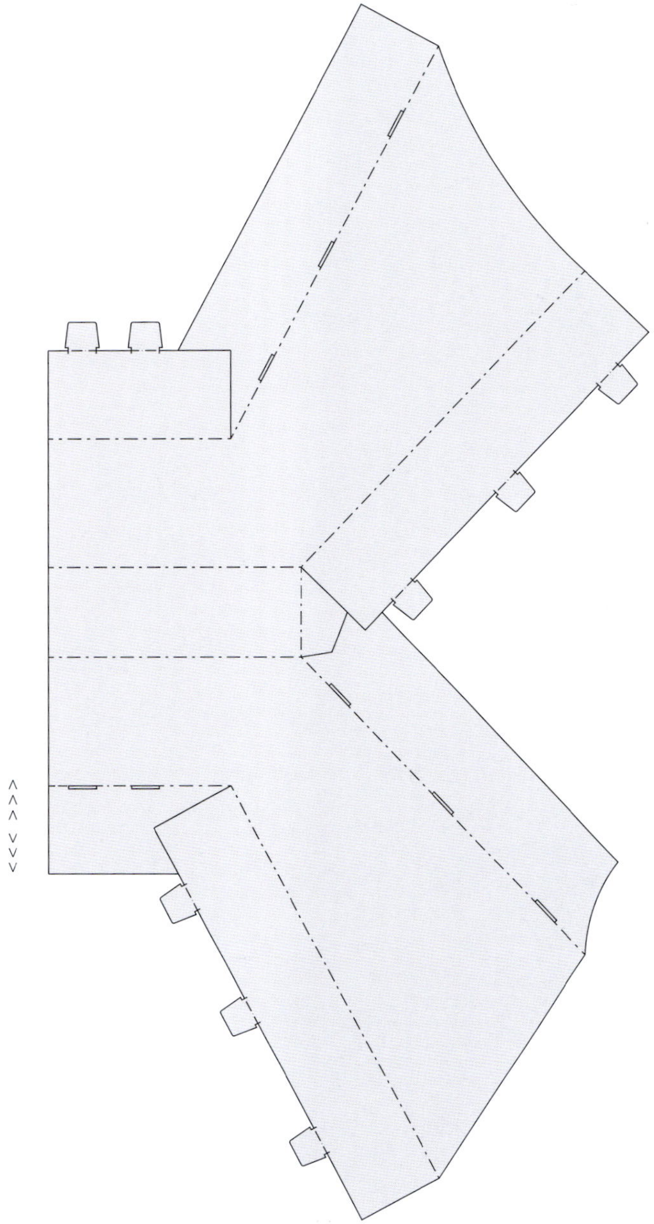

The expertise of the packaging designer can often be employed in industries beyond the traditional boundaries of packaging. This innovative design requires the knowledge of working with corrugated board and carton board to derive practical and environmental benefits that former designs and materials could not deliver. Previously, a metal-exhaust duct had been used in a chemical-extraction process, but the duct, which required daily cleaning, wasted time, money, and resources. The use of a corrugated-board duct that can be recycled saves time and prevents the need for chemical cleaning.

BOOK-COVER CARTON

The anticipation and delight of opening the cover of a storybook for the first time and revealing the contents by turning each page is something every child remembers. This design achieves a similar sense of expectation by incorporating a large flap fixed to the left edge of the carton's main face that acts as a front cover. The internal faces provide an excellent platform for product information and graphic design.

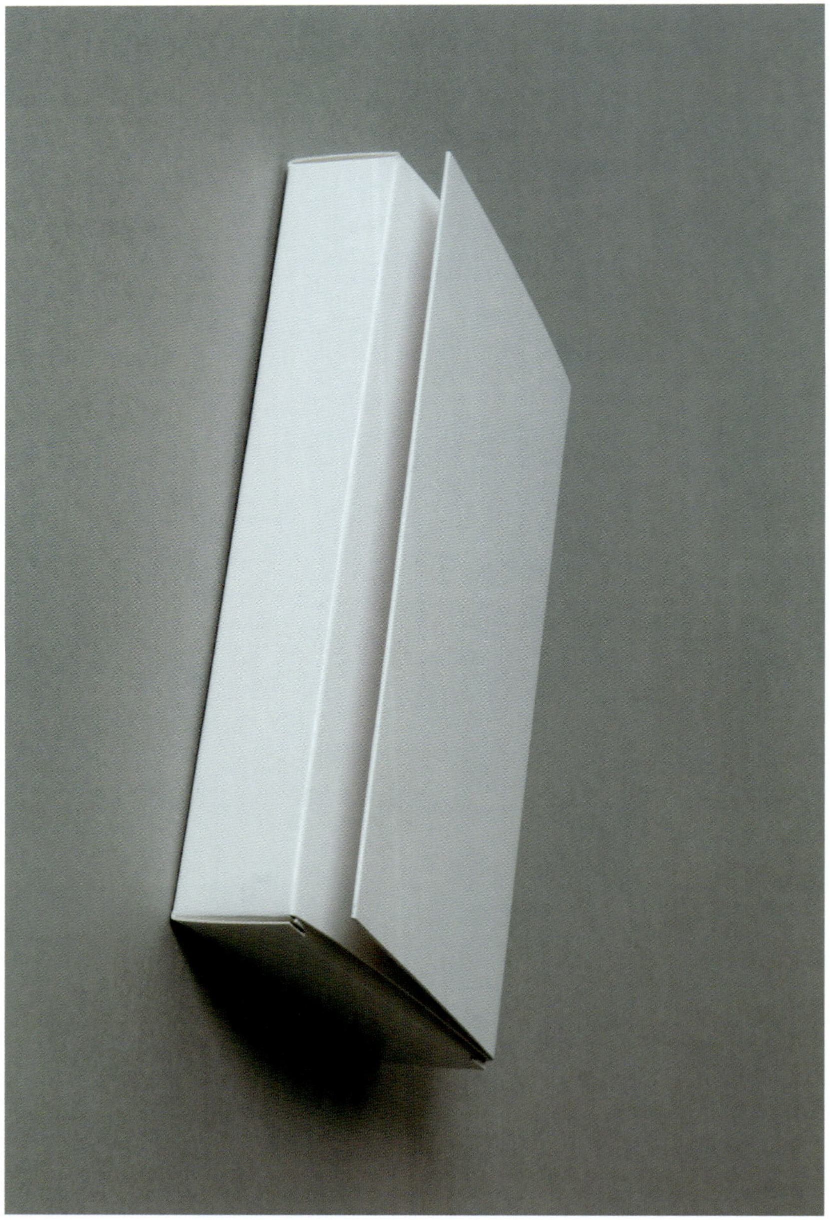

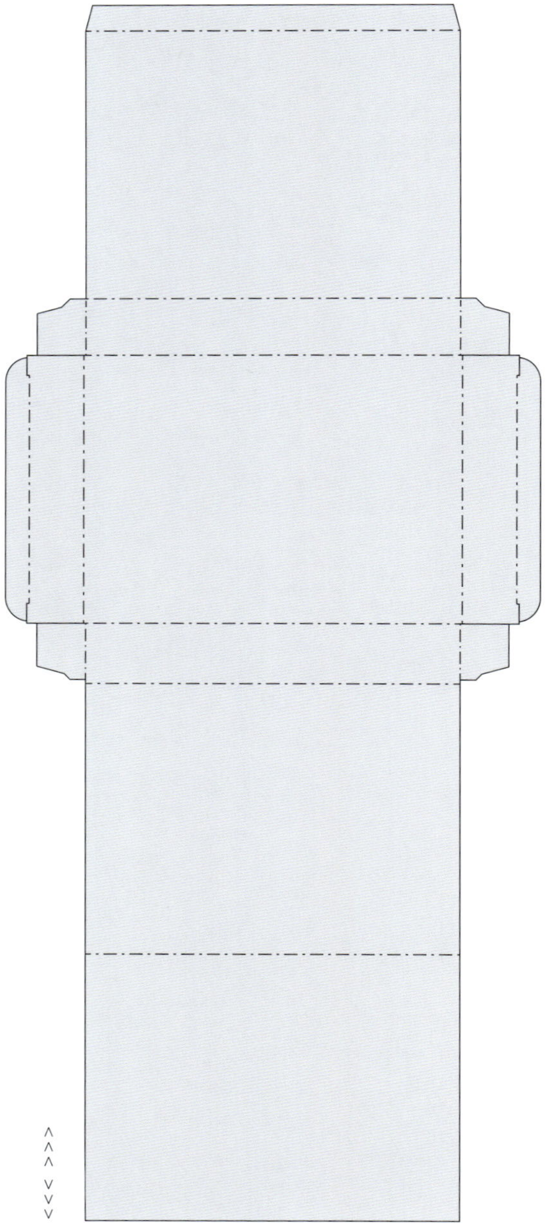

CMCS

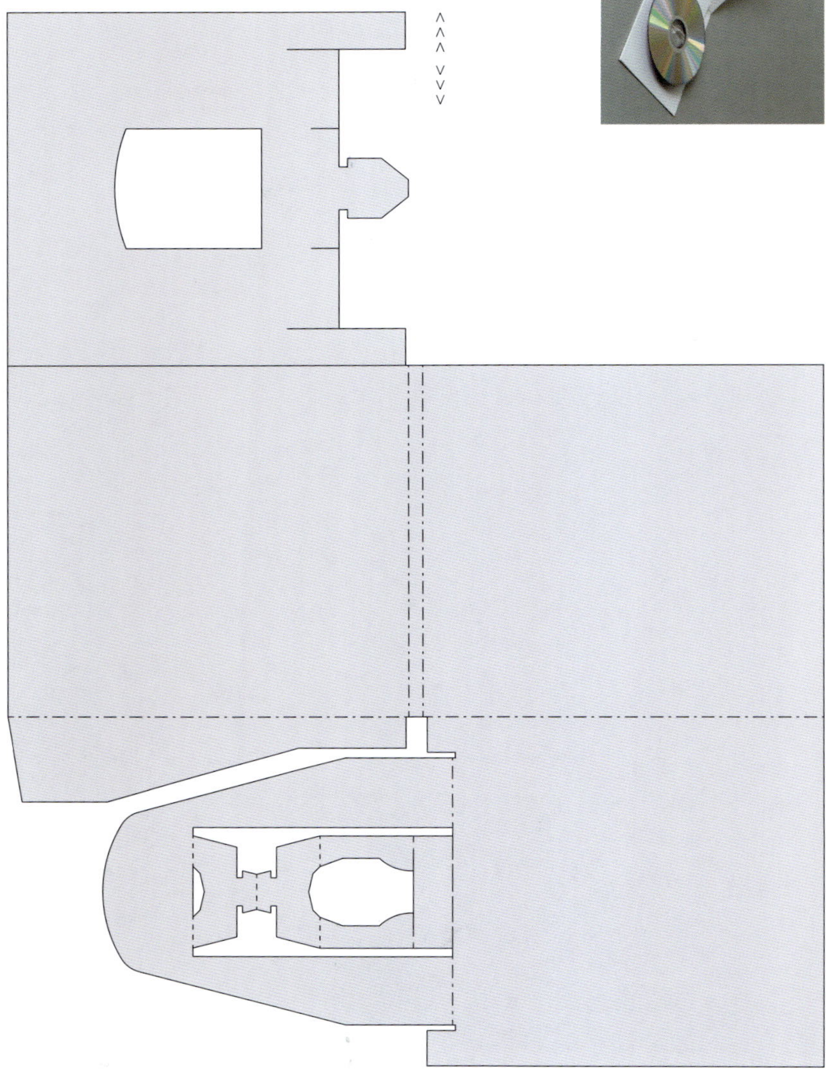

DiscLev (Schausberger)

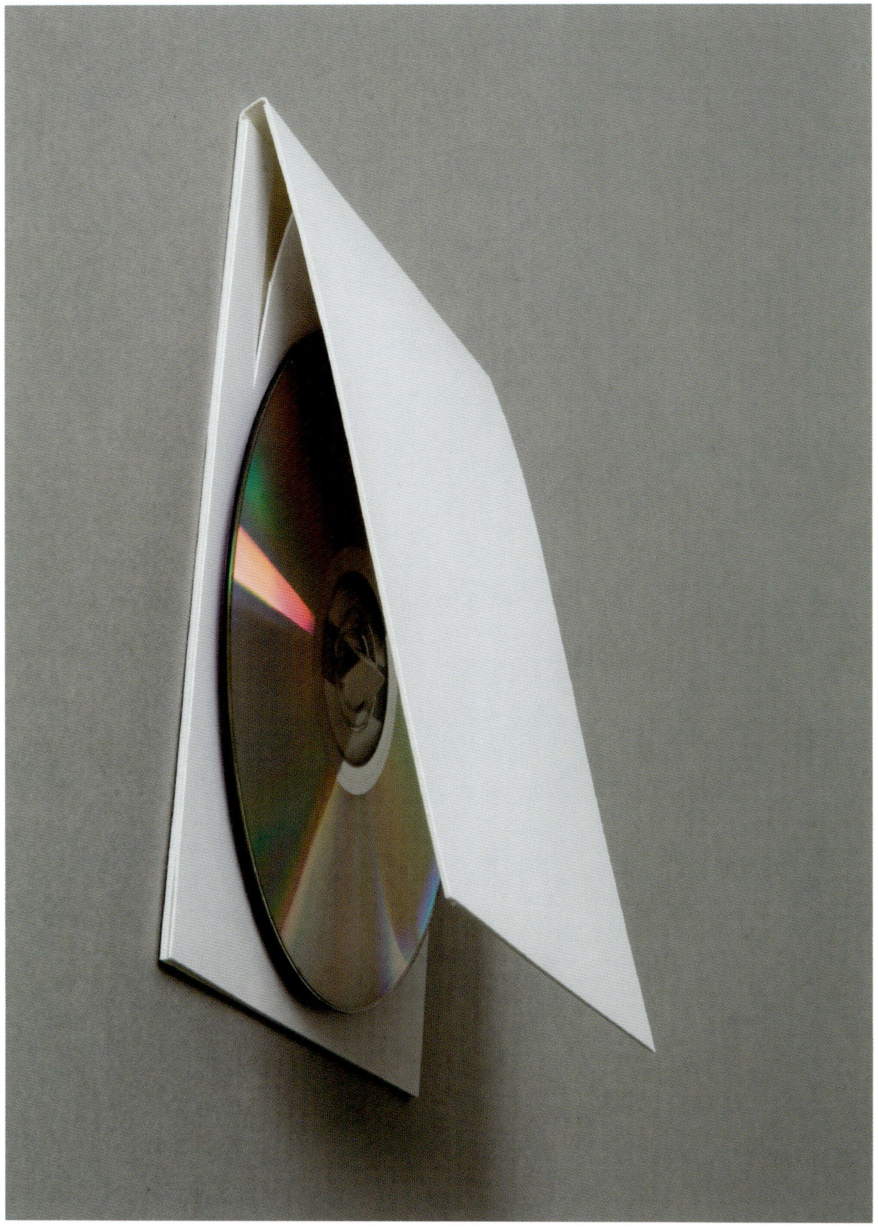

COMPACT-DISC CASE

Dynamic, innovative, and intelligent design has created an exceptional case for containing and presenting compact discs. The case is delivered flat-packed, but after it has been folded, the locking mechanism is activated and hereafter, the levitation device rises to present the compact disc each time the front of the case is opened. This highly effective design solution demonstrates the versatility and excellent potential of carton board in markets traditionally dominated by other materials.

DISPLAY PACK

Sometimes the pack and the product are indistinguishable. In this instance, the product is the package and the package is the product. Two glued components, a piece of paper and a folded piece of carton board, provide a means of information display. The wallet protects the display sheet inside, which automatically unfolds through a series of creases in the paper when the wallet is opened.

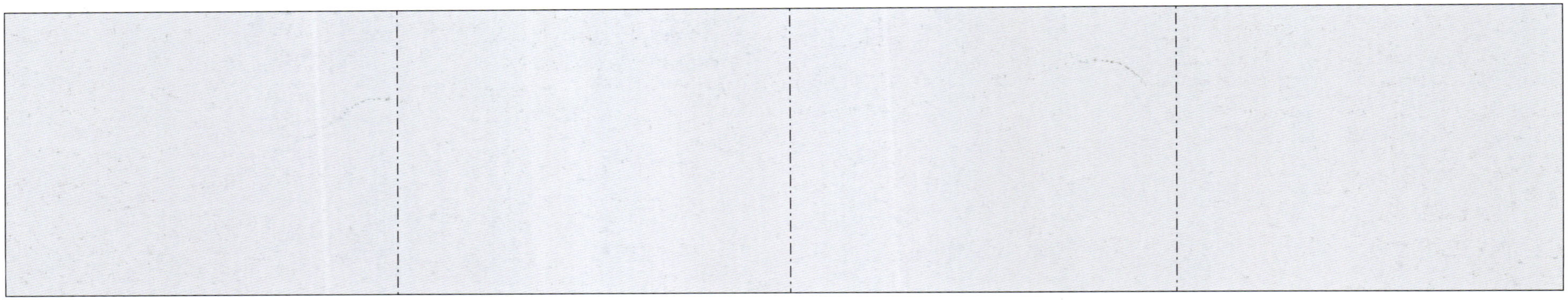

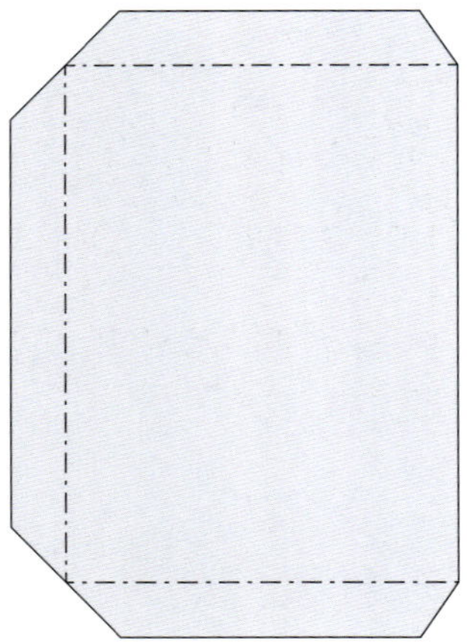

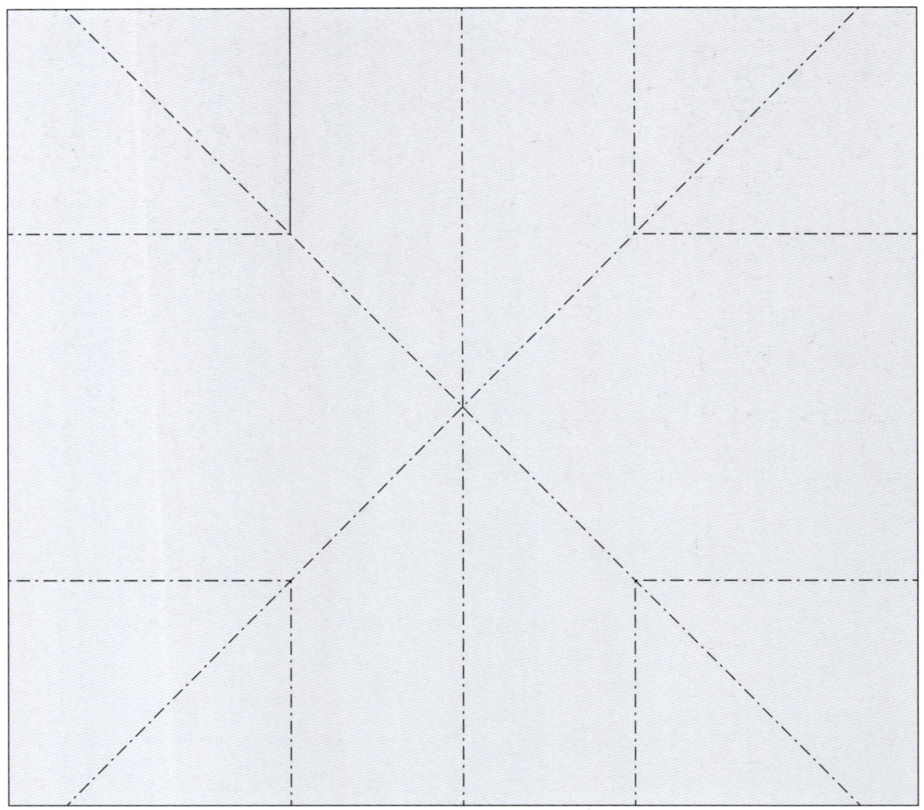

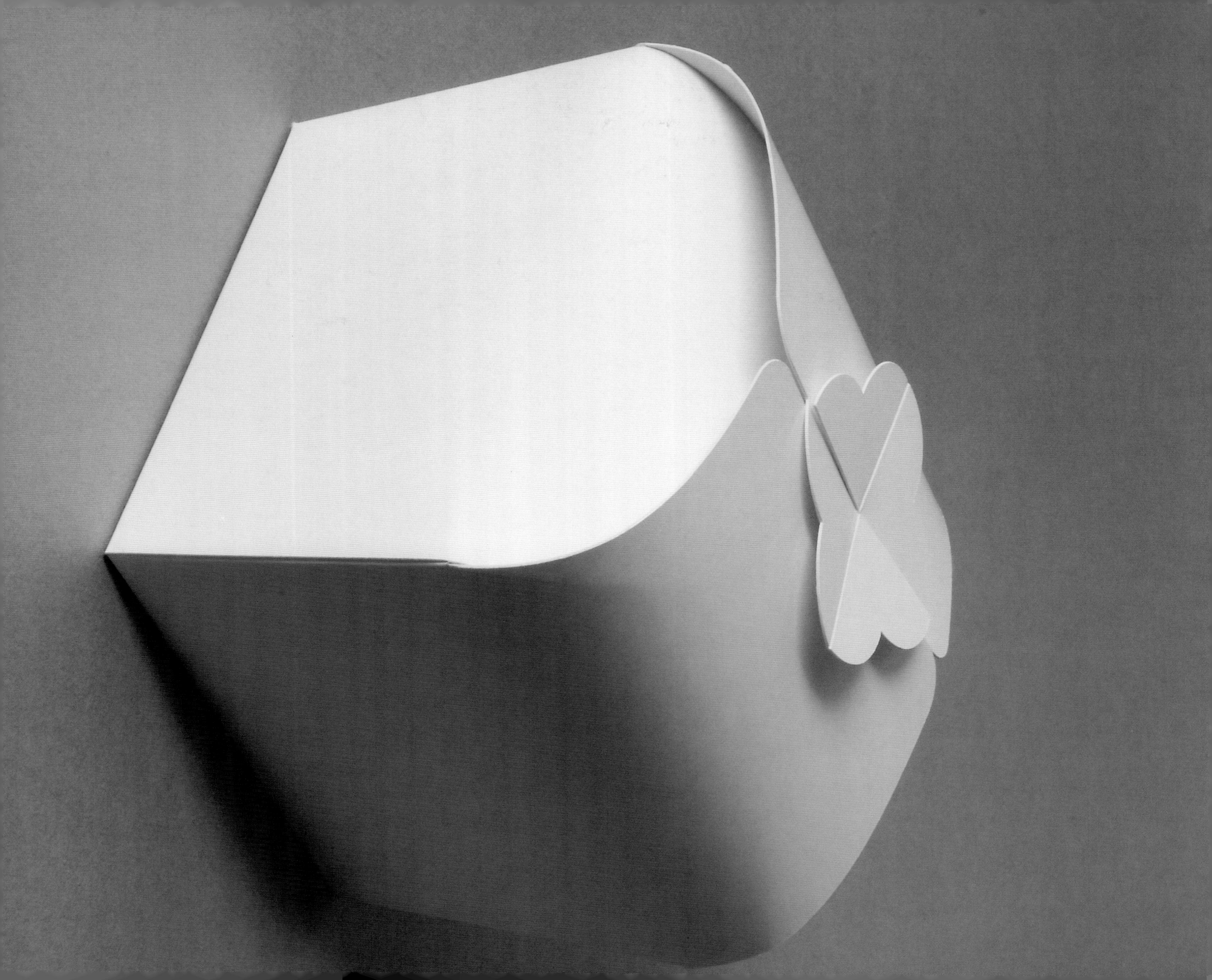

MR Design

<<< >>>

DOUBLE-ARCED TOPPED CARTON

A standard interlocking device provides the basis for this striking carton, whose tapered sides are gracefully rounded toward the top of the carton to form an interesting closure. The crash-lock base provides easy construction and the simple locking device on top of the carton makes this container ideal for take-out or novelty products.

GARTER CARTON

Packaging does not always have to contain a product completely. Sometimes, it serves as a presentation device, displaying the product rather than concealing it. This simple band serves as a structural device on which certain products, such as textiles, can be hung, while also providing a suitable point-of-sale vehicle that does not distract attention from the product. The carton is made from a single piece of board and requires no gluing.

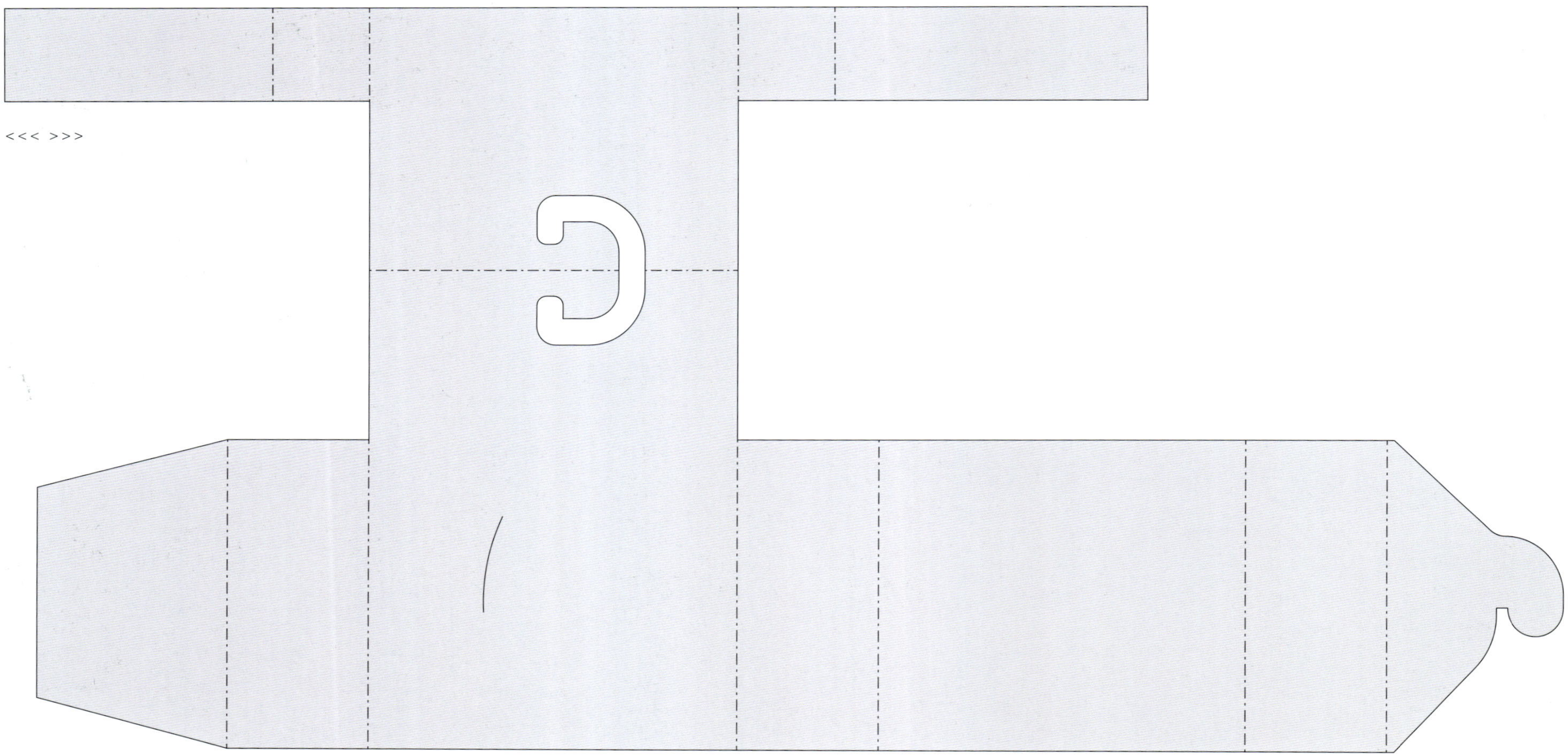

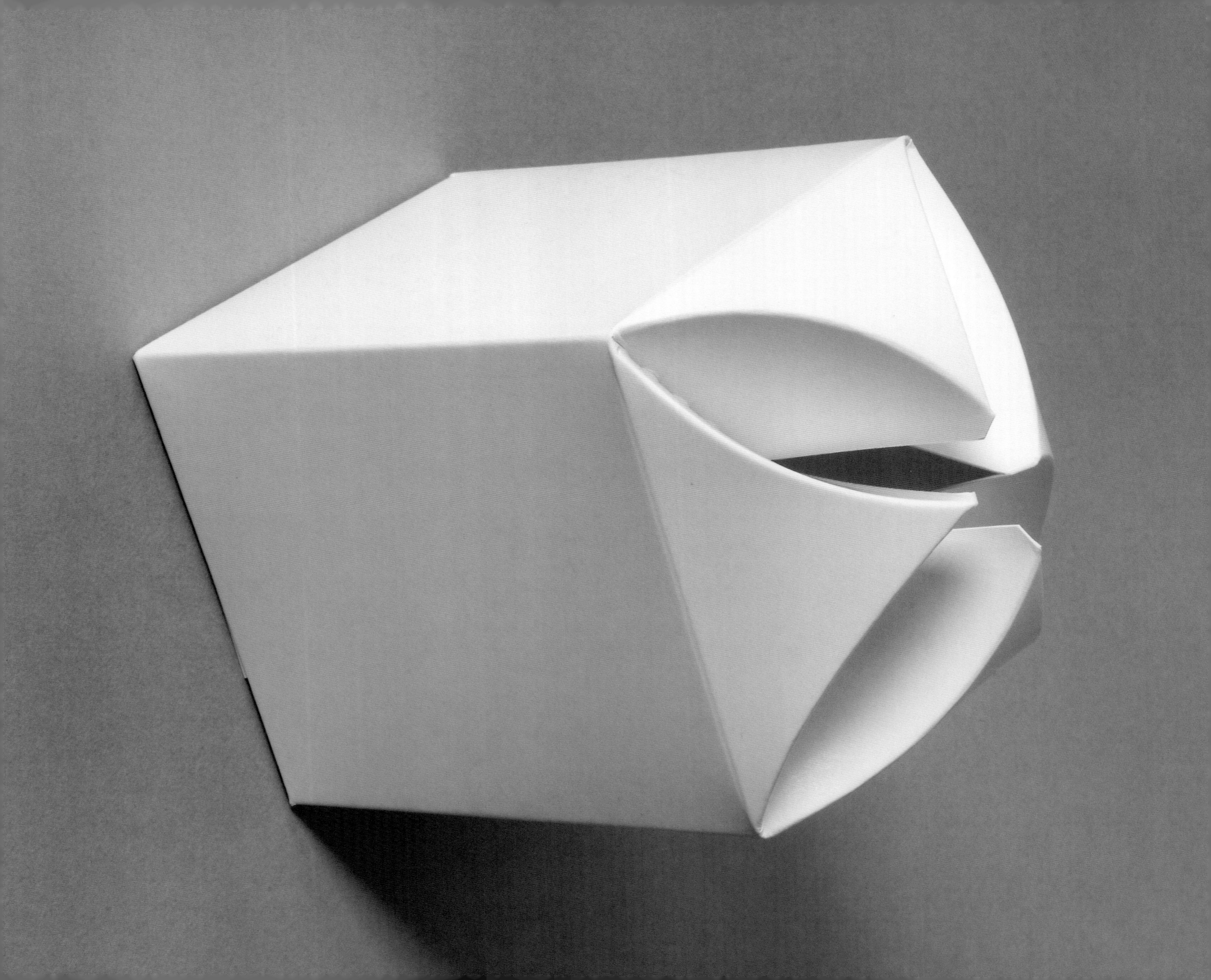

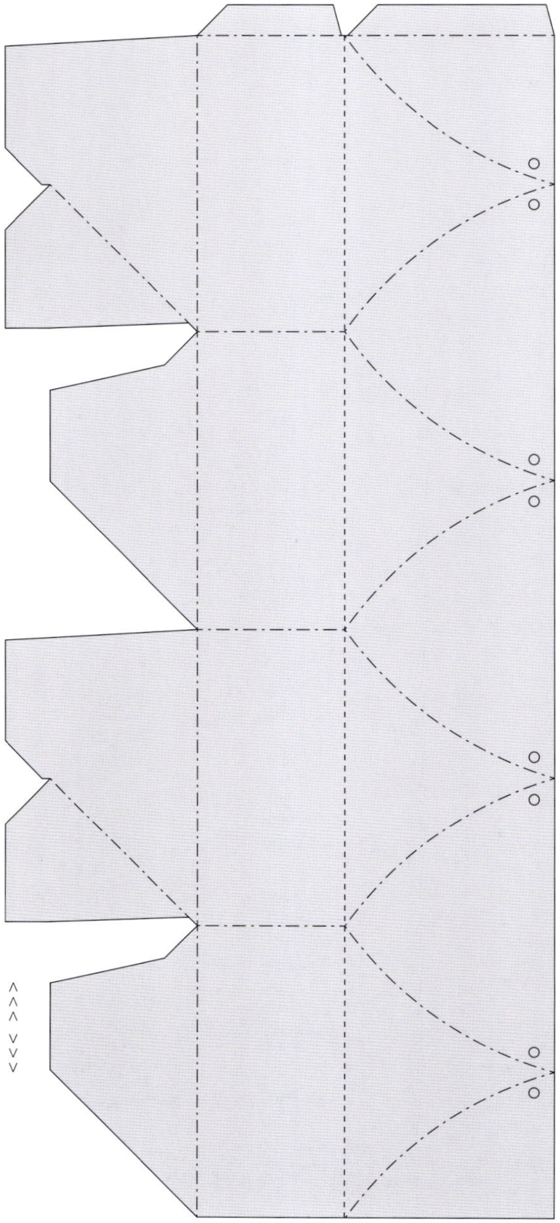

MR Design

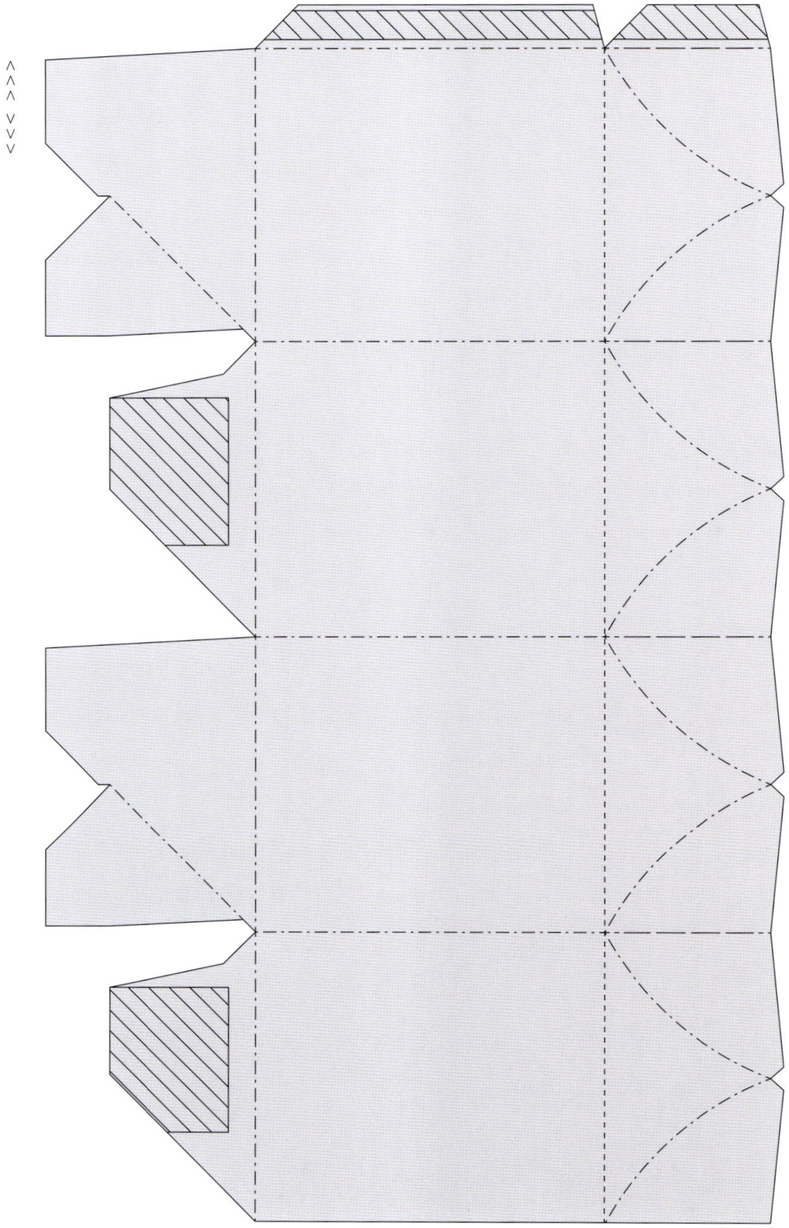

FLORAL-TOP CARTON

The lid of this carton offers a striking characteristic to an otherwise standard carton with crash-lock base. The four-sided floral design is achieved by the folds in the lid that, when compressed, fold naturally into the center of the carton and hold in place. More secure sealing could be achieved by using an adhesive sticker in the center of the floral design or by passing a ribbon through holes at the apex of each folded side.

GABLED-ROOF CARTON

Minor alterations to standard pack designs can produce idiosyncratic results. Here, the front face of a standard tuck-flap carton has been lowered to give the carton a sloping top surface. This has been further enhanced by the inclusion of a triangular extension on the front of the carton that forms a gable. This addition is both visually interesting and structurally important, as it forms a locking device at the back edge of the carton to secure the contents within the package.

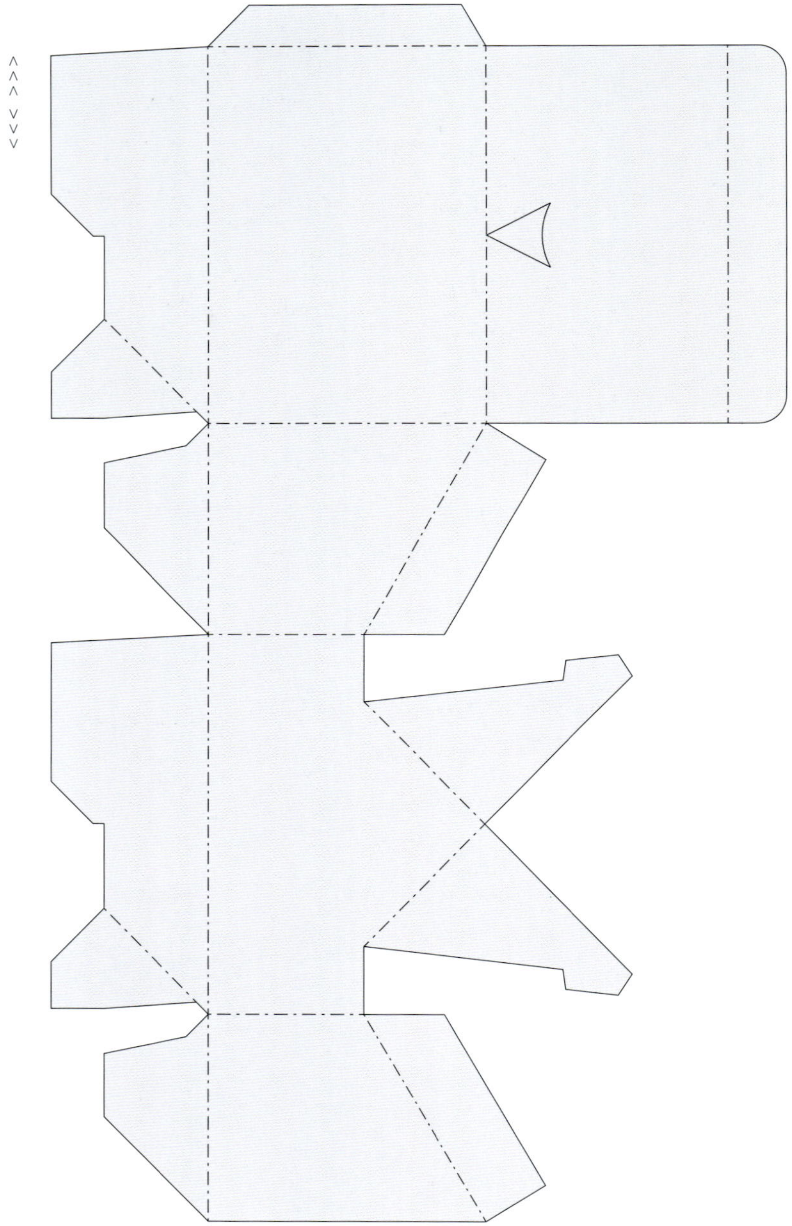

MR Design

MR Design

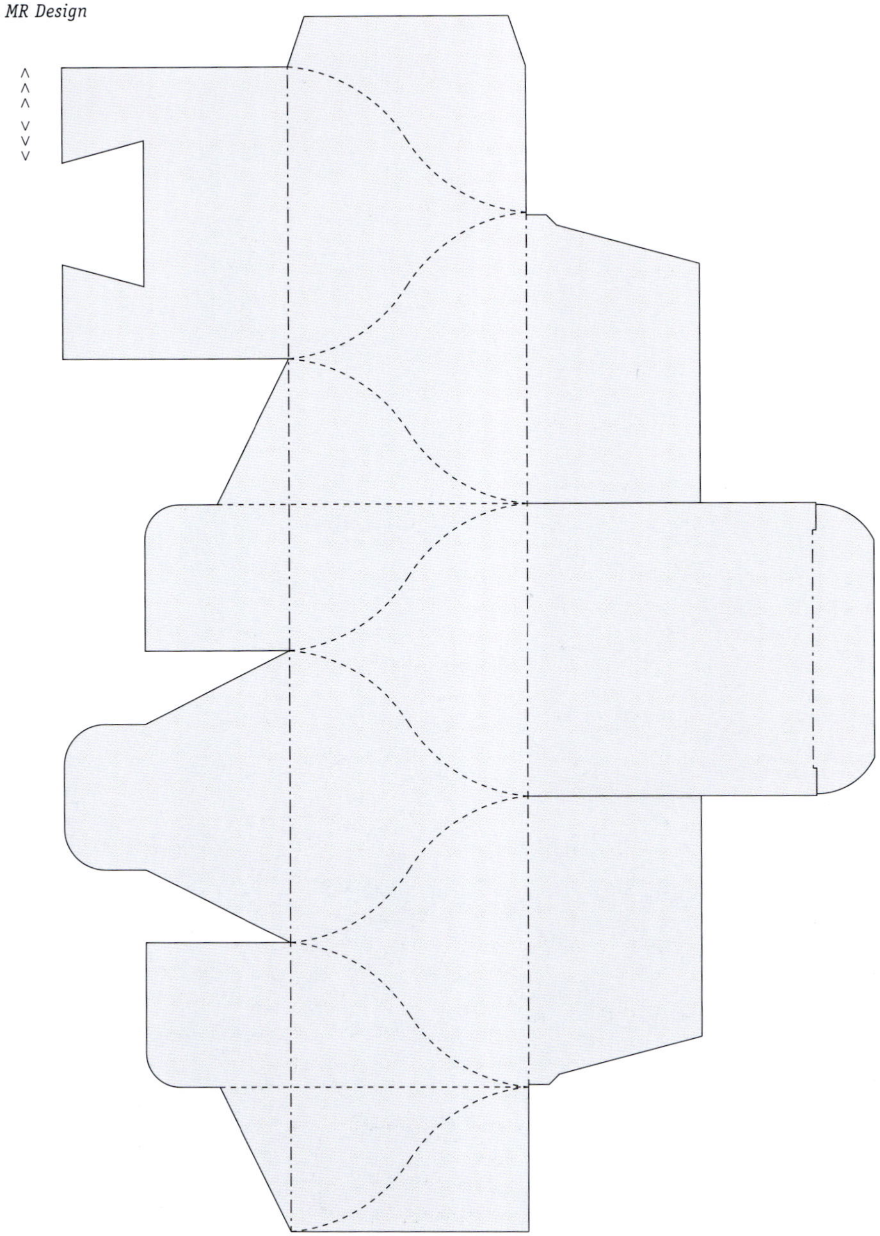

OGEE-ARCH CARTON

The refined but complex curves of an ogee arch have been cleverly incorporated into this container, whose four-sided crash-lock base and tab-locked top are deliberately misaligned. This allows the apex of one arch to meet at the corner of one end of the container and its base to converge at two corners at the other end of the container, while the arch's symmetrical form enables each adjacent arch to be inverted with its neighbor.

FOUR-SIDED PILLOW PACK

A slight adaptation to the design of a standard pillow pack through the incorporation of the arced ends on all four sides of the carton produces a container with a high degree of visual appeal and structural integrity, and which is also easy to erect. This pack can be used on its own or stacked on top of one another and held together with a supporting band to create a visually interesting towered structure.

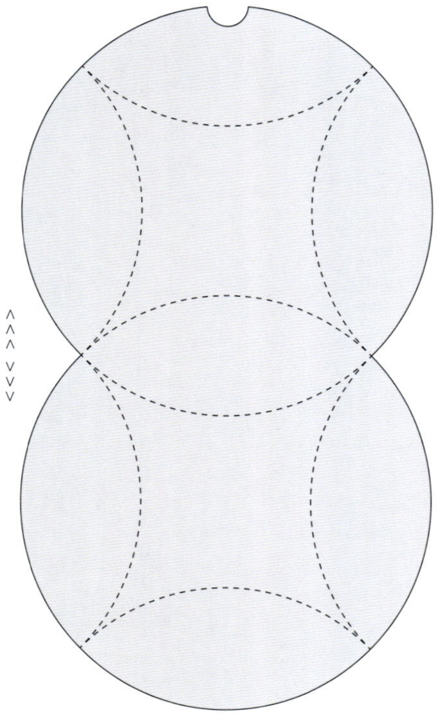

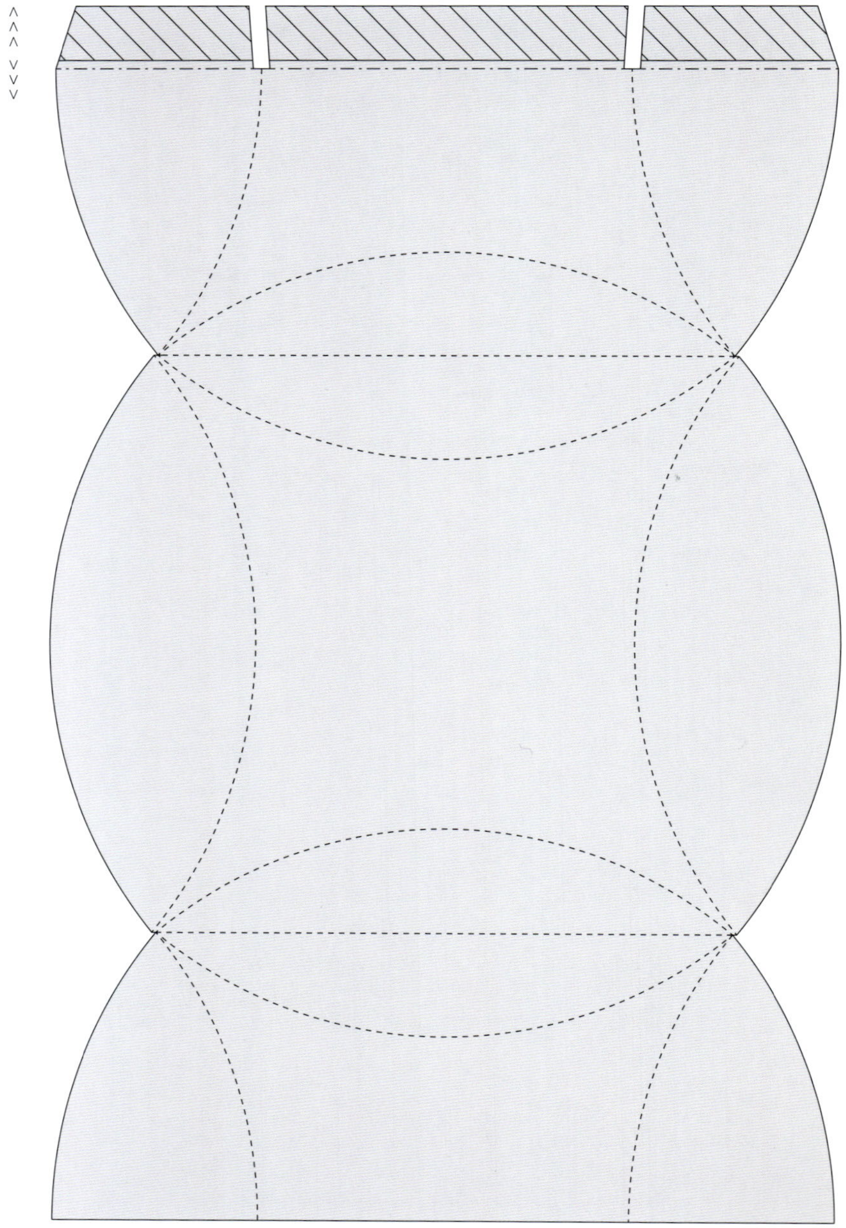

MR Design

MR Design

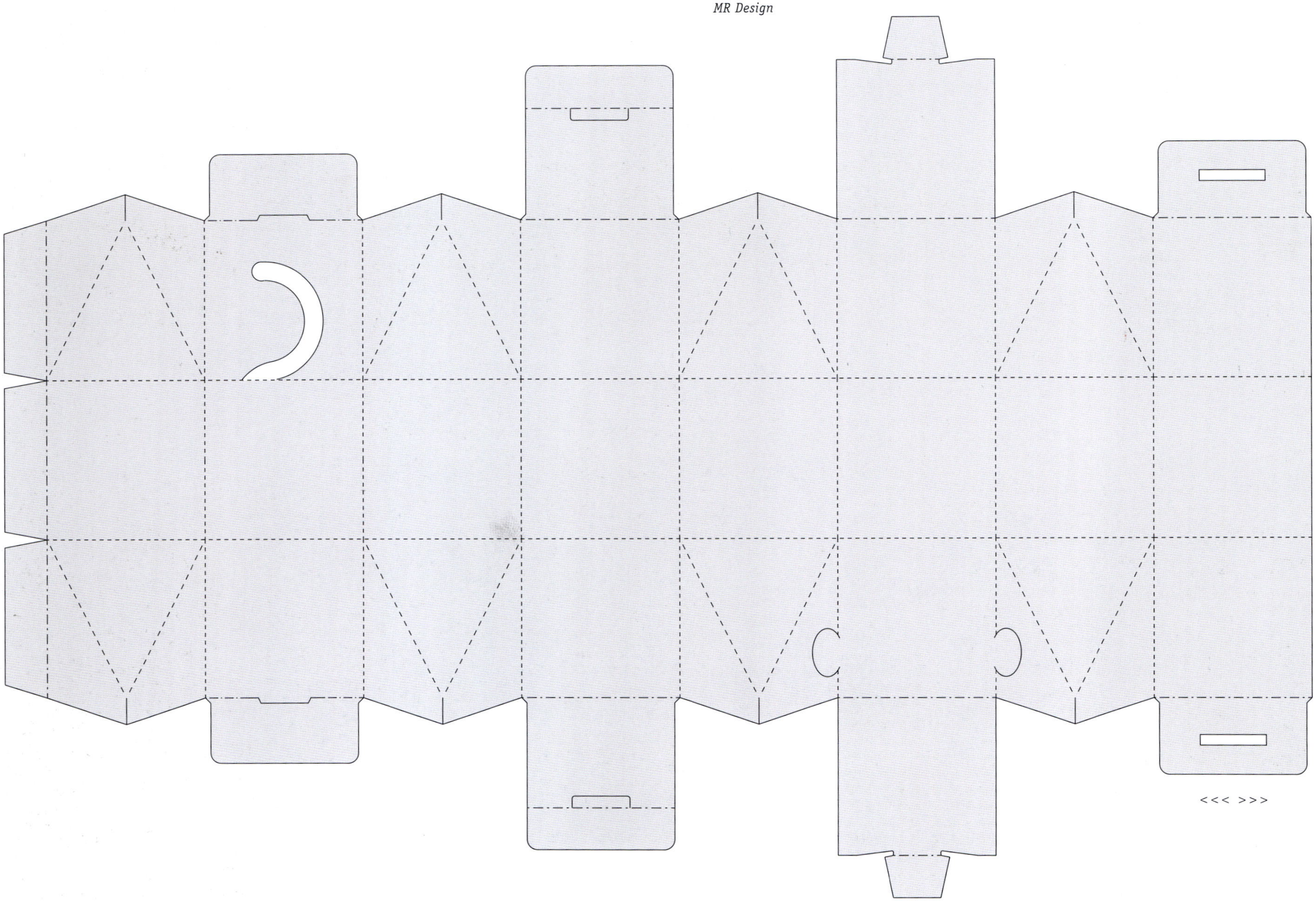

ADAPTED BALL CARTON

An economical way of producing interesting or visually appealing carton designs can be achieved by employing minor adjustments to the design or detailing of common carton designs. This example illustrates the familiar spherical carton, which has been wittily adapted into an animated form by die-cutting ears and a curly tail into its upper surface.

MR Design

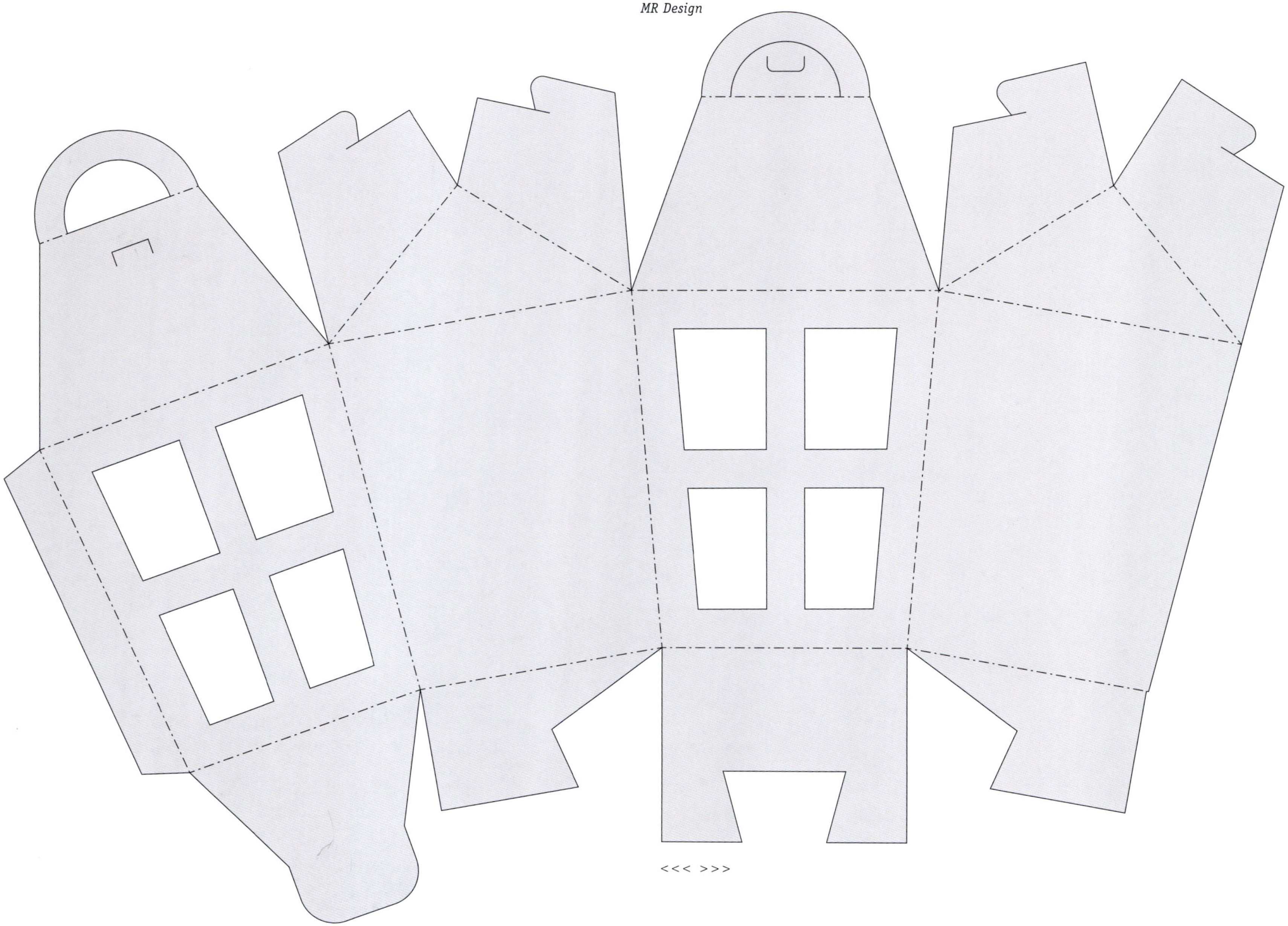

LANTERN CARTON

Die-cutting sections from the surfaces of standard cartons can be a cost-effective way of producing dramatic visual results. Here, the windows stamped from the sides of this carton provide an instantly recognizable form that has been further augmented by the gable-topped lid, which incorporates the handle device and simple locking mechanism.

GIFT PACK WITH HANDLE

There are many ways to incorporate a handle into a package to allow it to be carried by the user. In this instance, the carton's tapered design allows the rectangular crash-lock base to gently narrow to form a pointed top, sufficiently deep to allow a handle to be cut into each side. The handle is given extra strength by the large flap that wraps over the top of the carton from the rear and locks at the front using a simple tab-lock.

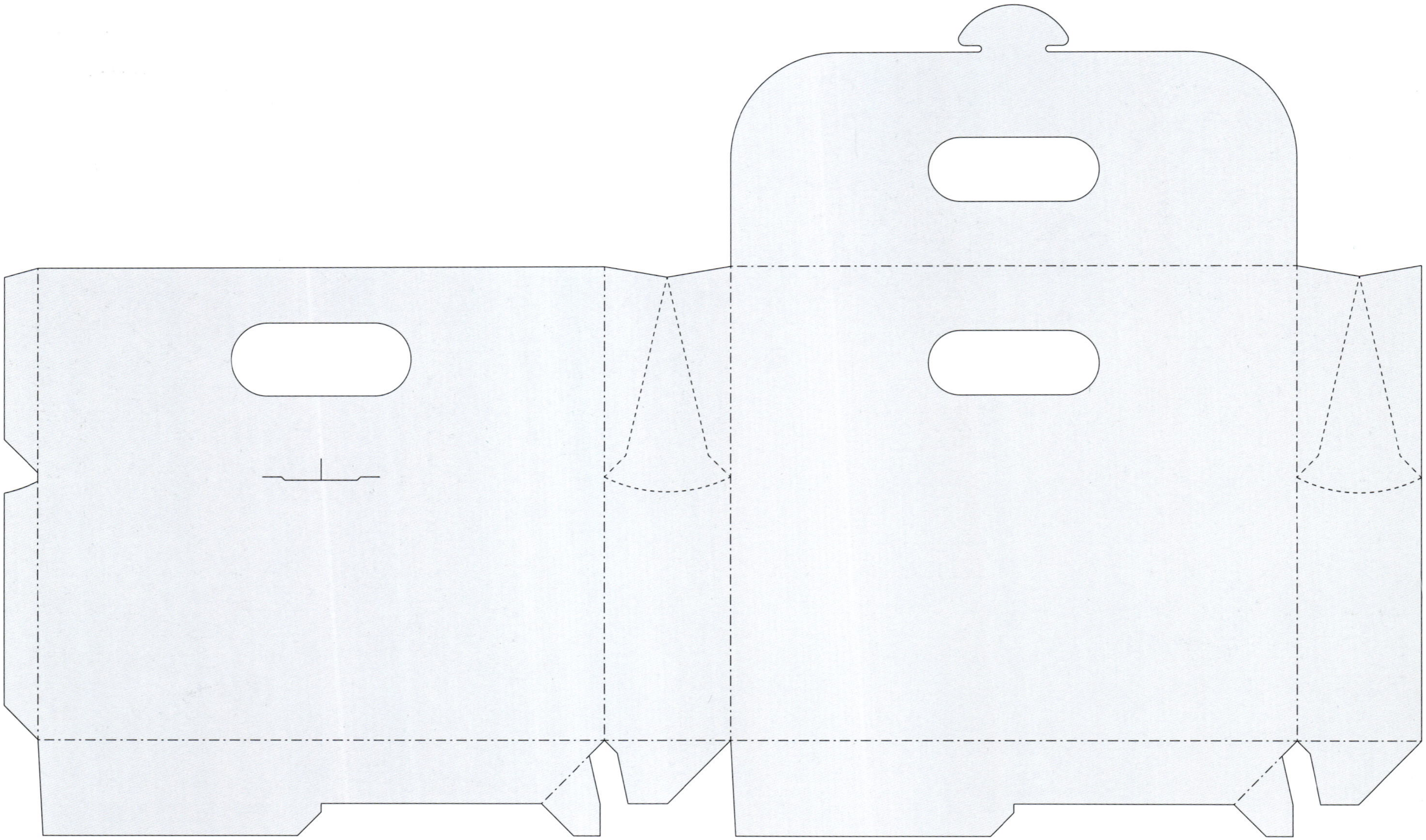

<<< >>>

Boxpak

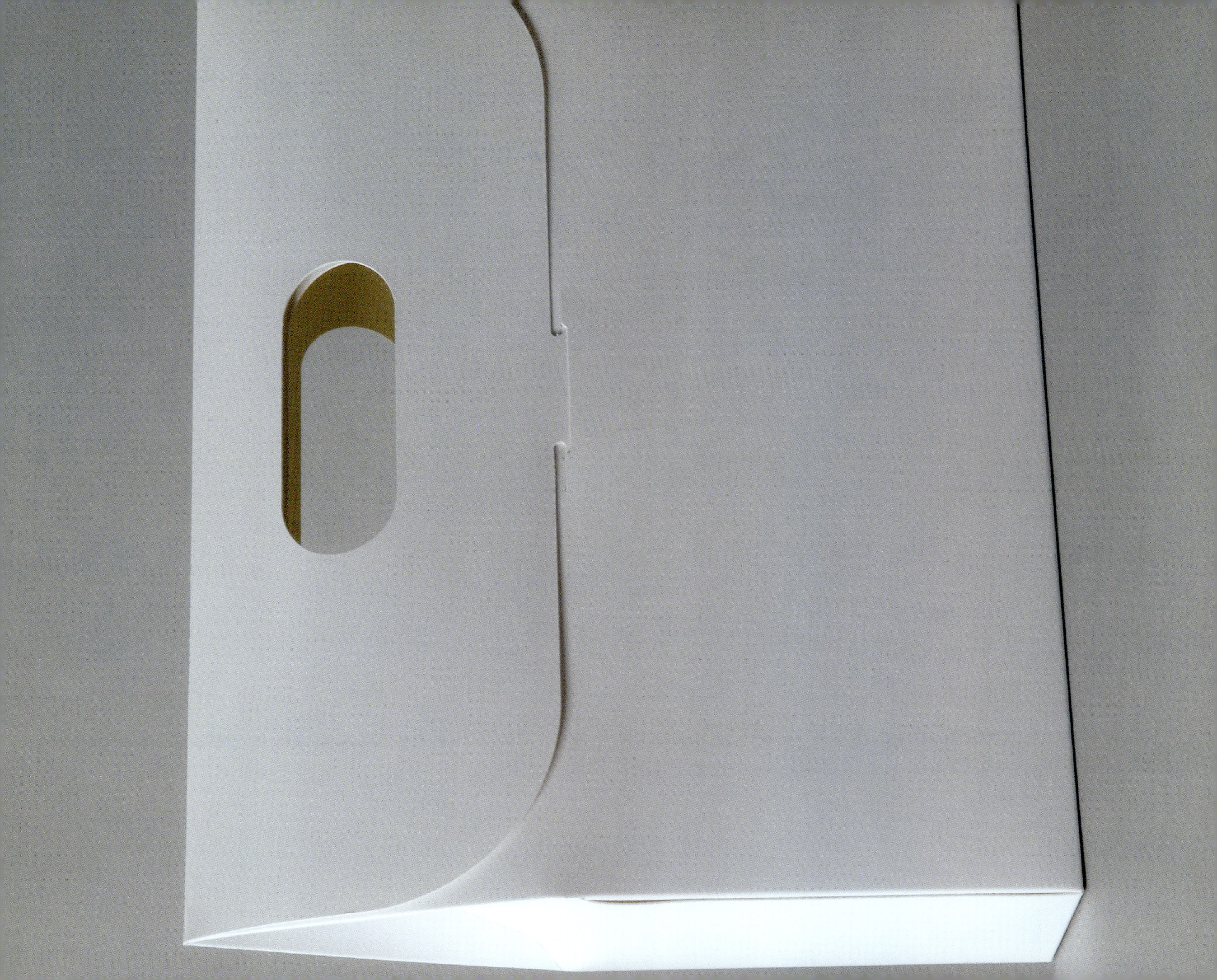

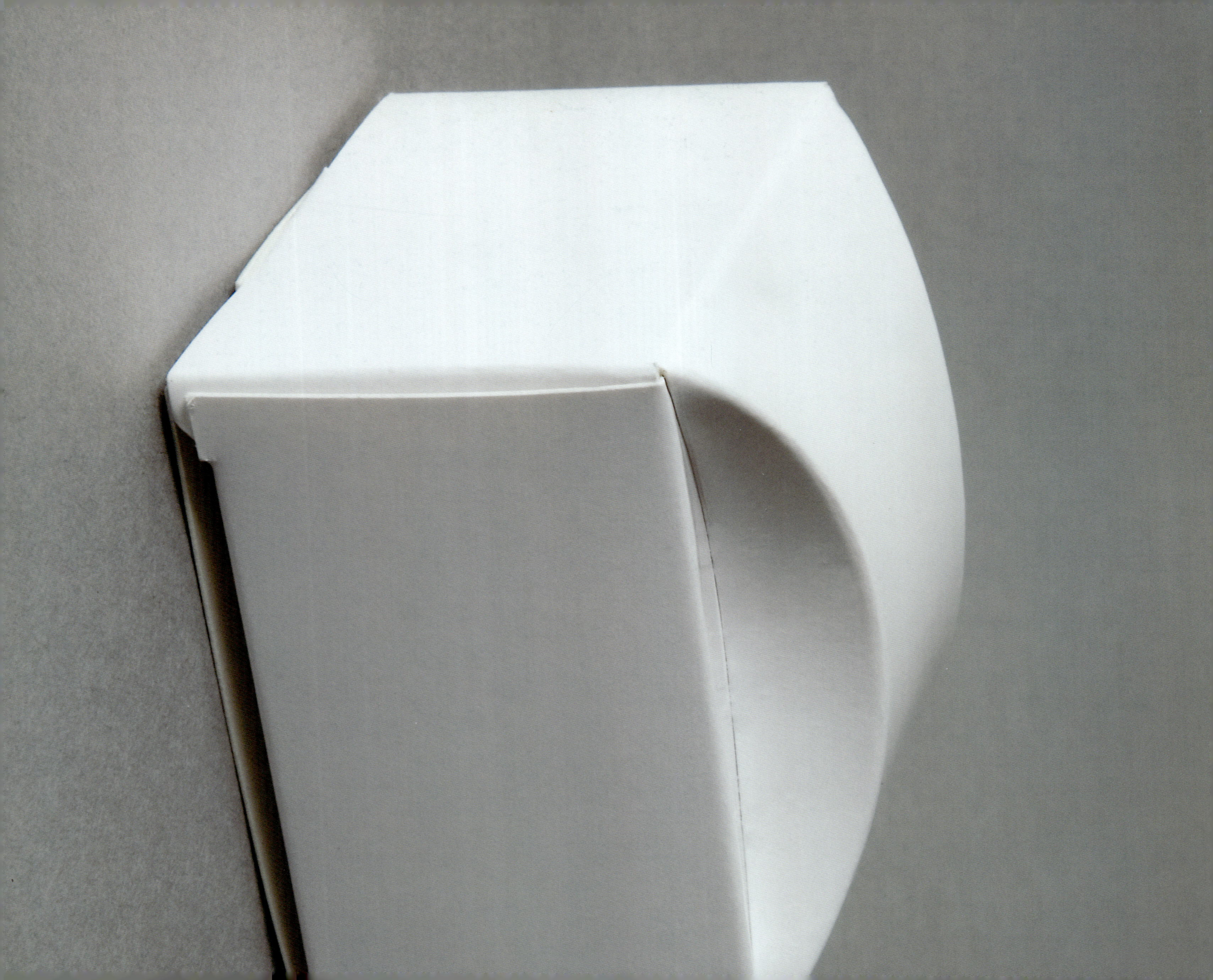

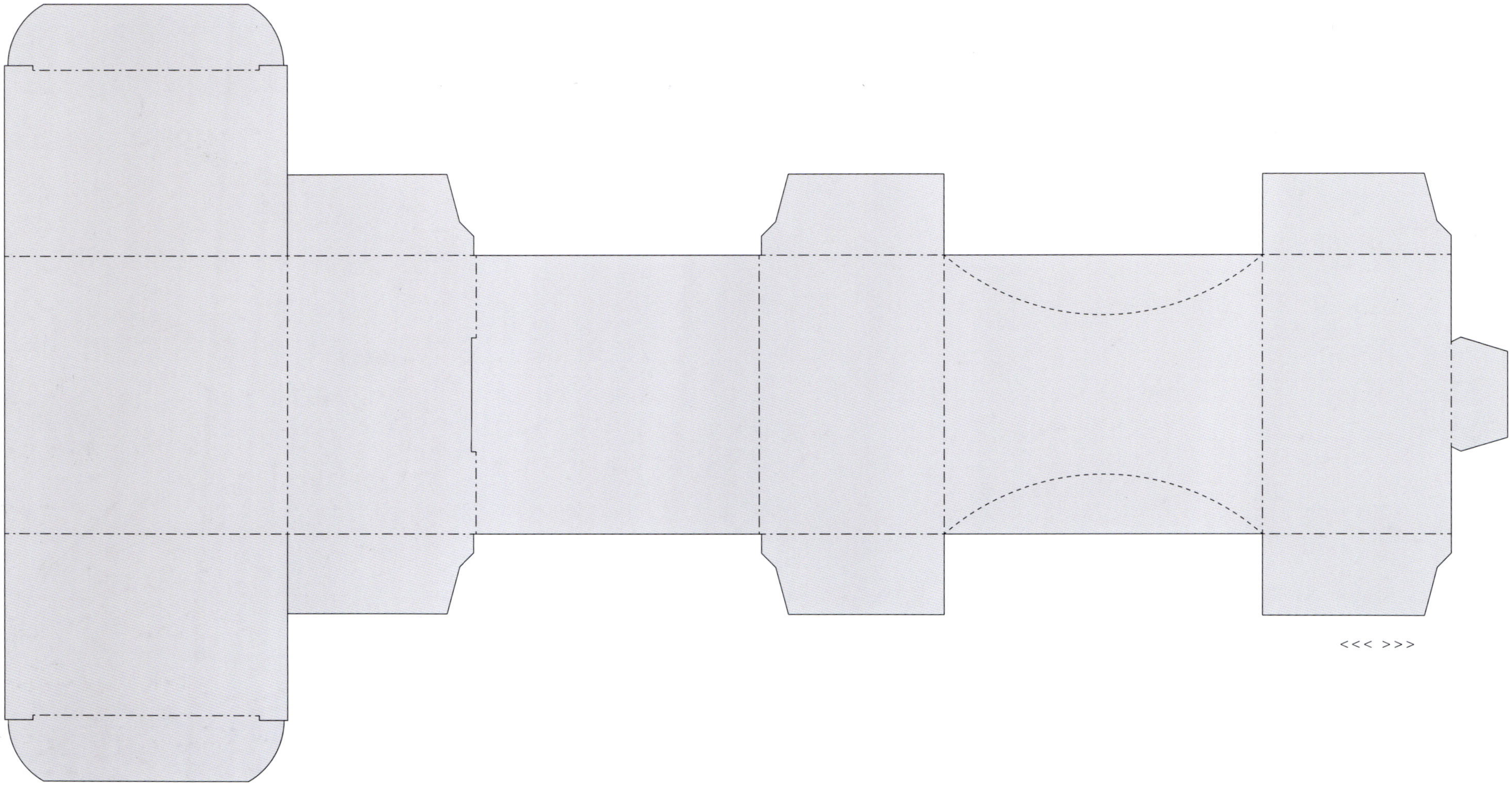

NOVELTY CASE

With a little adaptation, a simple cubed tuck-end carton can become a deceptively eye-catching design. Here, two extra sides have been added to a standard die-cut container to provide a novelty crown formed by arced folds that create a stylish contour to the top surface of an otherwise simple cubed form. The pack requires no glue in construction and is sealed using a tab-lock.

CONTOURED NOVELTY CARTON

Complex and interesting surface finishes can be applied to regular cartons with the addition of extra sides on the template, which act as an outer skin on one or more faces of the carton. Here, an interesting contoured effect has been applied to the top and front face of the carton, complementing the carton's rectilinear form underneath. The additional material needed to achieve this sort of effect increases the price of the carton, making it particularly popular for gift or novelty items with a relatively high added value.

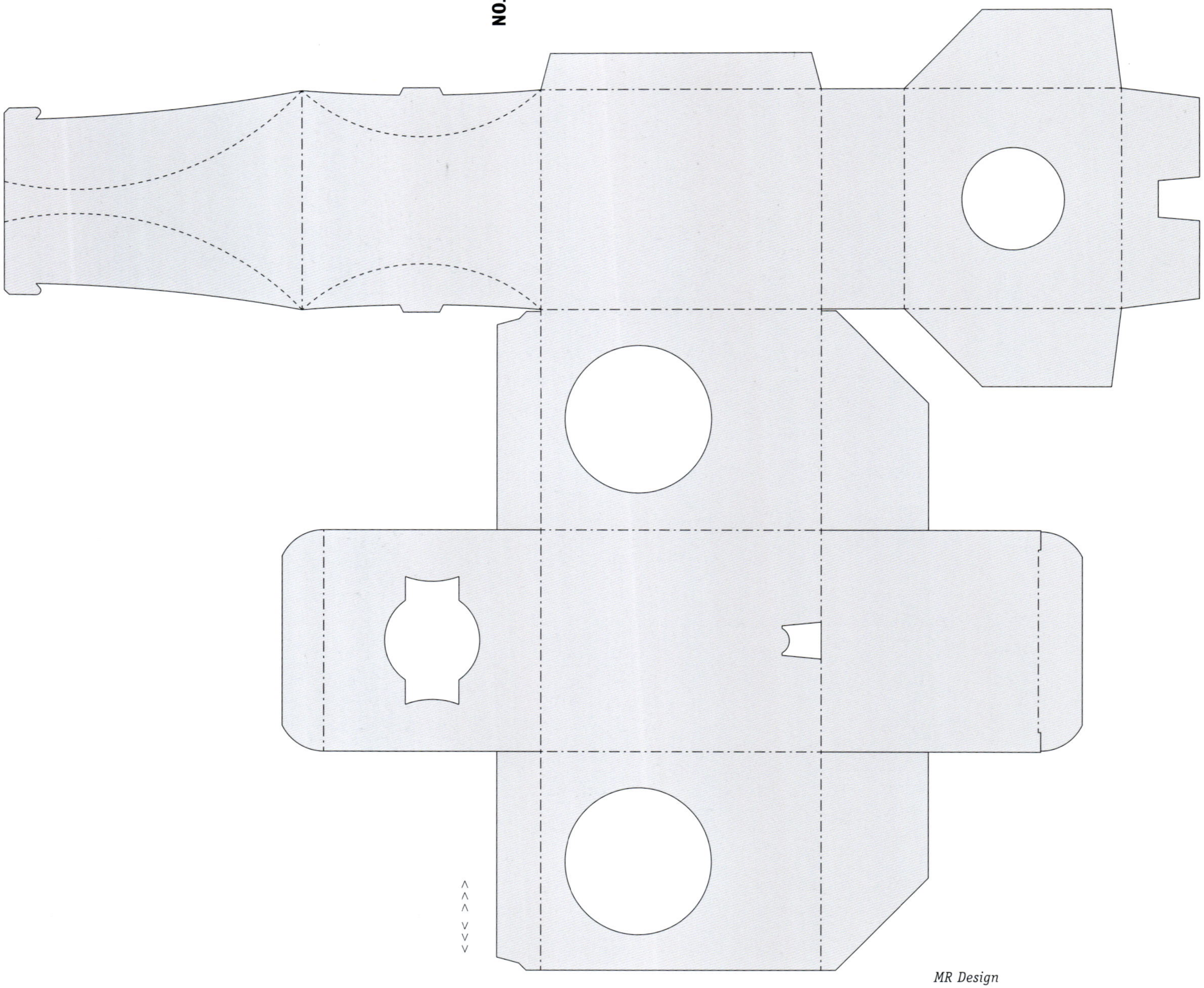

MR Design

HEXAGONAL BOX

This hexagonal container provides a sturdy packaging solution that is easy to construct and is visually appealing. The base is constructed using a series of unglued flaps that fold neatly on the underside of the container to provide a strong structure, while the inside is finished by flaps on each side of the hexagon, folding internally to give the interior of the box the same surface finish as the exterior. The lid is constructed in a similar manner, using minimal glue and requiring only simple internal folds to achieve a strong structure that fits over the base.

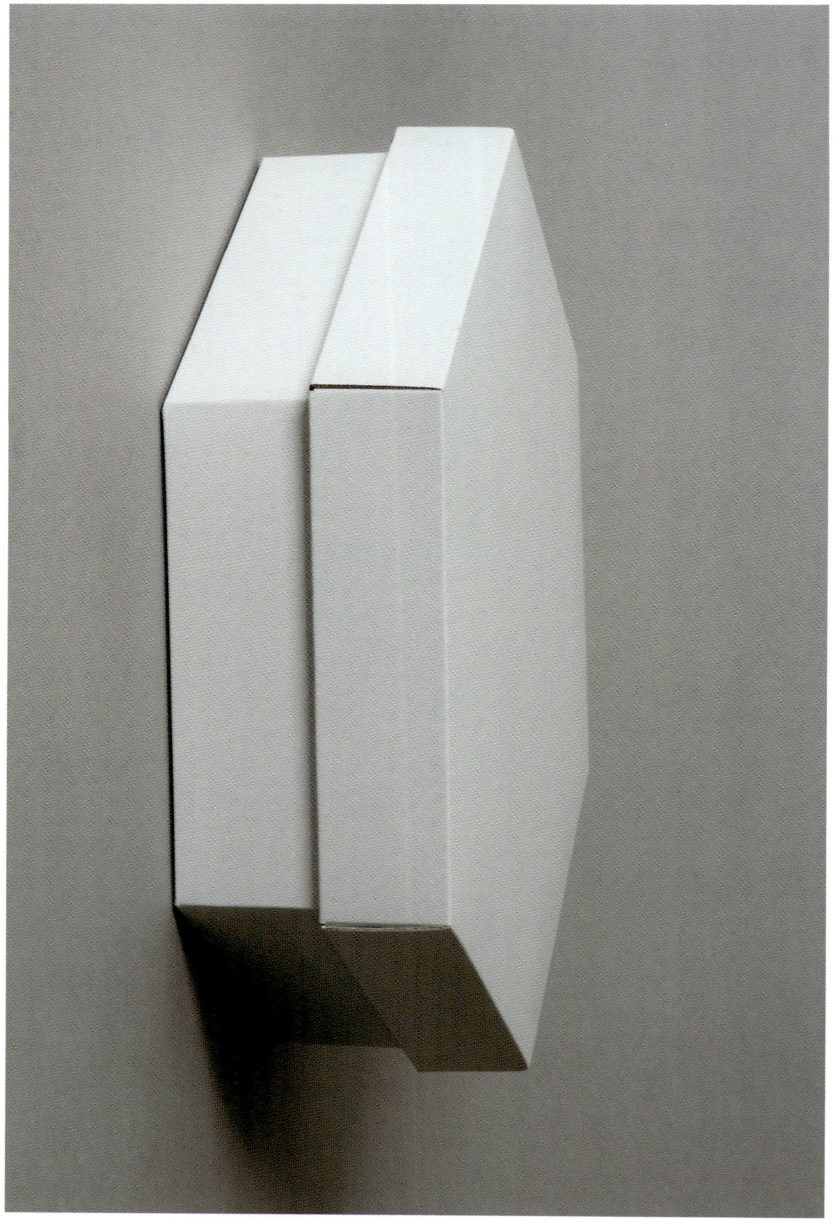

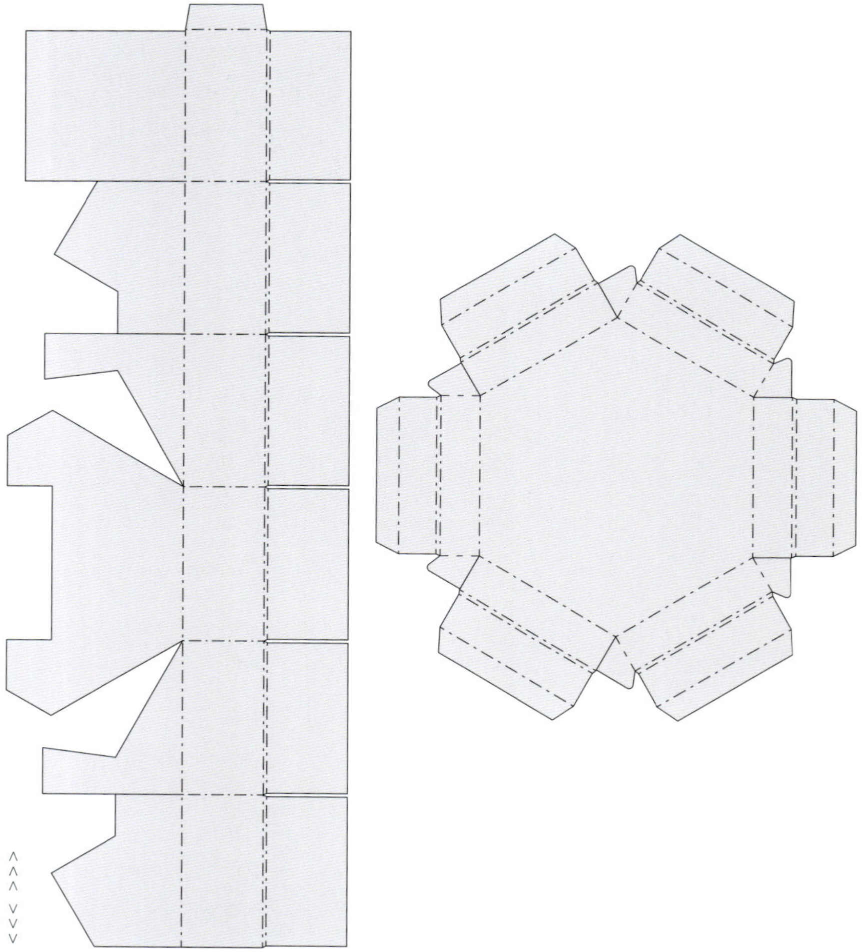

Wedgwood

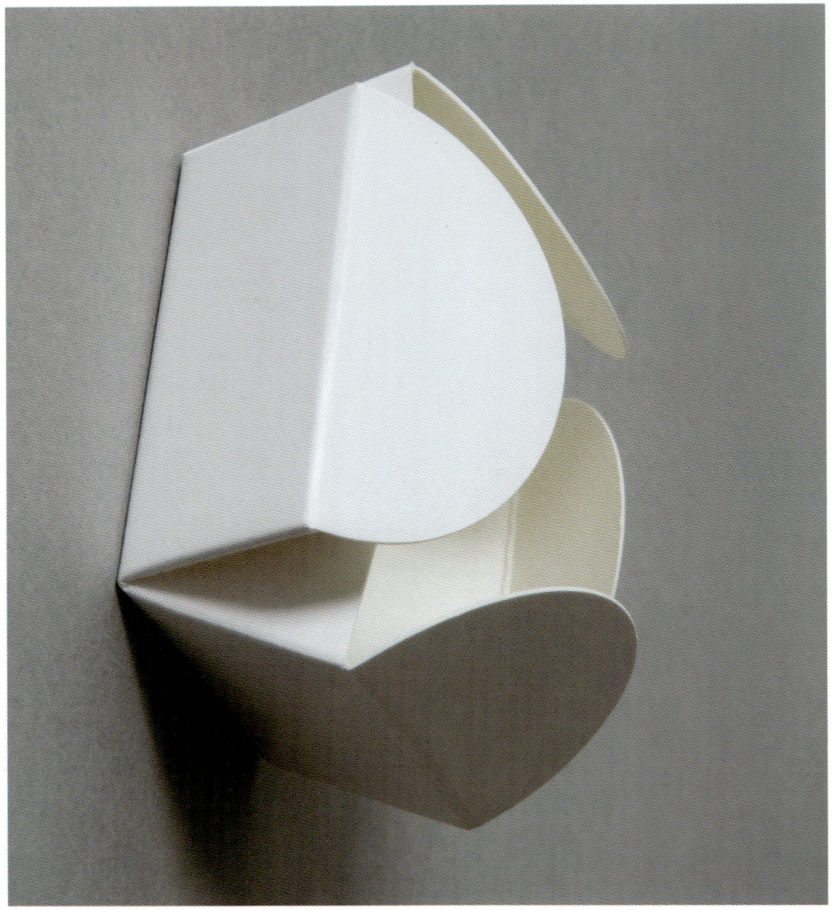

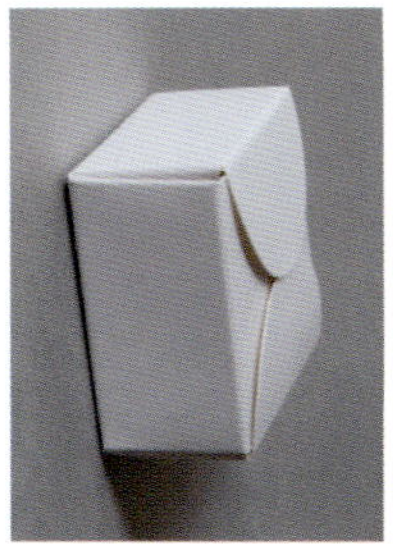

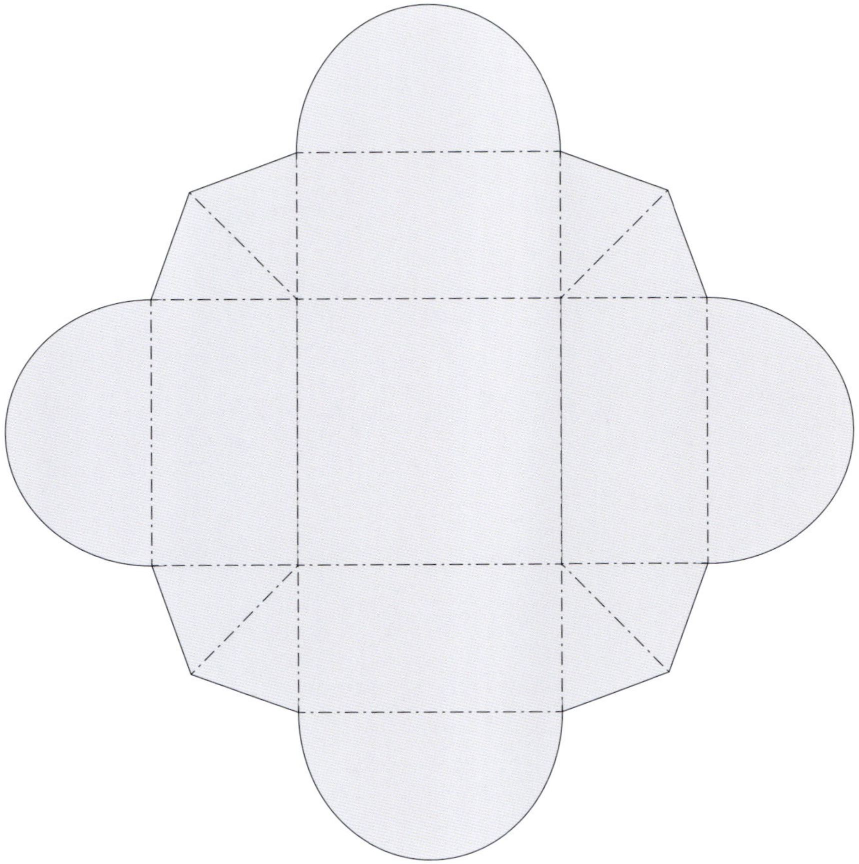

WEB-CORNERED CUBE

Simple cuboid shapes are often made from a cruciform template, with tuck-flaps at the top and base to seal the carton. Here, this is achieved by an unusual quadrangular template. The corners are webbed to create the sides, and semicircular tabs that fold to meet in the middle form the base, which is sealed by gluing. An advantage of the quadrangular template is that it opens out to form a large surface area that can serve a functional purpose during the preparation or consumption of the product, such as a plate from which to eat take-out food.

DOUBLE CUP HOLDER

Folded corrugated or carton board can provide packaging functionality without comprising the entire package. Being a cheap, recyclable, and sustainable material, paperboard is an ideal complementary material that can be used concurrently with other materials to form an effective packaging device. This simple sheet of corrugated board has been designed to provide an alternative to the double or quadruple paper-pulp take-out drinks holders. Lighter and cheaper to produce, this device is designed to be either a standalone piece of packaging or an insert in a take-out carry bag.

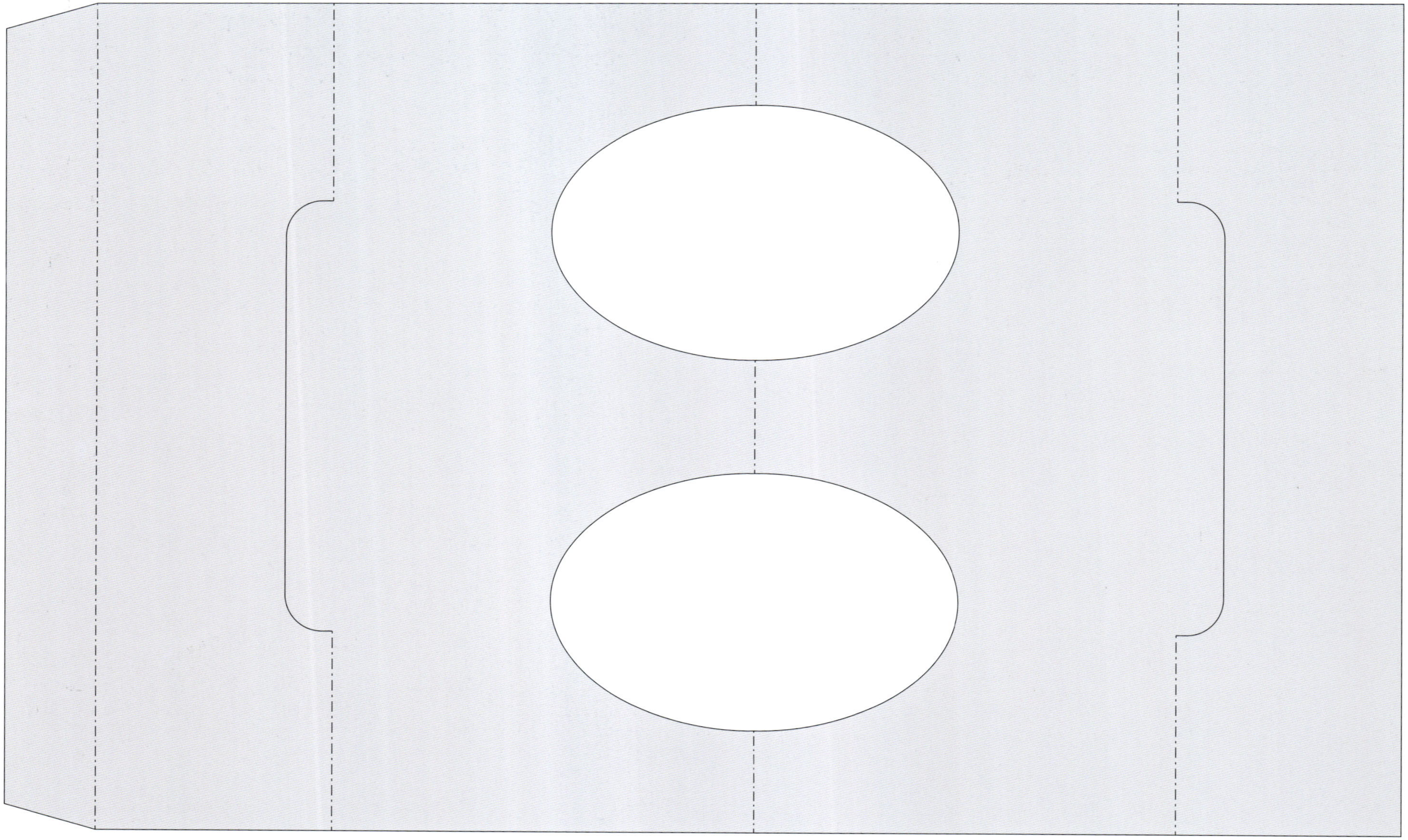

Rokov Design

<<< >>>

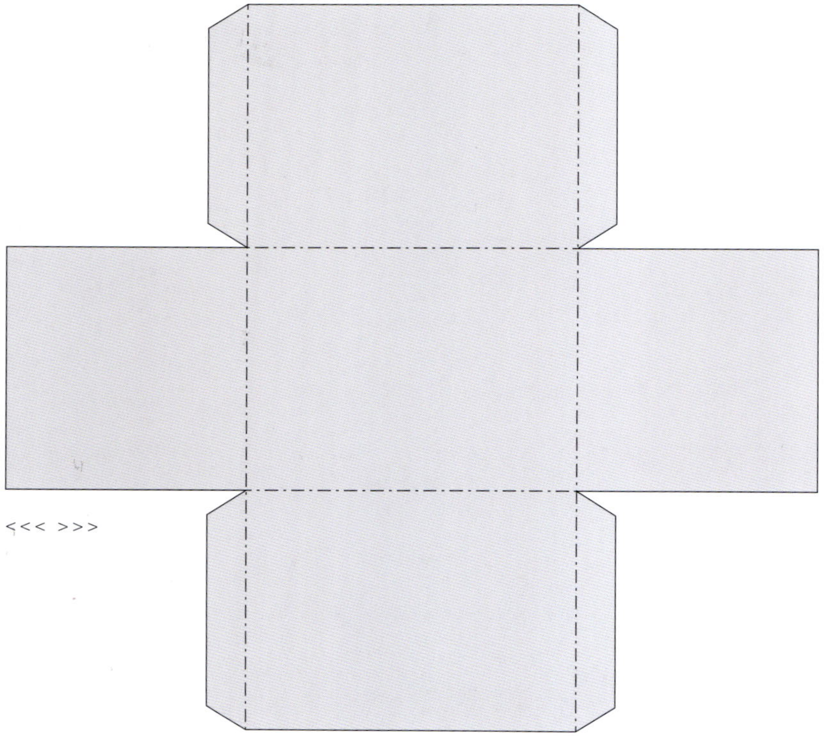

<<< >>>

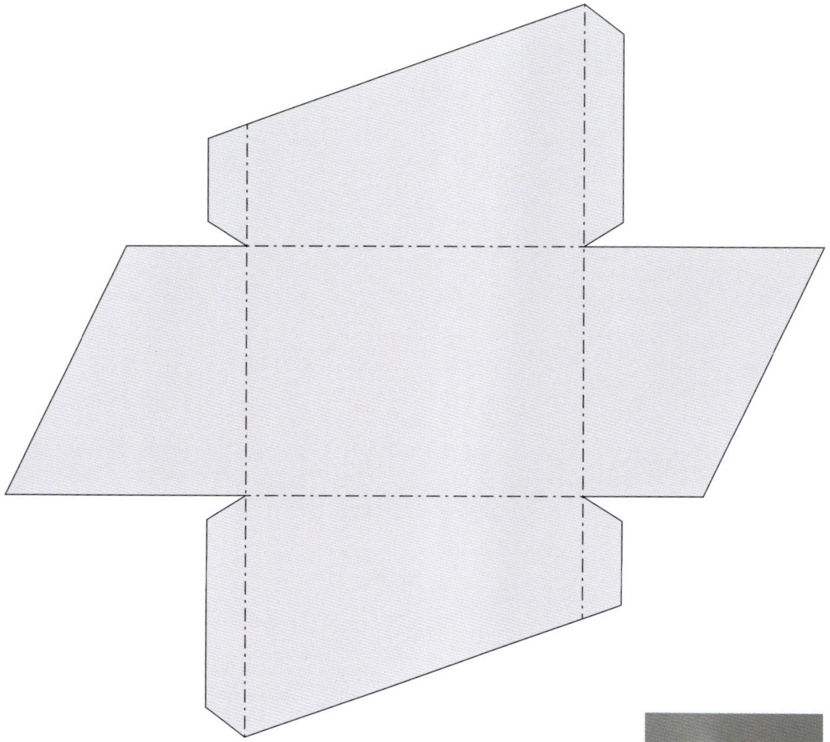

TWO-PIECE INTERCONNECTING CARTONS

Using one or more components in a package allows for greater flexibility in structural and graphic design, as well as choice of materials. The series of designs shown here and on pages 236 and 237 use two interconnecting cartons that fit one inside the other. Subtle design detailing in the cut-away ends of the outer sleeve expose the internal container, lending the package a unique character that can be replicated in a number of different package shapes and forms. The contrast between the inner and outer sleeves can be accentuated through printed graphic design or material texture.

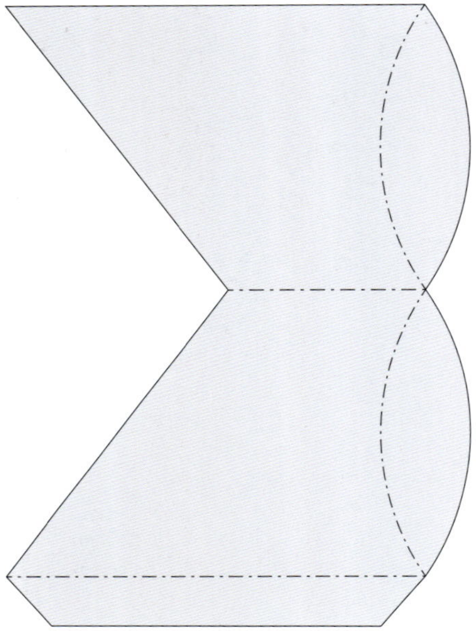

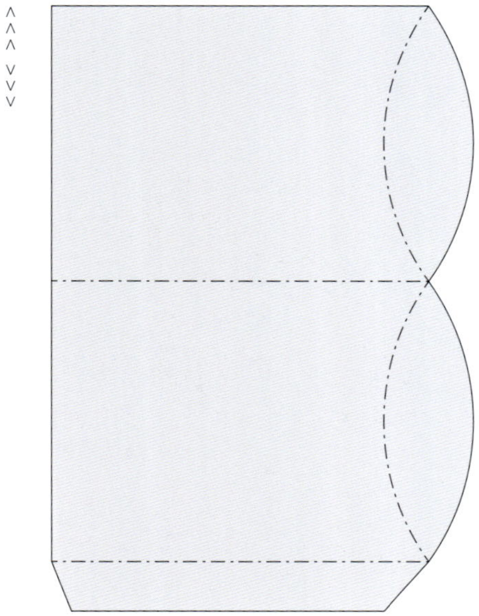

Sophie Cullingford

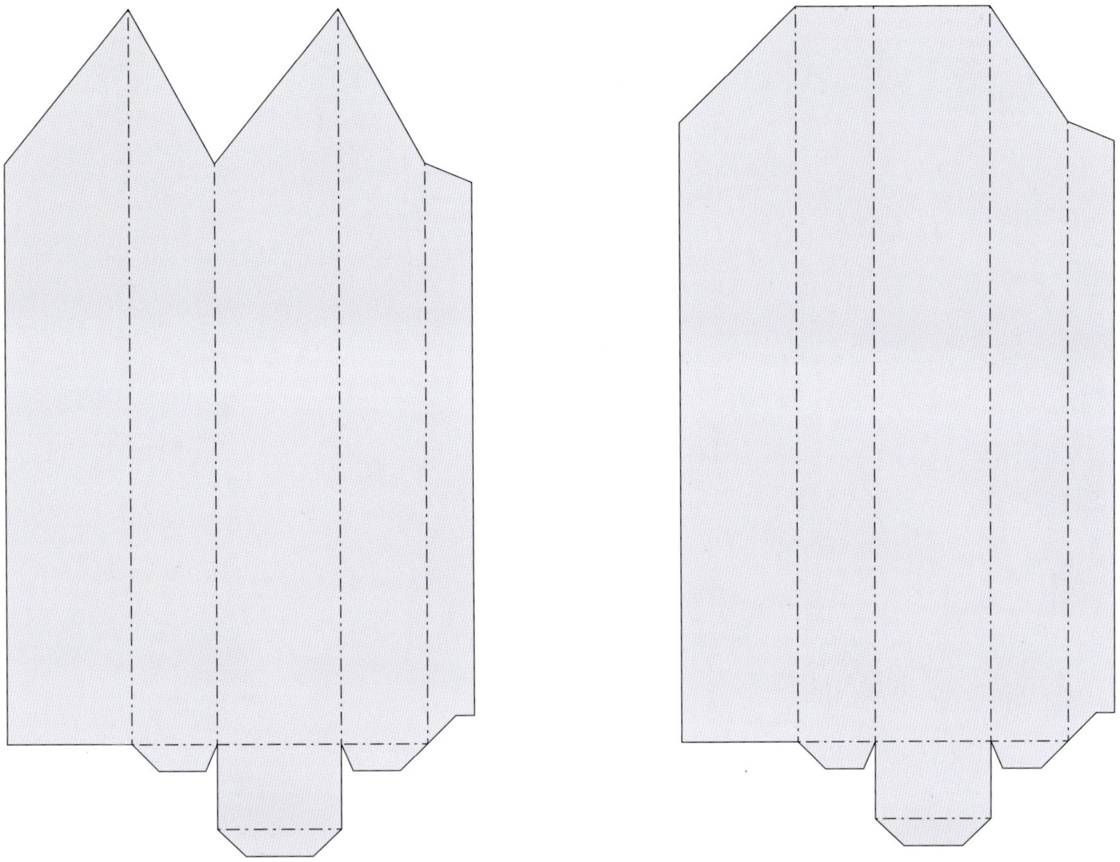

<<< >>>

Sophie Cullingford

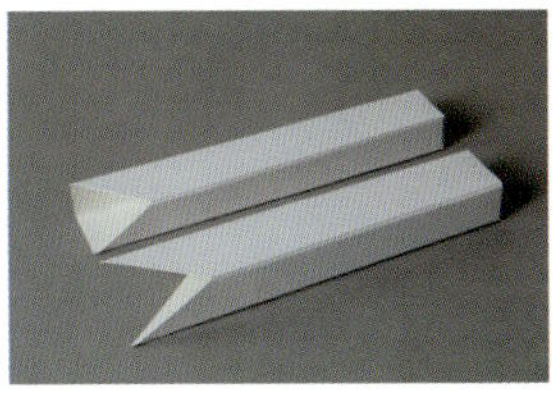

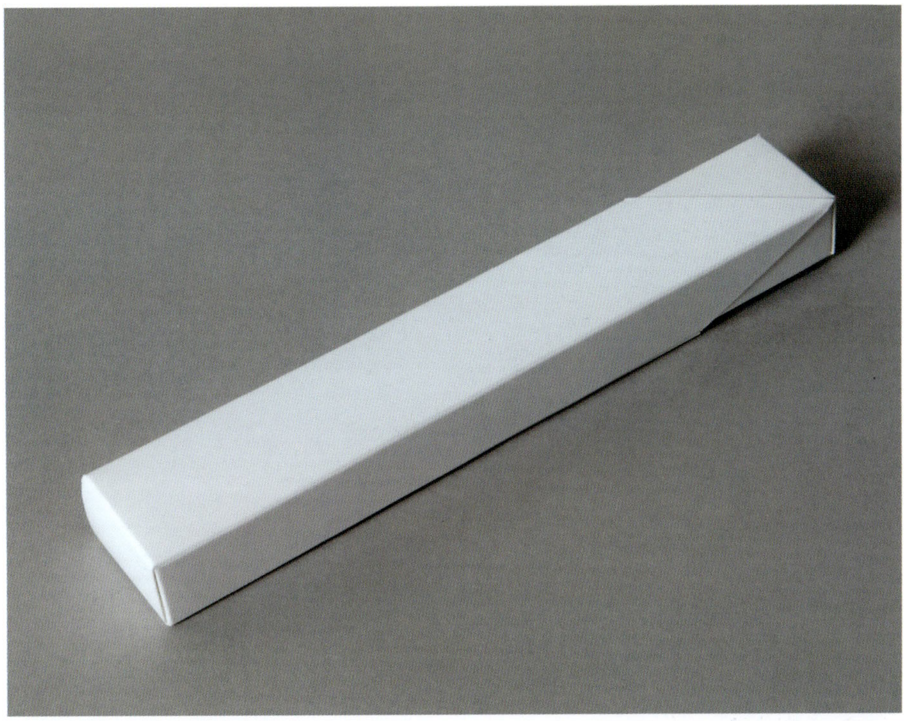

Karen Slade

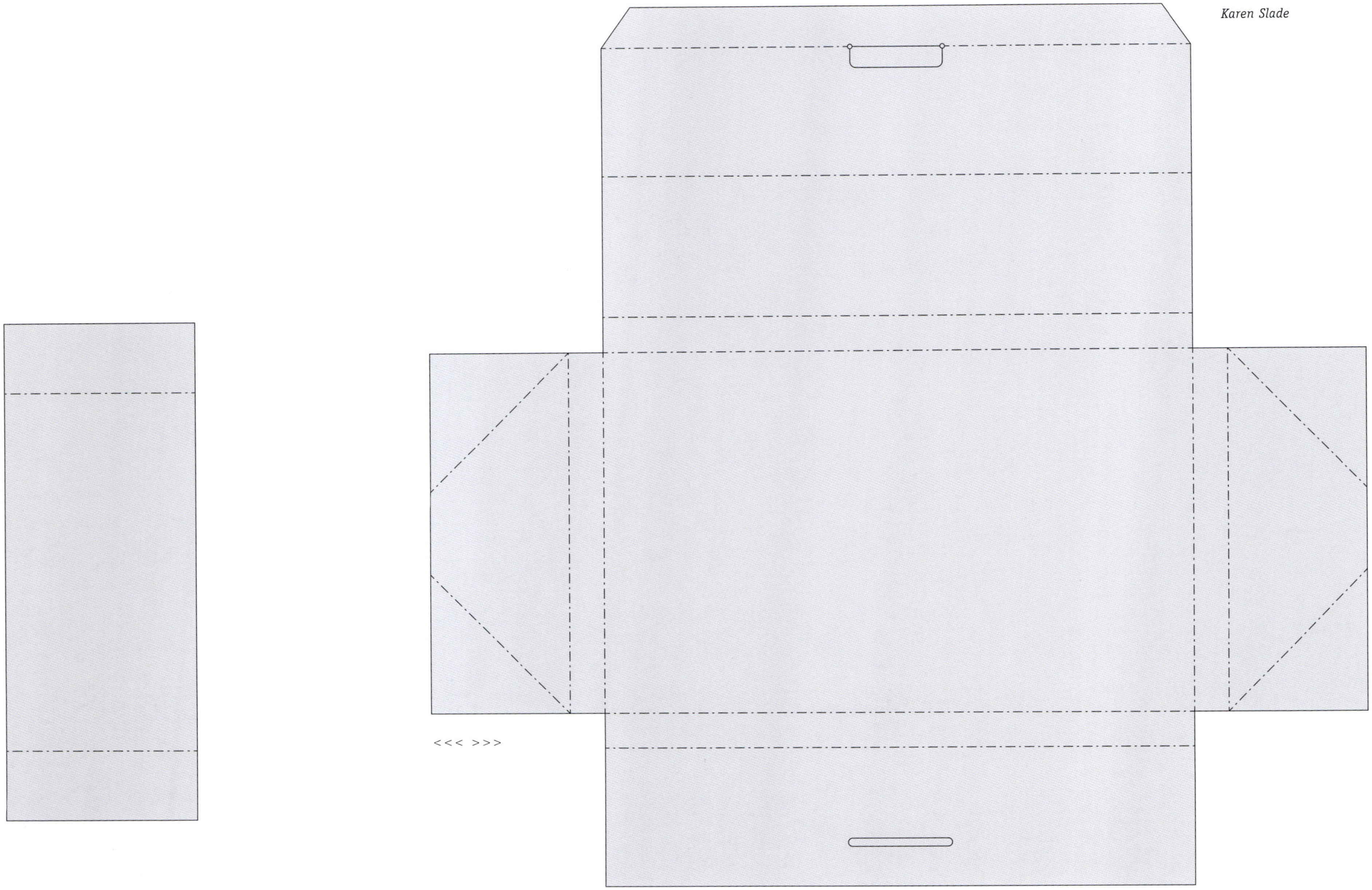

Bulk packing of small, fragile items causes particular problems for packaging designers. The effective separation and safe storage of many individual products, perhaps of varying sizes and shapes, in a single container, often requires complex packaging solutions that must provide easy construction, limited use of materials, and exceptional performance. This solution (also shown on pages 240 and 241) uses a series of inserts that fold to provide individual compartments within a larger outer container that can be reused indefinitely. Each compartment can be varied in size using a different configuration of inserts that keep the products safely housed and separated from one another during transit and distribution.

MULTIPARTITIONED TRAY

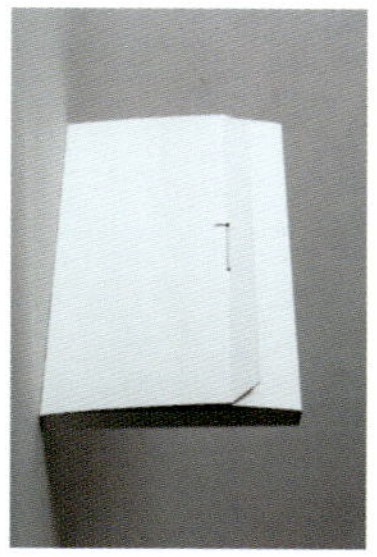

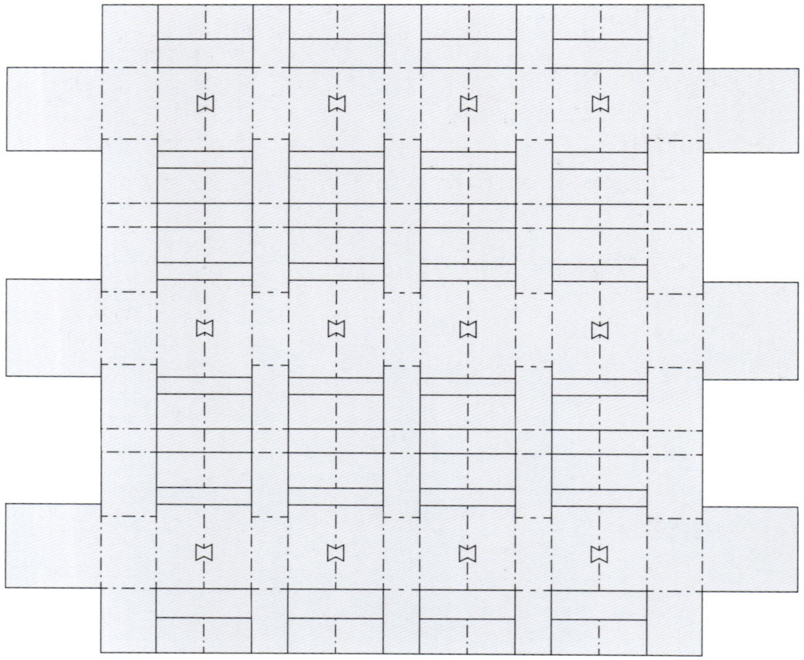

<<< >>>

Karen Slade

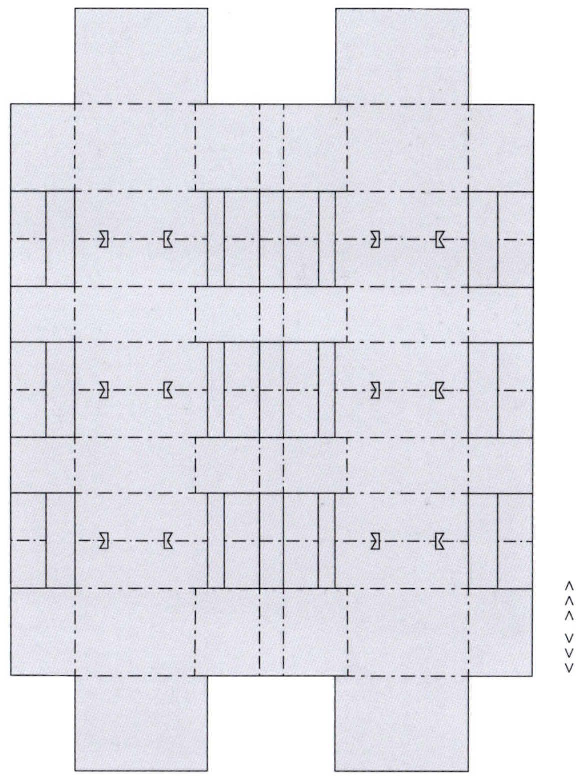

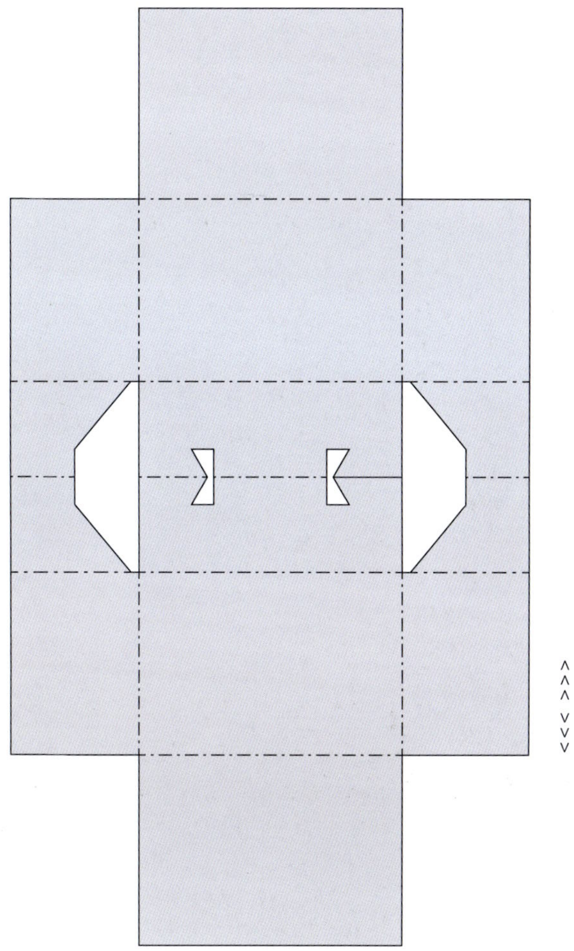

Karen Slade

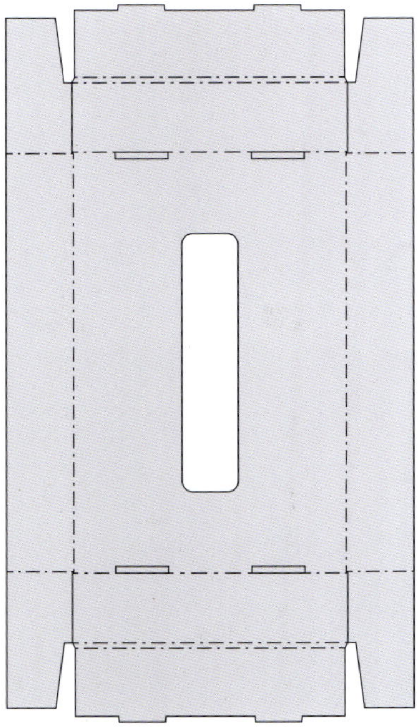

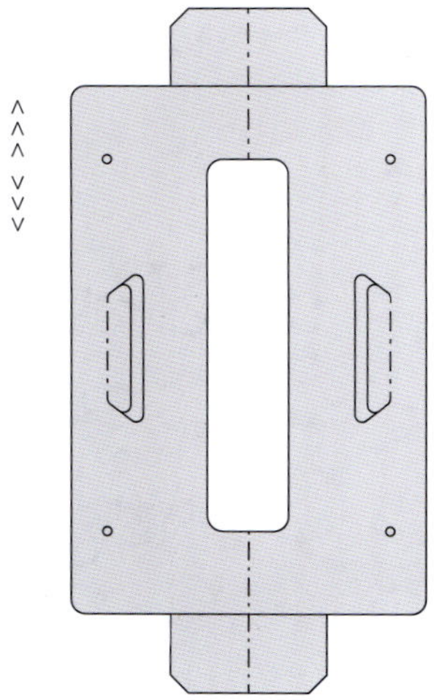

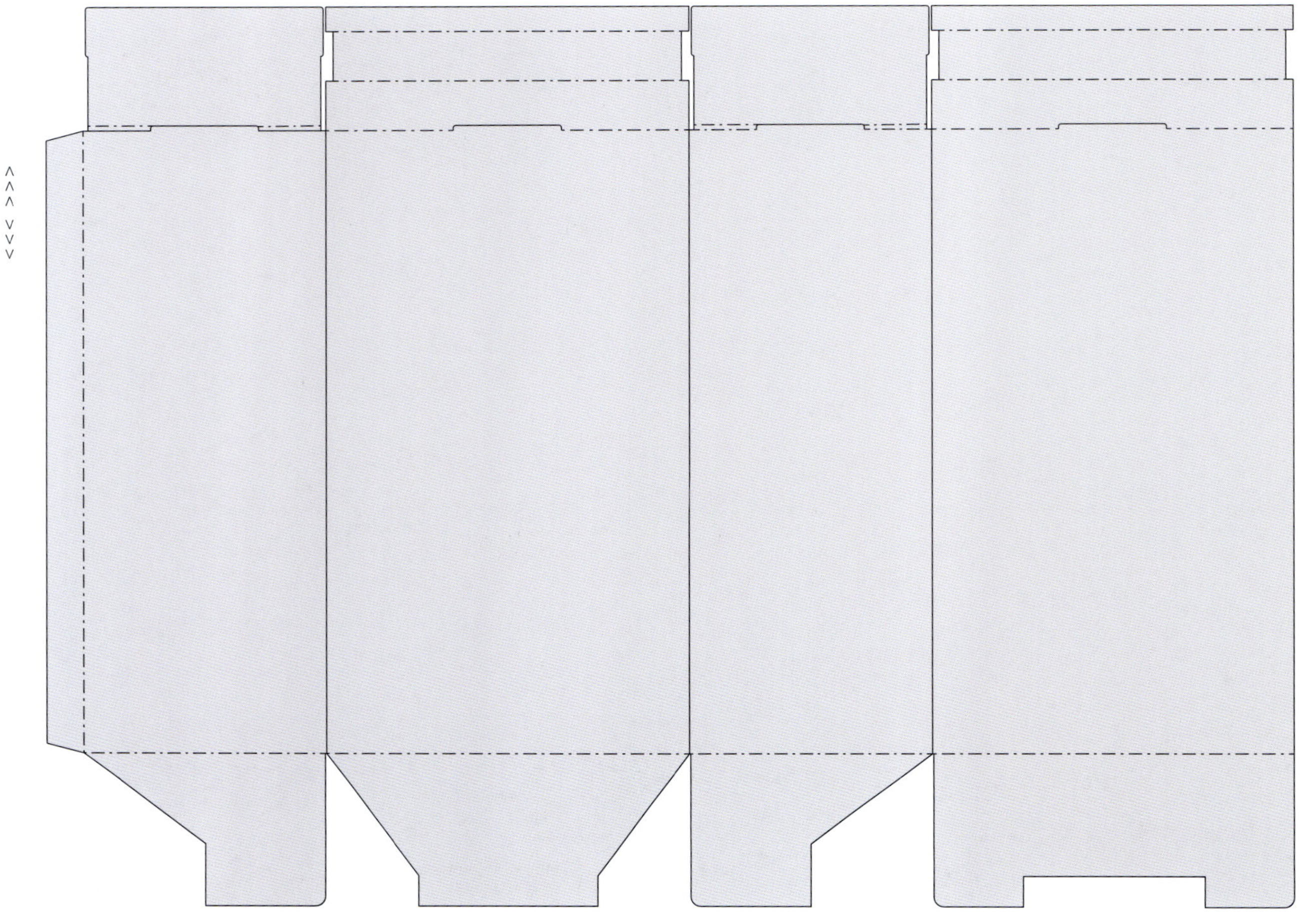

CORRUGATED WASTE CONTAINER

The unique properties of corrugated board make it an excellent material to use in packaging situations that demand reuse, a long life, or biodegradability. This waste box uses these properties effectively in a container that is designed as a temporary repository for refuse. Its sturdy structure allows it to be extensively reused, and the material allows it to be recycled or composted at the end of its life.

ANGLE-SIDED CARTON

Perpendicular angles invariably provide the most efficient use of space in packaging design, allowing individual units to be packed side by side in transit and on display without any loss in volume. This take-out food carton, which is delivered flat-packed and erected at the point-of-sale, is not constrained by these factors and so can be an irregular shape. The small base and a large lid give the carton a distinctive angular appearance.

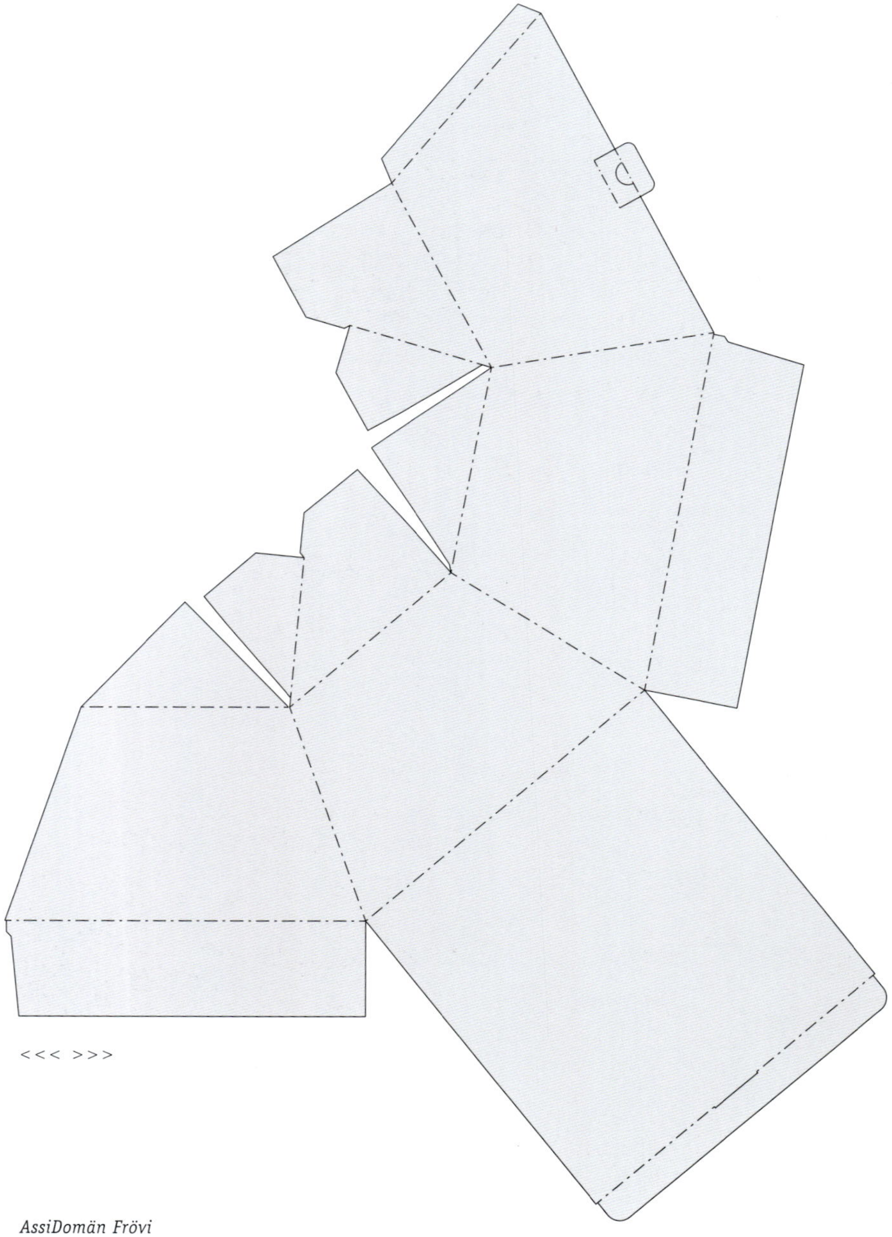

<<< >>>

AssiDomän Frövi

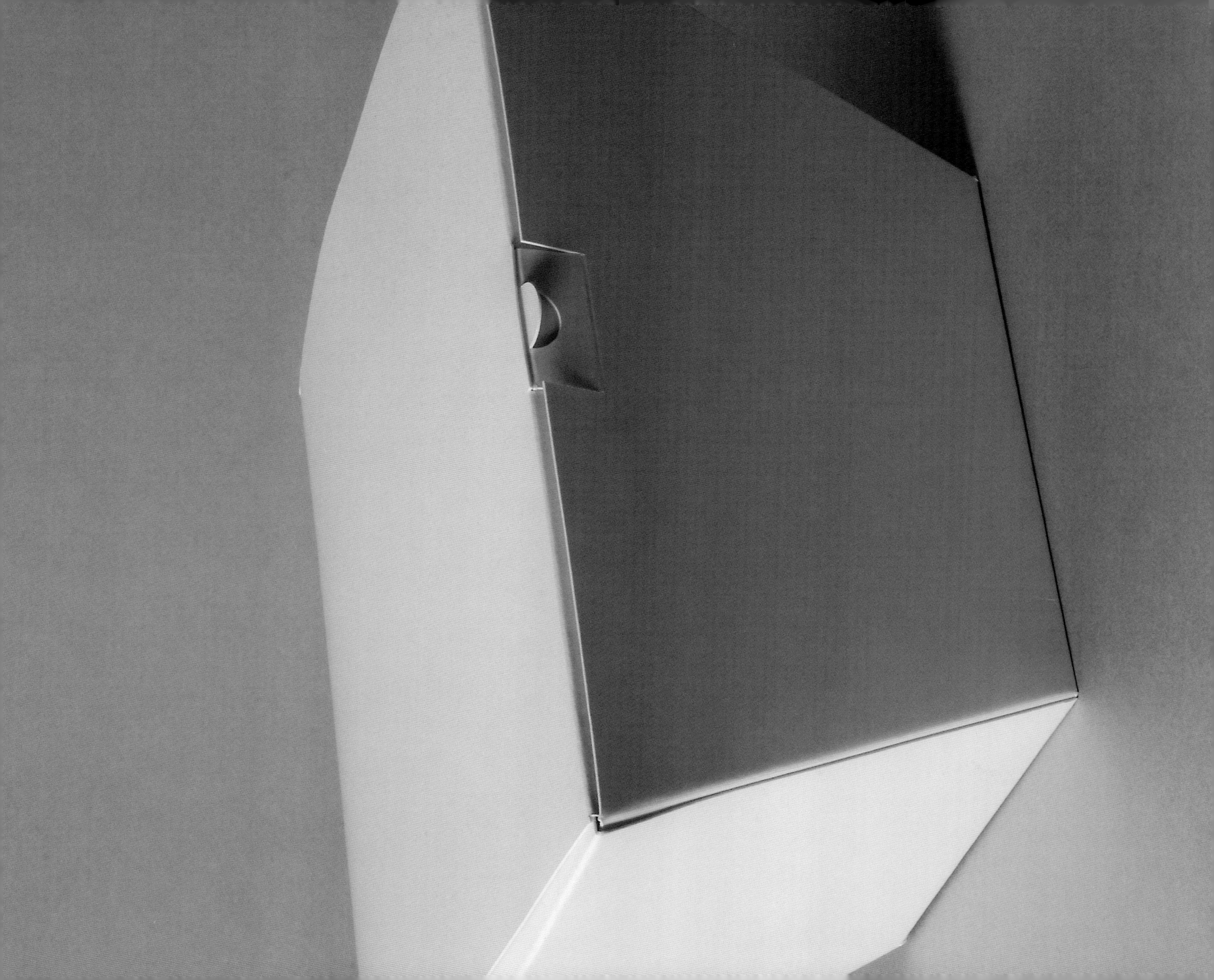

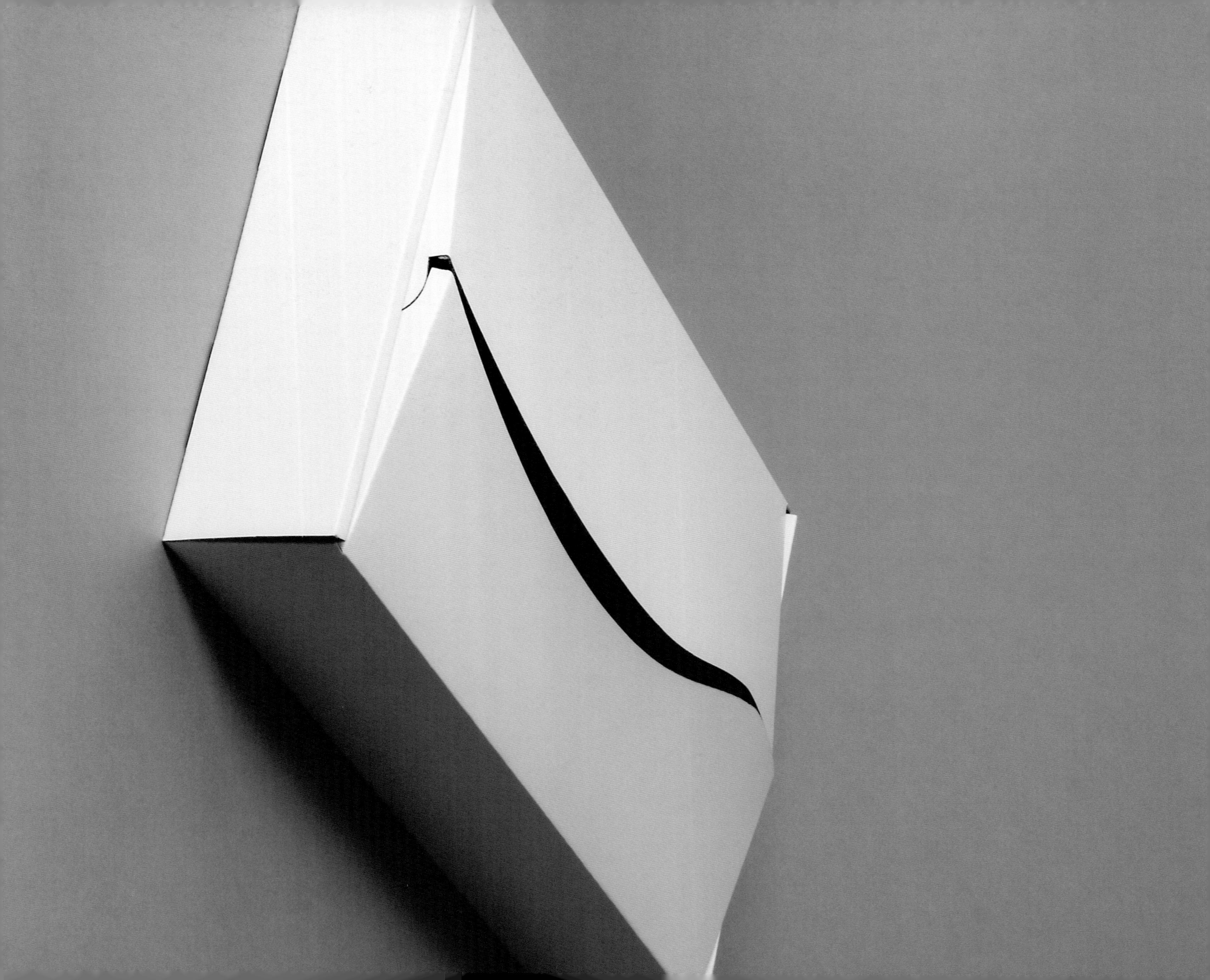

Arden Packaging

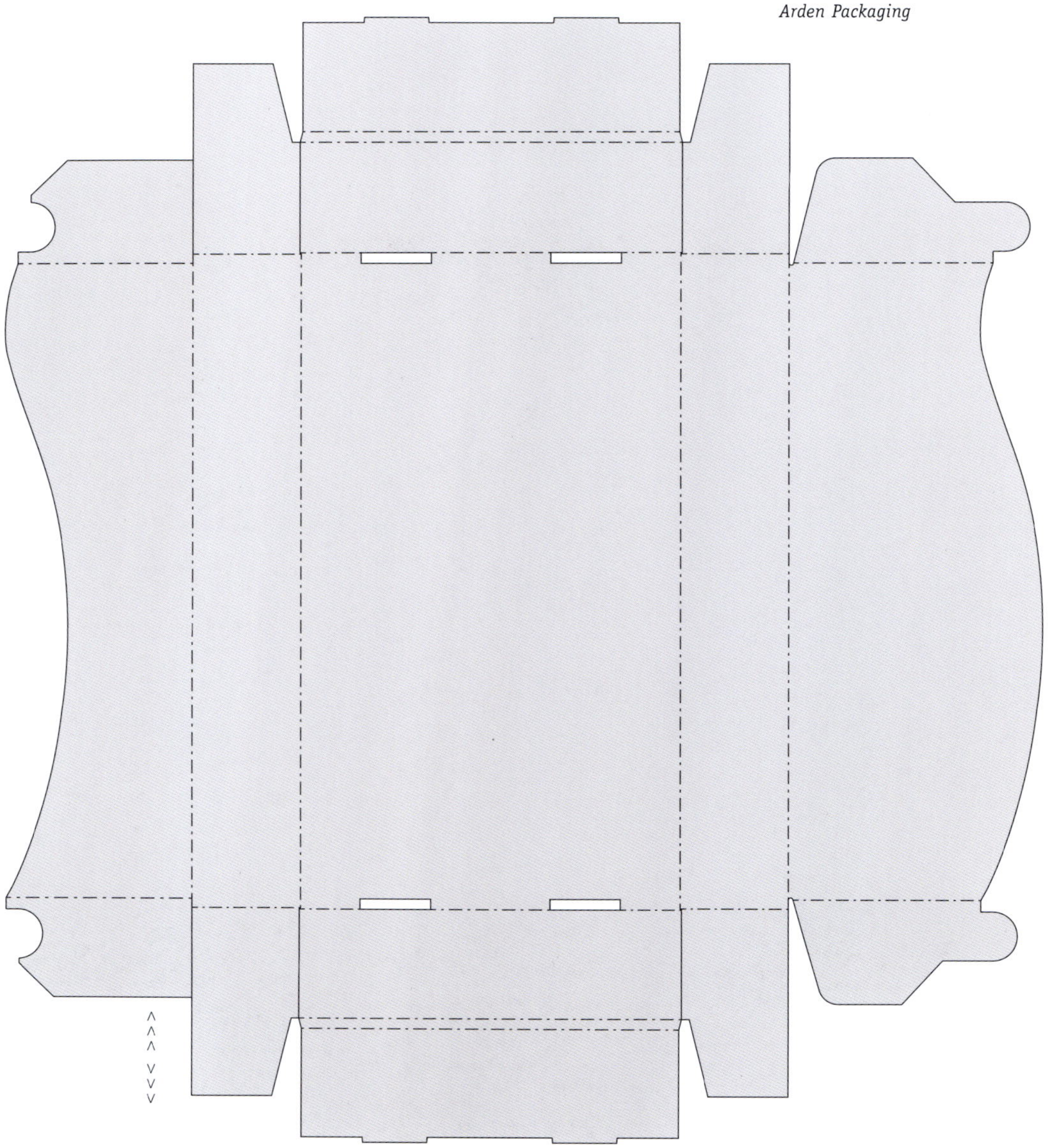

CORRUGATED PRESENTATION CASE

An unprecedented rise in internet shopping in recent years has led to an increased demand in packaging designed to withstand the rigors of the postal system. Often, this requires packs that are both structurally tough and visually appealing. This presentation case is made from corrugated board to exploit its exceptional structural characteristics, while a gentle curve cut in its top creates an attractive opening. The case contains a series of partitioned compartments in which to safely house individual products such as wine, cheese, or chocolates, and could be easily adapted or resized to cater for different product requirements.

CORRUGATED CONES

Sustainable sources of materials have key environmental benefits over nonsustainable sources, which will become increasingly expensive as supplies become scarce. The designer of this corrugated cone has used considerable ingenuity to create a product that replaces the plastic cone, which is deemed expensive and inefficient to store. The flat-pack design is made from a single piece of board and is easily erected, using waste material from the cutting process to further strengthen the constructed product. Such products provide an important example of the diverse application of the packaging designer's skills, just as other areas of design increasingly rely on the resourcefulness of those working effectively with sustainable materials.

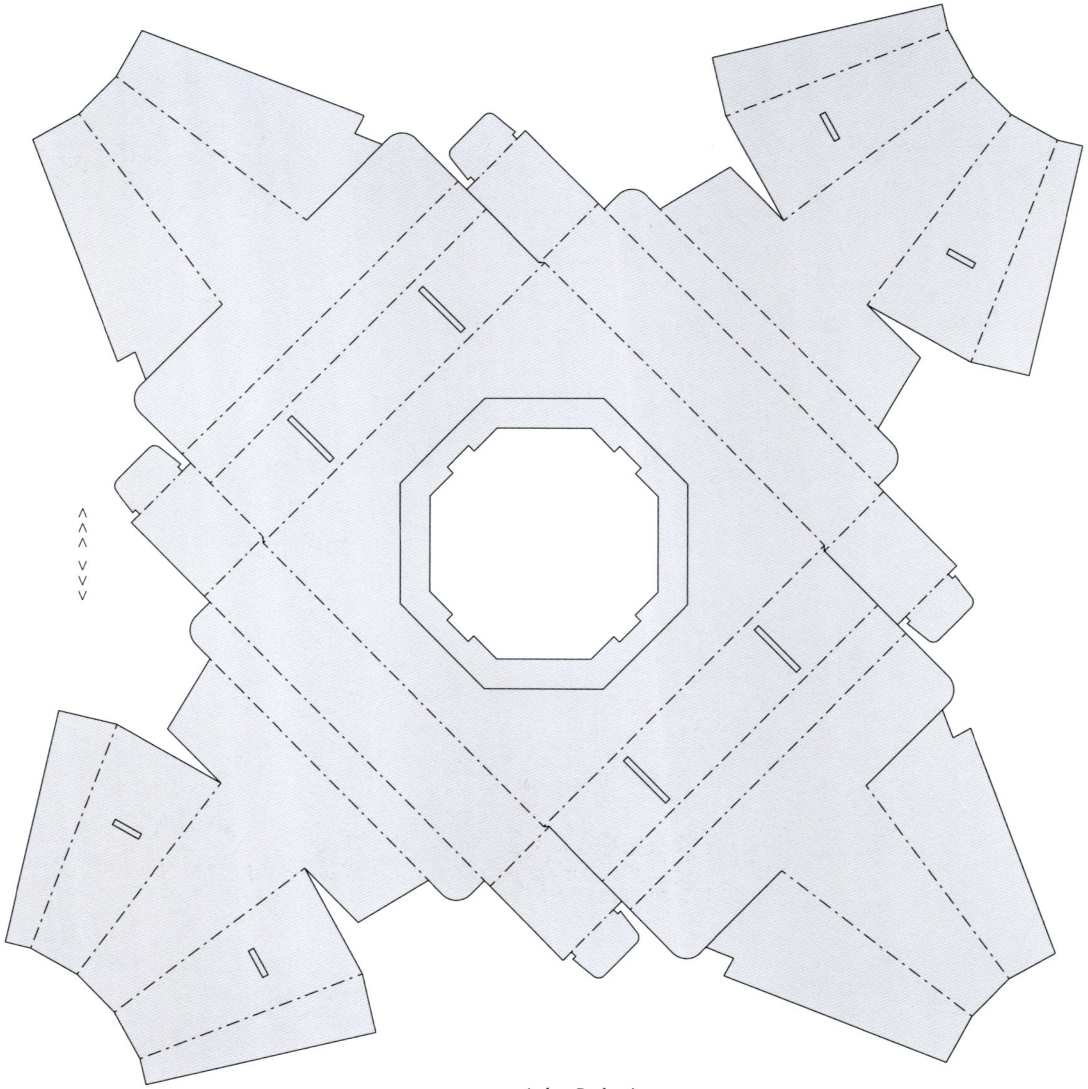

Arden Packaging

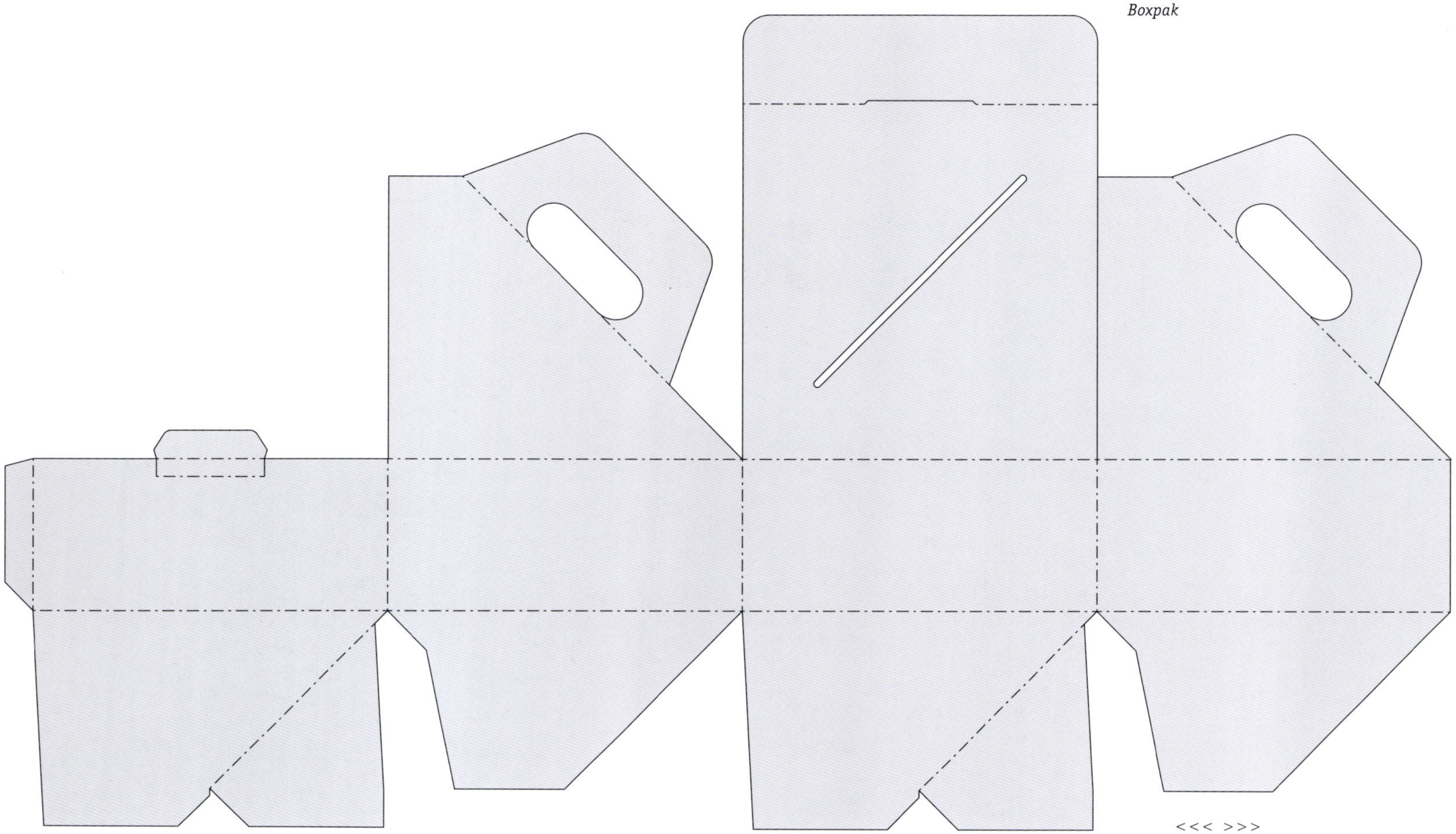

Boxpak

DIAGONAL-HANDLED CARRY CARTON

The top of this simple crash-lock carton has been designed to include a handle that transforms the pack into a carry case. The two-piece handle inserts through a slit in a flap that forms the lid of the carton diagonally across its surface, providing a visual modification that distinguishes it from other types of conventional carry cartons.

DISPENSING CARTON

Cartons often have to perform functionally throughout the life of the product, not just in transit and at the point-of-sale. This carton offers a simple example of a design that allows multiple products to be dispensed individually at different times, between which the carton can be resealed using a tab-lock. Here, a cut-away section inside the carton gestures to the user the method of dispensing the product, though the uncomplicated design could be adapted easily to allow the product to be dispensed using alternative methods, such as perforated sections or tear-away strips.

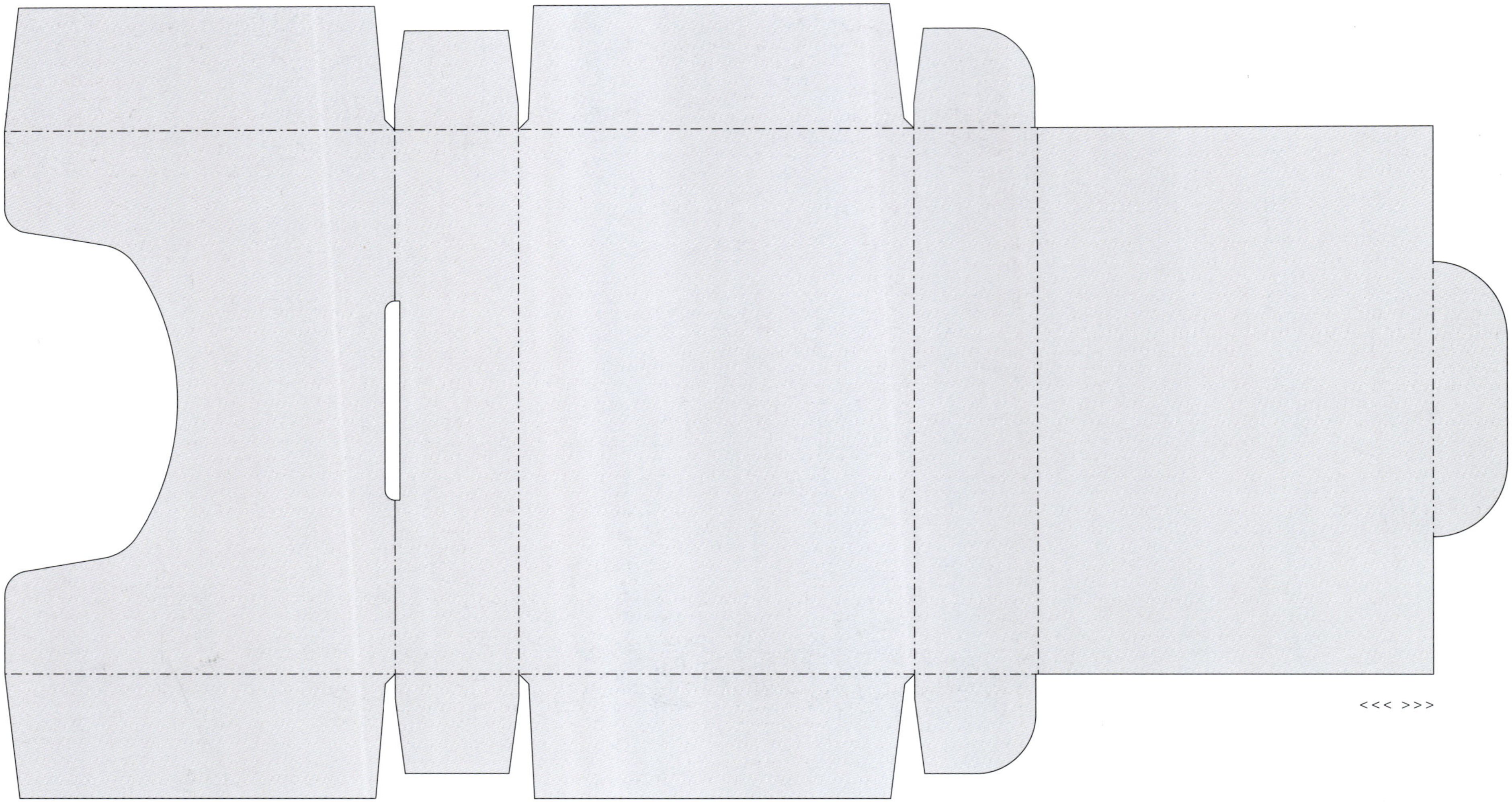

Benson Box

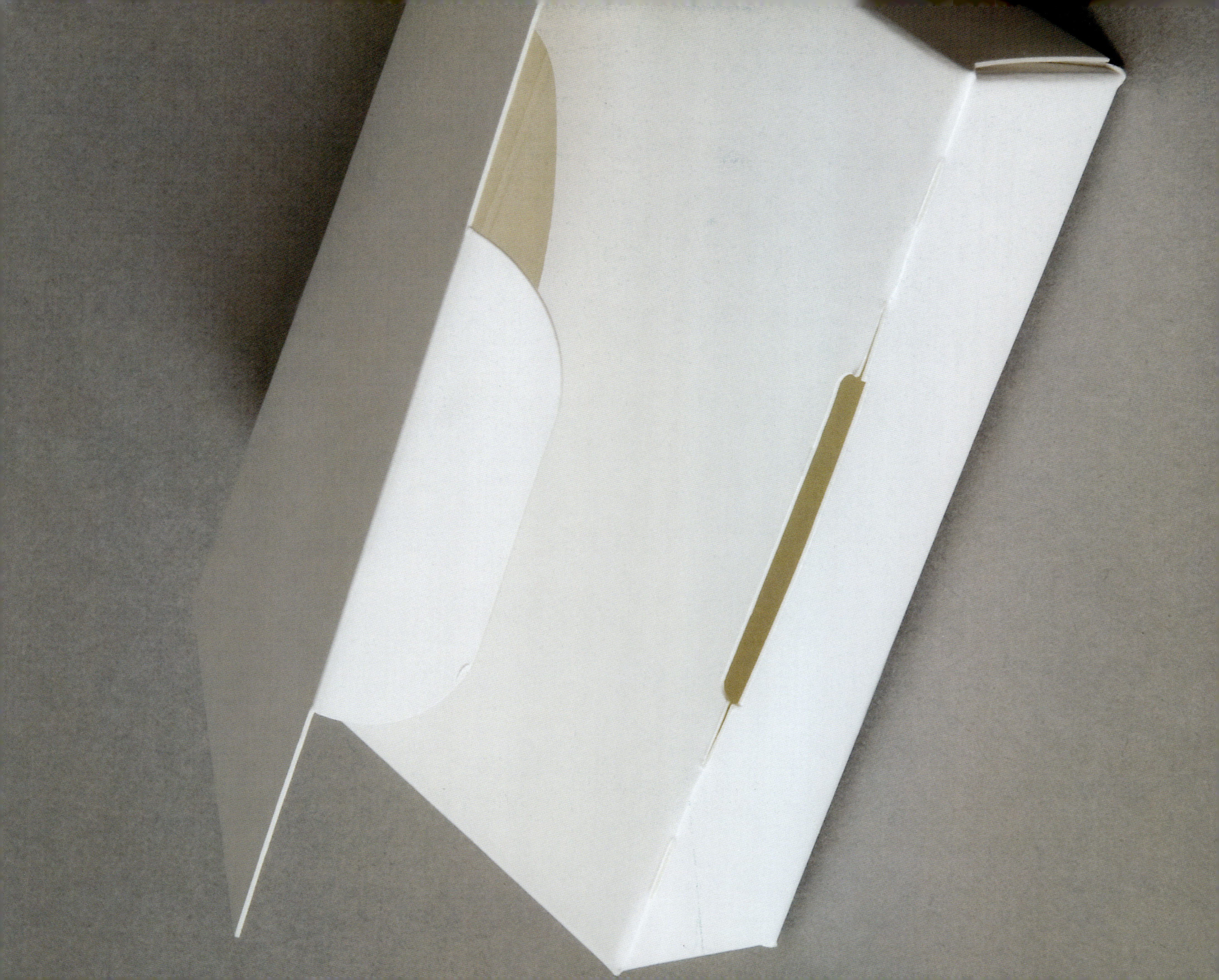

SEGREGATED TRAY

A useful carton to support two or more bottles, or similar upstanding items securely, this carton also provides a flat surface for point-of-sale graphics or promotions. The height of the carton will be determined by the height and composition of the product being contained, and the print on the product must not be obscured by the carton. The support arm across the carton provides a useful divide to separate the products and prevent them from clashing against each other, as well as giving a degree of rigidity to the carton.

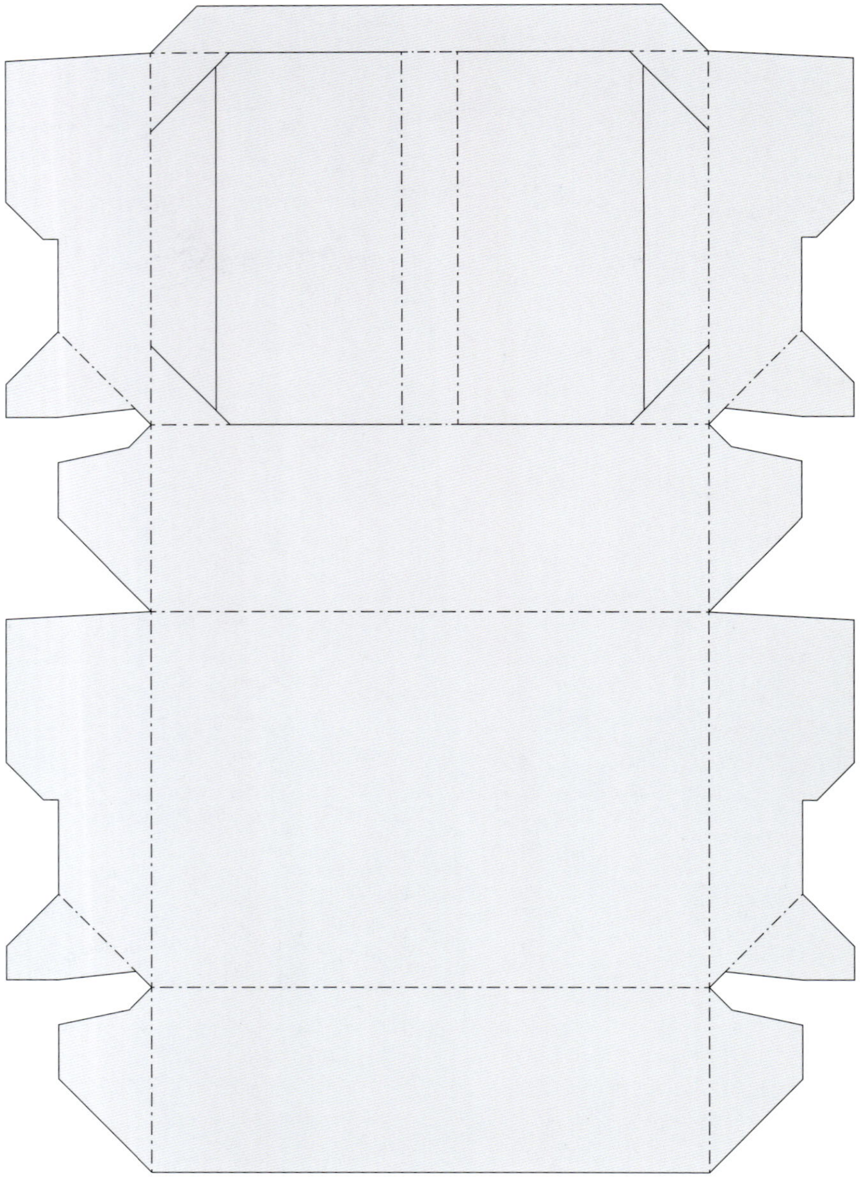

FOUR-SPOT-GLUED CORNER-TUCK CARTON

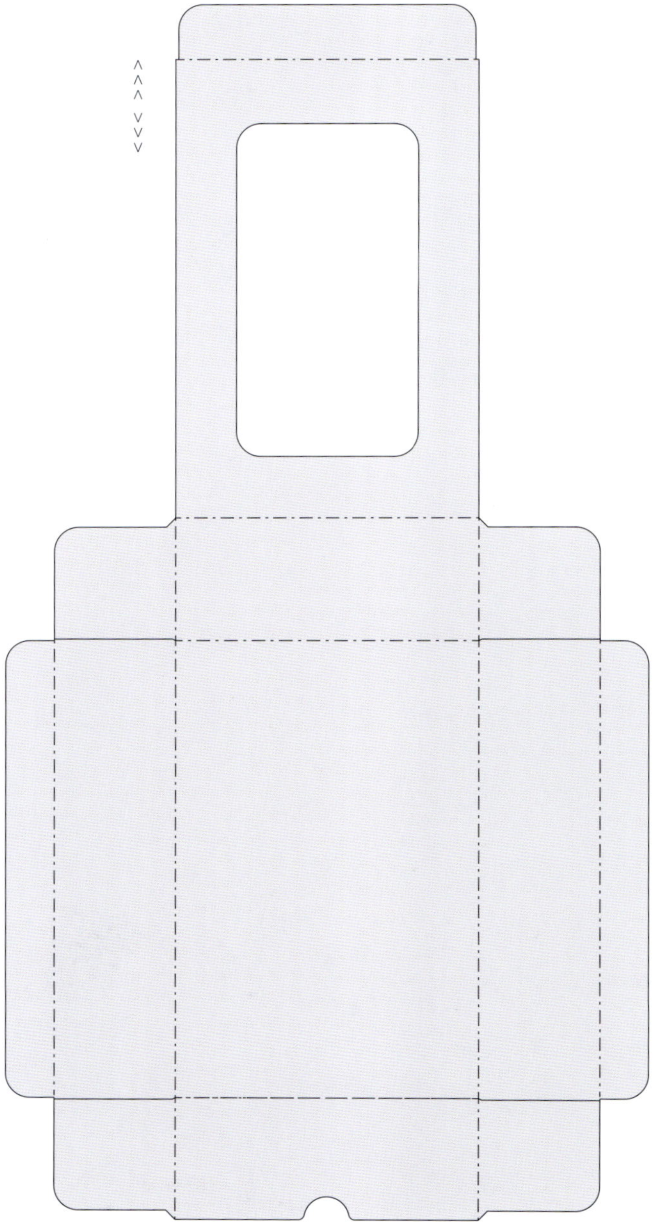

The diminished stength of this carton caused by the large window display makes it suited to the storage of light products such as clothing. The inner flaps on the side of the carton prevent the product from becoming trapped or escaping out from under the lid. The use of a window in the lid provides the consumer with a visual reference to the product.

SUB-DIVIDED CARTON

A carton with a standard, reversible tuck base and top flaps. The three-part compartment is partitioned inside using cut-out sections. Clever use of these die-cut sections provides variation in an otherwise simple design without using additional material or component parts. This design is useful for holding delicate accompaniments to a main product. The cut-outs in the lid flaps might clasp the neck of a bottle.

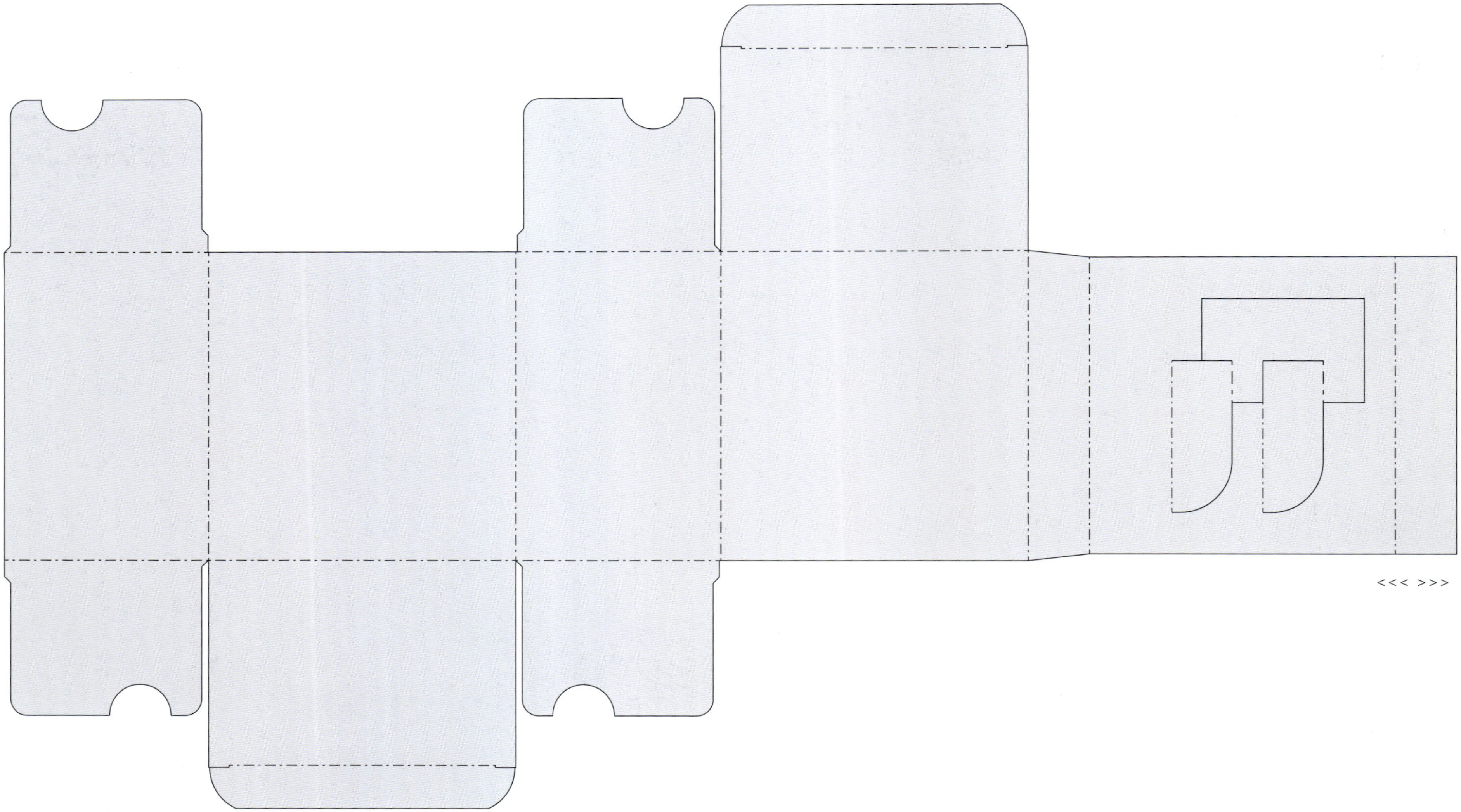

<<< >>>

GCM Boxmore

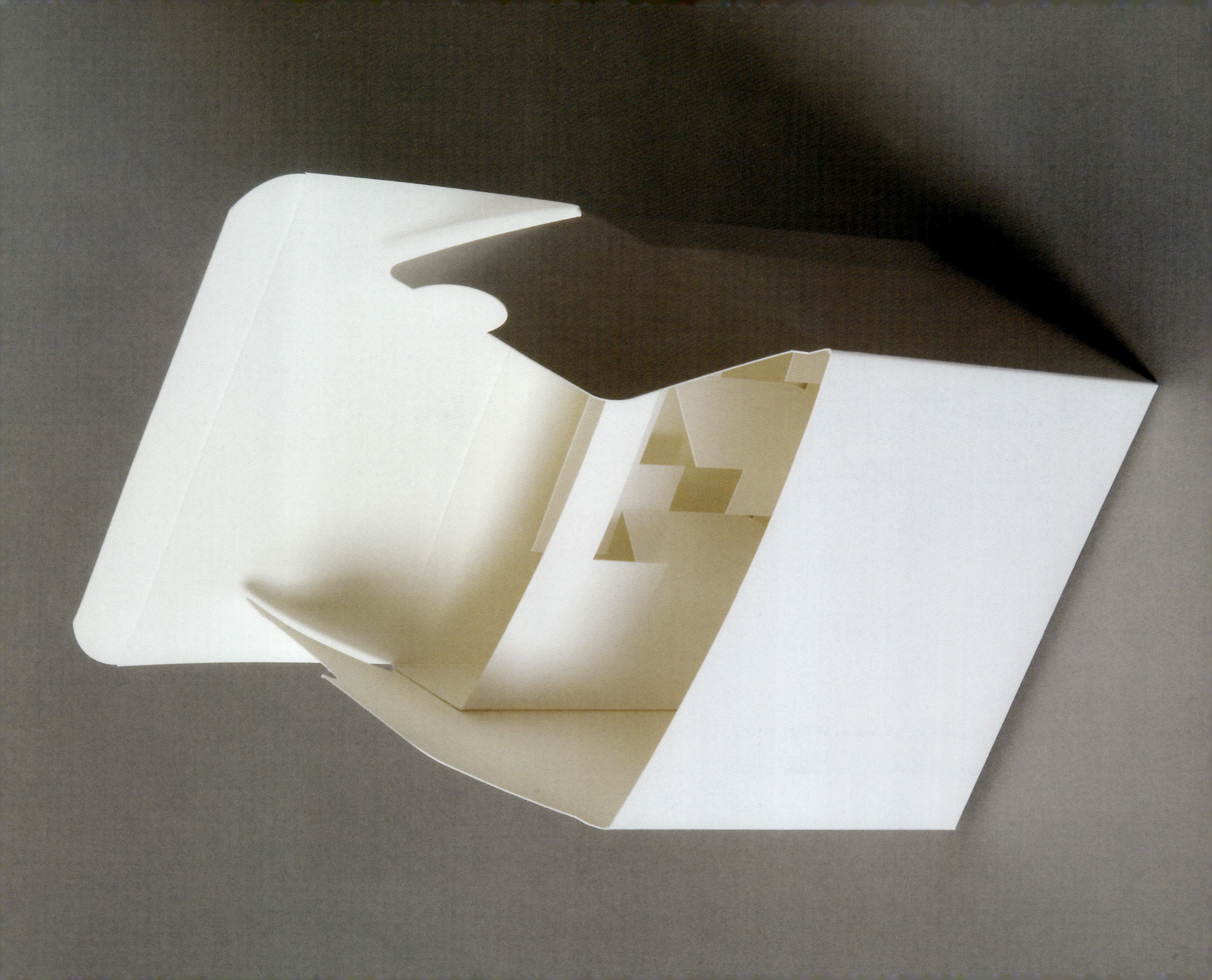

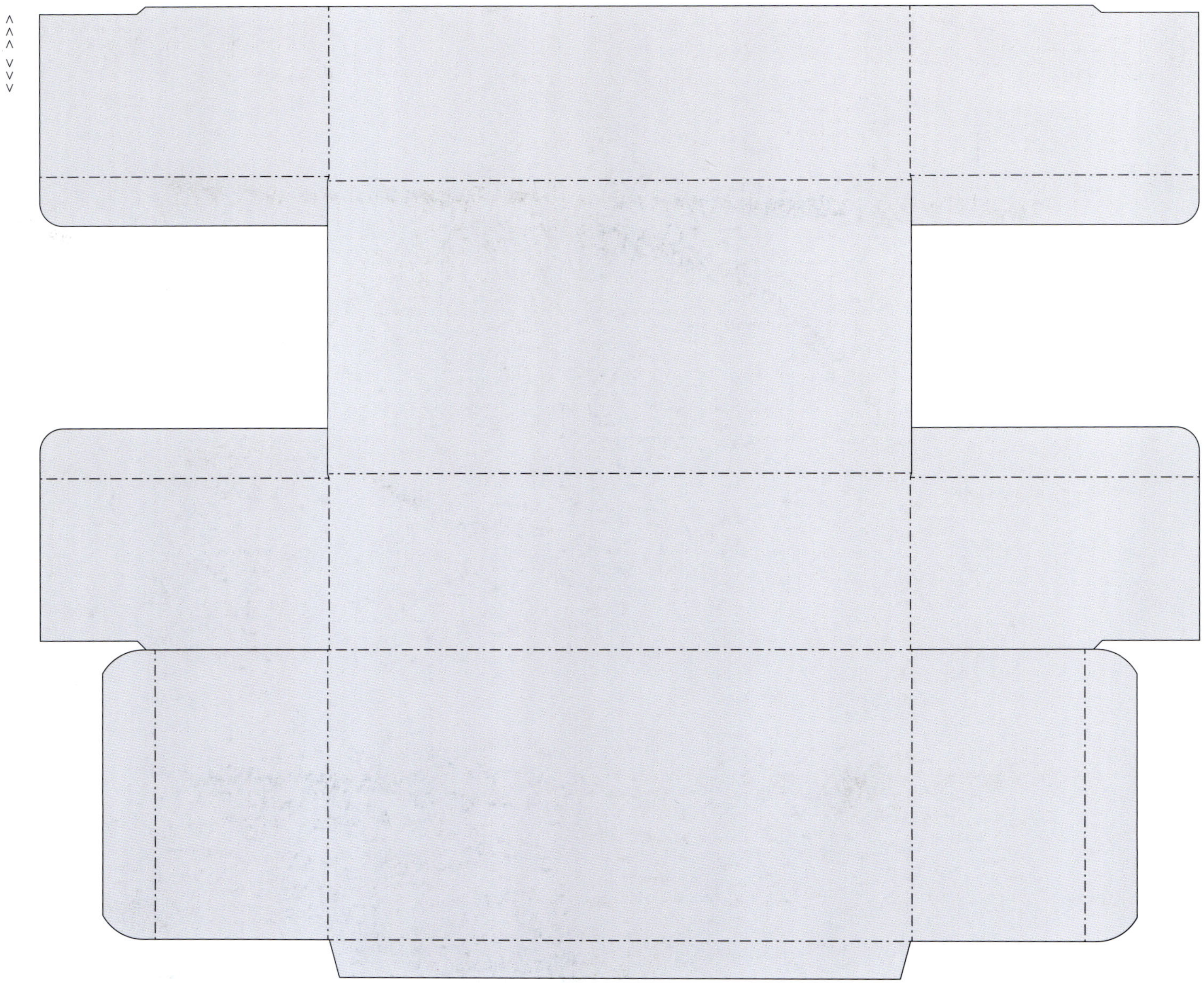

FULL OVERLAP-TUCK CARTON

An example of how a straight-tuck carton can be given added value by having fully overlapping flaps at the access to the contents. It also provides a barrier to the product on which graphics or instructions could be applied.

Contact Details

Contributing Companies

Acket Drukkerij-Kartonnage BV
Gasstraat Oost 13
5349 AH Oss
The Netherlands
T: +31 (0)412 699799
E: info@acket.nl
W: www.acket.nl

Alcan Packaging
Alcan Inc.
1188 Sherbrooke Street West
Montreal
Quebec
H3A 3G2
Canada
T: +1 514 848 8000
F: +1 514 848 8115
E: communications.packaging@alcan.com
W: www.alcanpackaging.com

Arden Packaging Ltd.
Whitefield Road
Bredbury
Stockport
Cheshire
SK6 2SJ
UK
T: +44 (0)161 494 6699
F: +44 (0)161 406 6273
E: info@ardenpackaging.co.uk
W: www.ardenpackaging.co.uk

AssiDomän Frövi
SE-718 80 Frövi
Sweden
T: +46 581 370 00
F: +46 581 310 67
E: info@frovi.com
W: www.frovi.com

Benson Box Co. Ltd.
Interlink Park
Bardon
Leicestershire
LE67 1PE
UK
T: +44 (0)1530 518200
F: +44 (0)1530 518222
E: solutions@bensonbox.co.uk
W: www.bensonbox.co.uk

Boxpak
65 Church Road
Newtownabbey
Belfast
Co. Antrim
Northern Ireland
BT36 7LR
UK
T: +44 (0)28 9036 5421
F: +44 (0)28 9086 6731
E: sales@boxpak.co.uk
W: www.boxpak.co.uk

CMCS Group plc
Unit 1
Kennet Road
Dartford
Kent
DA1 4QN
UK
T: +44 (0)20 8308 5000
F: +44 (0)20 8308 5005
E: sales@cmcs.co.uk
W: www.cmcs.co.uk

CPC Packaging (UK) Ltd.Knapps LaneBristolBS5
7UNUKT: +44 (0)117 951 6751F: +44 (0)117 935
4038E: enquiries@cpcpackaging.co.uk
W: www.cpcpackaging.co.uk

Grobox Gardens Ltd.
Spring Avenue
Gee Cross
Hyde
Cheshire
SK14 5LT
UK
T: +44 (0)161 366 9462
E: customerservices@groboxgardens.co.uk.

Marchmont Packaging Ltd.
Cloverhill Industrial Park
Clondalkin
Dublin 22
Ireland
T: +353 1 4130200
F: +353 1 4130299
E: sales@marchmont.com
W: www.marchmont.com

Mayr-Melnhof Packaging
Brahmsplatz 6
A-1040 Wien
Austria
T: +43 (0)1 502 570
F: +43 (0)1502 57250
E: office@mm-packaging.com
W: www.mm-packaging.com

Model PrimePac AG
Auenstrasse 5
CH-9434 Au (SG)
Switzerland
T: +41 (0)71 747 91 91
F: +41 (0)71 747 91 00

Model PrimePac SA
Route de Neuchâtel 8
CH-1032 Romanel-sur-Lausanne
Switzerland
T: +41 (0)21 731 93 19
F: +41 (0)21 731 93 10
E: info@modelprimepac.com
W: www.modelgroup.com

MR Designs
Unit 6
Lower Farm
High Street
Irchester
Northamptonshire
NN29 7AB
UK
T: +44 (0)1933 410016
F: +44 (0)1933 419929
E: mrdesigns@btconnect.com

Rob. Leunis & Chapman
Andertensche Wiese 18-25
D-30169, Hannover
Germany
T: +49 (0)5 11 / 164 99 0
F: +49 (0)5 11 / 164 99 45
E: info@rlc-packaging.com
W: www.leunisman.de

Rapid Action Packaging (RAP) UK
107 Mortlake High Street
London
SW14 8HQ
UK
T: +44 (0)20 8392 8320
F: +44 (0)20 8392 8321
E: info@rapuk.com
W: www.rapuk.com

Rokov Design
T: +44 (0)870 774 4343
E: info@rokov.biz
W: www.rokov.biz

Ernst Schausberger & Co. GmbH
Heidestraße 19
A-4623 Gunskirchen
Austria
T: +43 7246 6493 353
F: +43 7246 6493 277
E: office@schausberger.com
W: www.schausberger.com

S Sheard Packaging & Sons Ltd.
Solar Works
Calder Street
Greetland
Halifax
West Yorkshire
HX4 8AQ
UK
T: +44 (0)1422 373649
F: +44 (0)1422 310090
E: sales@sheard.co.uk
W: www.sheard-packaging.co.uk

Stora Enso
Office address:
World Trade Center
Klarabergsviadukten 70, C4
P.O. Box 70395
SE-107 24 Stockholm
Sweden
T: +46 (0)1046 46 000
F: +46 8 10 60 20

Stora Enso International Office
9 South Street
London
W1K 2XA
UK
T: +44 (0)20 7016 3100
F: +44 (0)20 7016 3200
W: www.storaenso.com

Contributing Universities

Blackpool and The Fylde College
T: +44 (0)1253 352352
F: +44 (0)1253 356127
E: visitors@blackpool.ac.uk
W: www.blackpool.ac.uk

Brunel University
T: +44 (0)1895 274000
F: +44 (0)1895 232806
E: admissions@brunel.ac.uk
W: www.brunel.ac.uk

Hull College
T: +44 (0)1482 329943
F: +44 (0)1482 598733
E: info@hull-college.ac.uk
W: www.hull-college.ac.uk

Kingston University
T: +44 (0)20 8547 2000
F: +44 (0)20 8547 7388
E: admissions-info@kingston.ac.uk.
W: www.kingston.ac.uk

New College Nottingham
T: +44 (0)115 910 0100
F: +44 (0)115 916 6242
E: enquiries@ncn.ac.uk
W: www.ncn.ac.uk

North Warwickshire College
T: +44 (0)24 7624 3154
E: the.college@nwhc.ac.uk
W: www.nwhc.ac.uk

Norwich School of Art and Design
T: +44 (0)1603 610561
F: +44 (0)1603 615728
E: info@nsad.ac.uk
W: www.nsad.ac.uk

Sheffield Hallam University
T: +44 (0)114 225 5555
F: +44 (0)114 225 2094
E: enquiries@shu.ac.uk
W: www.shu.ac.uk

South East Derbyshire College
T: +44 (0)1332 590500
F: +44 (0)1332 294861
E: enquiries-admissions@derby.ac.uk
W: https://ulib.derby.ac.uk

Suffolk College
T: +44 (0)1473 255885
F: +44 (0)1473 230054
E: info@suffolk.ac.uk
W: www.suffolk.ac.uk

Swindon College
T: +44 (0)1793 491591
E: Admissions@swindon-college.ac.uk

University of Teesside
T: +44 (0)1642 384228
F: +44 (0)1642 384201
E: registry@tees.ac.uk
W: www.tees.ac.uk

University Centre Yeovil
T: +44 (0)1935 845454
F: +44 (0)1935 415483
E: ucy@yeovil.ac.uk
W: www.ucy.ac.uk

Associations, Institutes, and Federations

Aluminium Federation
Broadway House
Calthorpe Road
Five Ways
Birmingham
B15 1TN
UK
T: +44 (0)121 456 1103
F: +44 (0)121 456 2274
E: alfed@alfed.org.uk
W: www.alfed.org.uk

Aluminum Foil Container Manufacturers Association
10 Vecilla Lane
Hot Springs Village
AR 71909
USA
T: +1 501 922 7425
F: +1 501 922 0383
W: www.afcma.org

Argentinean Institute of Packaging
Av. Jujuy 425 (1083)
Buenos Aires
Argentina
T: +54 11 4957 0350
F: +54 11 4956 1368
E: instituto@envase.org
W: www.packaging.com.ar

Asian Packaging Federation
SPG Media Private Ltd.
2nd Floor
Challa Chambers
Raj Bhavan Road
Somajiguda
Hyderabad 500082
India
T: +91 40 55837363
F: +91 40 23318066
E: info@spg-global.com

The Association of European Producers of Steel for Packaging (APEAL)
Avenue Louise, 89
BE-1050
Brussels
Belgium
T: +32 (2)537 91 51
F: +32 (2)537 86 49
E: info@apeal.be
W: www.apeal.org

Association of Plastics Manufacturers
PlasticsEurope
Avenue E van Nieuwenhuyse 4
Box 3
B-1160 Brussels
Belgium
T: +32 (2)675 32 97
F: +32 (2)675 39 35
E: info@plasticseurope.org
W: www.plasticseurope.org

Australian Institute of Packaging
P.O. Box 1059
Chatswood
NSW 2057
Australia
E: info@aipack.com.au
W: www.aipack.com.au

Belgian Packaging Institute
Z.1. Research Park 280
B-1731 Zellik
Belgium
T: +32 (2)464 02 10
F: +32 (2)464 02 39
E: packaging@ibebvi.be
W: www.ibebvi.be

British Adhesives and Sealants Association
5 Alderson Road
Worksop
Nottinghamshire
S80 1UZ
UK
T: +44 (0)1909 480888
F: +44 (0)1909 473834
E: secretary@basaonline.org
W: www.basaonline.org

British Aerosol Manufacturers Association
Kings Buildings
Smith Square
London
SW1P 3JJ
UK
T: +44 (0)20 7828 5111
F: +44 (0)20 7834 8436
E: enquiries@bama.co.uk
W: www.bama.co.uk

British Box and Packaging Association
64 High Street
Kirkintilloch
Glasgow
G66 1PR
UK
T: +44 (0)141 777 7272
F: +44 (0)141 777 7747
E: BPA@natpack.org.uk
W: www.boxpackaging.org.uk

British Fibreboard Packaging Association
2 Saxon Court
Northampton
N1 1ST
Northampton
UK
T: +44 (0)1604 21002

British Glass
9 Churchill Way
Chapeltown, Sheffield
S35 2PY
UK
T: +44 (0)114 290 1850
F: +44 (0)114 290 1851
E: info@britglass.co.uk
W: www.britglass.org.uk

The British Plastics Federation
6 Bath Place
Rivington Street
London
EC2A 3JE
UK
T: +44 (0)20 7457 5000
F: +44 (0)20 7457 5045
E: bpf@bpf.co.uk
W: www.bpf.co.uk

British Printing Industries Federation
Farringdon Point
29-35 Farringdon Road
London
EC1M 3JF
UK
T: +44 (0)870 240 4085
F: +44 (0)20 7405 7784
W: www.britishprint.com

British Soft Drinks Association
20-22 Stukeley Street
London
WC2B 5LR
UK
T: +44 (0)20 7430 0356
F: +44 (0)20 7831 6014
E: bsda@britishsoftdrinks.com
W: www.britishsoftdrinks.com

The Canadian Corrugated Case Association
45 Sheppard Ave. E.
Suite # 900
North York
Ontario M2N 5W9
Canada
T: +1 416 590 9471
F: +1 416 590 9812
E: ccca_@cccassociation.com
W: www.cccassociation.com

Chartered Society of Designers
1 Cedar Court
Royal Oak Yard
Bermondsey Street
London
SE1 3GA
UK
T: +44 (0)207 357 8088
F: +44 (0)207 407 9878
E: info@csd.org.uk
W: www.csd.org.uk

China Packaging Federation
No. 46 Dongsi Xidaji
Dongcheng District
Beijing
100711
China
T: +86 10 65285035
F: +86 10 65285117
W: www.cpta.org.cn

Confederation of Paper Industries
1 Rivenhall Road
Swindon
Wiltshire
SN5 7BD
UK
T: +44 (0)1793 889600
F: +44 (0)1793 878700
E: cpi@paper.org.uk
W: www.paper.org.uk

Corrugated Packaging Council
25 Northwest Point Boulevard
Suite 510
Elk Grove Village
IL 60007
USA
T: +1 847 364 9600
F: +1 847 364 9639
W: www.corrugated.org

Design Council
34 Bow Street
London
WC2E 7DL
UK
T: +44 (0)20 7420 5200
F: +44 (0)20 7420 5300
E: info@designcouncil.org.uk
W: www.design-council.org.uk

The European Carton Makers Association (ECMA)
P.O. Box 85612
NL 2508 CH
The Hague
The Netherlands
T: +31 (0)70 312 39 11
F: +31 (0)70 363 63 48
E: mail@ecma.org
W: www.ecma.org

The European Design Centre
Havensingel 1
5211 TX's-Hertogenbosch
Postbus 1860
5200 BW's-Hertogenbosch
The Netherlands
T: + 31 (0)73 6104000
F: + 31 (0)73 6104060
E: Info@edc.nl
W: www.edc.nl

European Federation of Corrugated Board
Manufacturers (FEFCO)
Avenue Louise 250
B-1050 Brussels
Brussels
Belgium
T: + 32 (2)646 40 70
F: + 32 (2)646 64 60
E: information@fefco.org
W: www.fefco.org

European Organization for Packaging and the Environment
Le Royal Tervuren
Avenue de l'Armée 6 Legerlaan
1040 Brussels
Belgium
T: +32 (2)736 3600
F: +32 (2)736 3521
E: packaging@europen.be
W: www.europen.be

European Plastics Converters
Avenue de Cortenbergh, 66
P.O. Box 4
1000 Brussels
Belgium
T: +32 (2)732 41 24
F: +32 (2)732 42 18
E: info@eupc.org
W: www.eupc.org

Finnish Packaging Association
Ritarikatu 3 h A
FIN 00170
Helsinki
Finland
T: +358 9 684 0340
F: +358 9 6840 3410
W: www.pakkaus.com

Flexible Packaging Association
971 Corporate Boulevard
Suite 403
Linthicum
MD 21090
USA
T: +1 410 694 0800
F: +1 410 694 0900
E: fpa@flexpack.org
W: www.flexpack.org

German Packaging Association
Deutsches Verpackungsinstitut e.V. (DVI)
Bund Deutscher Verpackungsingenieure e.V. (BDVI)
Hohenzollernstraße 12
14163 Berlin
Germany
T: +49 (0)30 80 49 858 0
F: +49 (0)30 80 49 858 18
E: info@verpackung.org
W: www.verpackung.org

Glass Packaging Institute
515 King Street
Suite 420
Alexandria
VA 22314
USA
T: +1 (703)684 6359
F: +1 (703)684 6048
E: info@gpi.org
W: www.gpi.org

Hellenic Packaging Institute
86-88 M. Antypa str.
Gr 163 46 Hellioupolis
Athens
Greece
T: +30 210 9982100
F: +30 210 9969100
E: info@hepo.gr
W: www.hepo.gr

Industry Council for Packaging and the Environment (INCPEN)
SoanePoint
6-8 Market Place
Reading
Berkshire
RG1 2EG
UK
T: +44 (0)118 925 5991
F: +44 (0)118 925 5993
E: info@incpen.org
W: www.incpen.org

Information Resource for the Packaging Industry
W: www.idspackaging.com

The Institute of Packaging/The Packaging Society
(A division of The Institute of Materials, Minerals and Mining)
Stamford Regional Office
Willoughby House
Broad Street
Stamford
Lincolnshire
PE9 1PB
UK
T: +44 (0)1780 759211
F: +44 (0)1780 759220
E: iop@pi2.org.uk
W: www.pi2.org.uk

The Institute of Packaging Professionals
1601 North Bond Street, Suite 101
Naperville
IL 60563
USA
T: +1 630 544 5050
F: +1 630 544 5055
E: info@iopp.org
W: www.iopp.org

International Molded Pulp Environmental
Packaging Association
1425 W. Mequon Road
Suite A
Mequon
WI 53092
USA
T: +1 262 241 0522
F: +1 262 241 3766
E: info@impepa.org
W: www.impepa.org

Italian Institute of Packaging
E: istituto@istitutoimballaggio.it
W: www.istitutoimballaggio.it

Japan Package Design Association
E: info@jpda.or.jp
W: www.jpda.or.jp

Japan Packaging Institute
10F Togeki Bldg. 4-1-1
Tsukiji
Chuo-ku
Tokyo
Japan
T: +81 03 3543 1189
F: +81 03-3543-8970
E: pack@jpi.or.jp
W: www.jpi.or.jp

Metal Packaging Manufacturers Association
SoanePoint
6-8 Market Place
Reading
Berkshire
RG1 2EG
UK
T: +44 (0)118 925 5520
F: +44 (0)118 925 5888
E: enquiries@mpma.org.uk
W: www.mpma.org.uk

Mexican Packaging Association
Homero # 538-101 Col. Chapultepec Morales
México D.F.
T: +1 55 5545 6258
F: +1 55 554 6258
E: amee@starnet.net.mx
W: www.amee.org.mx

Netherlands Packaging Centre
P.O. Box 164
NL-2800
AD Gouda
The Netherlands
T: +31 (0)182 512411
F: +31 (0)182 512769
E: info@nvc.nl
W: www.nvc.nl

The Packaging Association of Canada
2255 Sheppard Avenue East
Suite E420
Toronto, Ontario M2J 4Y1
CanadaT: +1 416 490 7860
F: +1 416 490 7844
E: info@pac.ca
W: www.pac.ca

Bureau du Québec
C.P. 43010
1859 René-Laennec
Succ. Vilamont
Laval, Québec
H7M 6A1
Canada
T: +1 514 990 0134
F: +1 450 668 2691
E: quebec@pac.ca
W: www.pac.ca

Packaging Digest
2000 Clearwater Drive
Oak Brook
IL 60523
USA
T: +1 630 288 8000
F: +1 630 288 8750
E: packagingdigest@reedbusiness.com
W: www.packagingdigest.com

Packaging Gateway
SPG Media Ltd.
Brunel House
55-57 North Wharf Road
London W2 1LA
UK
T: +44 (0)207 915 9957
F: +44 (0)207 915 9958
E: packaging@spgmedia.com
W: www.packaging-gateway.com

The Paperboard Packaging Council
201 N. Union Street
Suite 220
Alexandria
VA 22314
USA
T: +1 703 836 3300
F: +1 703 836 3290
E: paperboardpackaging@ppcnet.org
W: www.ppcnet.org

Pro Carton
250 Avenue Louise
B-1050 Brussels
Belgium
T: +32 (0)2 640 4955
F: +32 (0)2 640 4959
E: info@procarton.com
W: www.procarton.com

Swedish Institute for Packaging and Logistics
STFI-Packforsk AB
Box 5604
SE-114 86 Stockholm
Sweden
T: +46 8 676 70 00
F: +46 8 411 55 18
E: info@stfi.se
W: www.stfi-packforsk.se

Swiss Packaging Institute
Schweizerisches Verpackungsinstitut SVI
Geschäftsstelle
Brückfeldstrasse 18
3000 Bern 9
Switzerland
T: +41 (0)31 302 30 03
F: +41 (0)31 302 30 47
E: info@svi-verpackung.ch
W: www.svi-verpackung.ch

Waste Watch
56-64 Leonard Street
London
EC2A 4JX
UK
T +44 (0)20 7549 0300
F: +44 (0)20 7549 0301
E: info@wastewatch.org.uk
W: www.wasteonline.org.uk

World Packaging Organisation
c/o STFI-Packforsk
Box 5604
S-114 86 Stockholm
Sweden
T: +46 8 676 70 78
F: +46 8 411 55 18
W: www.packaging-technology.com/wpo

World Resource Foundation
Bridge House
High Street
Tonbridge
Kent
TN9 1DP
UK
T: +44 (0)1732 368 333
F: +44 (0)1732 368 337
E: wrf@wrf.org.uk
W: www.wrf.org.uk

Useful Reading

Calver, Giles. *What is Packaging Design?* RotoVision, 2004.

Cliff, Stafford. *50 Trade Secrets of Great Design: Packaging*. Rockport, 2002.

Denison, Edward and Cawthray, Richard. *Packaging Prototypes*. RotoVision, 1999.

Fishel, Catherine. *Design Secrets: Packaging*. Rockport, 2003.

Hargreaves, Ben. *Successul Food Packaging Design*. RotoVision, 2006.

King Gordon, Stacey. *Packaging Makeovers: Graphic Redesign for Market Change*. Rockport, 2005.

Opie, Robert. *Packaging Source Book*. Macdonald & Co. Ltd., 1989.

Rivers, Charlotte. *CD-Art: Innovation in CD Packaging Design*. RotoVision, 2003.

Rivers, Charlotte. *DVD-Art: Innovation in DVD Packaging Design*. RotoVision, 2005.

Roth, Laszlo and George L. Wybenga. *Packaging Designers' Book of Patterns*. Van Nosrand Reinhold, 1991.

Index